Brief Notes in Advanced DSP

Fourier Analysis with MATLAB®

Brief Notes in Advanced DSP

Fourier Analysis with MATLAB®

Artyom M. Grigoryan
Merughan M. Grigoryan

CRC Press
Taylor & Francis Group
Boca Raton   London   New York

CRC Press is an imprint of the
Taylor & Francis Group, an **informa** business

CRC Press
Taylor & Francis Group
6000 Broken Sound Parkway NW, Suite 300
Boca Raton, FL 33487-2742

First issued in paperback 2017

ISBN 13: 978-1-138-11774-7 (pbk)
ISBN 13: 978-1-4398-0137-6 (hbk)

Library of Congress Cataloging-in-Publication Data

Grigoryan, Artyom M.
    Brief notes in advanced DSP : Fourier analysis with MATLAB / Artyom M. Grigoryan, Merughan Grigoryan.
        p. cm.
    Includes bibliographical references and index.
    ISBN 978-1-4398-0137-6 (hardcover : alk. paper)
    1. Signal processing--Digital techniques--Mathematics. 2. Fourier analysis. 3. MATLAB. I. Grigoryan, Merughan. II. Title.

TK5102.9.G75 2009
621.382'2--dc22   2009000681

Visit the Taylor & Francis Web site at
http://www.taylorandfrancis.com

and the CRC Press Web site at
http://www.crcpress.com

Contents

Biography

Artyom M. Grigoryan received his M.S. degrees in mathematics from Yerevan State University (YSU), Armenia, USSR, in 1978, in imaging science from Moscow Institute of Physics and Technology, USSR, in 1980, and in electrical engineering from Texas A&M University, USA, in 1999. He received a Ph.D. degree in mathematics and physics from YSU in 1990. In 1990 to 1996, he was a senior researcher with the Department of Signal and Image Processing at the Institute for Informatics and Automation Problems of the National Academy Science of Republic of Armenia and Yerevan State University. From 1996 to 2000, he was a research engineer with the Department of Electrical Engineering, Texas A&M University. In December 2000, he joined the Department of Electrical Engineering, University of Texas at San Antonio, where he is currently an associate professor. Dr. Grigoryan is the author of *Multidimensional Discrete Unitary Transforms: Representation, Partitioning and Algorithms,* Marcel Dekker, 2003. He has authored many papers specializing in the theory and application of fast one- and multi-dimensional unitary transforms, integer Fourier transforms, paired transform, wavelets and unitary heap transforms, design of robust linear and nonlinear filters, image enhancement, image filtration, computerized tomography, and processing biomedical images.

Merughan M. Grigoryan received his M.S. degree in physics from Yerevan State University, Armenia, USSR, in 1979, and worked as a postdoctoral research associate from 1979 to 1981 on the dispersion of ultrashort impulses in the Department of Radio-Physics and Electronics at YSU. From 1982 to 1995, he was working as a senior research engineer at different science institutes, such as: All-Union Scientific Associations "Astro," "Neitron," and Scientific Research Institute of Non-Ferrous Metals (USSR), on topics including electronics, signal and image processing, and acoustic emission. He is currently conducting private research on the following topics: theory and application of quantum mechanics in signal processing, differential equations, Hadamard matrices, Haar transformation, fast integer unitary transformations, theory and methods of the fast unitary transforms generated by signals, and methods of encoding in cryptography.

Preface

Many interesting topics are studied in digital signal and image processing, and one of them is the theory and application of Fourier analysis. The Fourier transformation is the most used tool when analyzing and solving problems in the framework of linear systems that describe and approximate different physical systems in practice. In digital signal processing (DSP), this transformation gives the push for developing other fast discrete transformations, such as the Hadamard, cosine, and Hartley transformations. Another transformation that is used in DSP, as well as in speech processing and communication, is the discrete Haar transformation. This is the first orthogonal transformation developed after the Fourier transformation, which had been used as a basic stone to build the wavelet theory for the continuous-time signals. The Haar transformation used to be considered a transformation, that does not relate to the Fourier transformation. However, in the mathematical structure of the Fourier transformation, there is a unitary transformation, which is called the paired transformation and which coincides with the discrete Haar transformation, up to a permutation. Such a similarity is only in the one-dimensional case; the discrete paired transformations exist in two- and multi-dimension cases as well, and they are not separable.

The main purpose of using unitary transformations is in analyzing and processing the coefficients of decomposition of the signal in the corresponding basis. Each basic wave or function of the transformation is a carrier of a specific frequency (which is true for the Fourier, cosine, Hartley, Haar, and other transformations). The signal is thus transferred from the original time domain to the frequency domain, where the signal is analyzed and an effective solution of a given problem can be found. As an example, we can mention the complex operation of the cyclic linear convolution in the time domain, which is referred to in the frequency domain as the operation of multiplication for the Fourier transformation. In general, the basic functions of the transformation may be generated by characteristics other than frequencies; even the unitary property of the transformation may not be required, only the invertibility.

We focus here mainly on the unitary transformations, which are the Fourier, Hadamard, Hartley, Haar, and cosine transformations. The fundamental properties of these transformations can be found in many books written on signal and image processing by L.R. Rabiner, B. Gold, M. Proakis, S.K. Mitra, R.C. Gonzalez, and others. In this concise book, we give readers popular notes in advanced digital signal processing, the main part of which has been formed from the lectures given in advanced graduate level signal processing classes at the Department of Electrical and Computer Engineering at the University of Texas at San Antonio. This collection of notes addresses many concepts of DSP and their applications, which are based on our research in Fourier analysis. We also present many interesting problems and concepts we have been working on these last years. Our goal is to help readers, graduate students, and engineers to use new forms and methods of signals and images in the frequency domain, as well as in the so-called frequency-and-time domain. These notes also will be useful for self-study since much of the material is quite advanced. Many codes are given to show how to implement the discussed ideas in practice. These codes will help readers to compose their own programs and to understand the given concepts well. Each chapter contains a list of problems that we suggest readers work on and solve. Not all of these problems are simple and the difficult ones are marked by asterisks. These problems require diligent work with pencil and computer. To help instructors in solving these problems, we wrote the *Solution Manual: Brief Notes in Advanced DSP: Fourier Analysis with MATLAB®**. This manual contains the answers and the computer-based solutions of almost all problems.

The following describes the organization of the book. There are six chapters that include the context of 21 lectures numbered 3, 4, 4, 3, 2, and 5 in Chapters 1 through 6, respectively. Chapter 1 covers the basic concept of the discrete Fourier transformation and its properties. A brief review of the necessary background material, including the concept of the *splitting* of the transform, is given. The properties of the *discrete paired transformation* that result in the effective splitting of the Fourier transform are described. The paired transformation is unitary and allows for representing discrete-time signals in the frequency-and-time domain. It is not the Haar transformation; the relation between the Haar and paired transformations is explained. We discuss here the fast Fourier transform based on the splitting by the paired transform, and describe examples of the 8- and 16-point DFTs in detail. MATLAB-based codes for computing DFT and discrete Haar transform are given.

In Chapter 2, the methods of the lifting schemes and integer transformations with control bits are described and applied for integer approximations of the DFT. The algorithms for the 8- and 16-point integer approximation of the DFT are given in detail, when the paired transform is used. The inverse integer transforms are also described. In the second part of the chapter, we introduce and discuss the interesting concept of the vector DFT, when operations in the complex space are transferred to the real space. This concept shows a simple way for constructing the integer transformations, which have structure similar to the Fourier transformation. As a particular case, we present the so-called elliptic DFT that is based on the square roots of the identity matrix (2×2), which are not the Givens rotations. Main proper-

---

*MATLAB is a registered trademark of The MathWorks, Inc. For product information, please contact: The MathWorks, Inc. 3 Apple Hill Drive, Natick, MA 01760-2098 USA. Tel: 508-647-7000. Fax: 508-647-7001. Email: info@mathworks.com. Web: www.mathworks.com.

ties of such transformations are given and different examples are described. Chapter 3 is devoted to the discrete cosine transforms (DCT). A method of Coxeter-type matrices for computing short DCT is introduced, and the method of paired transforms for splitting the DCT is described in detail. Integer approximations of DCT are described by methods of lifting schemes, control bits, and nonlinear equations through the canonical representation of the DCT. In Chapter 4, we discuss the concept of the discrete Hadamard transformation (DHdT) and the paired transform method for calculating the DHdT. The mixed Hadamard as well as Fourier transformations are introduced, and the square roots and roots of high order of these transforms are described. Our attempt to generalize the discrete Hadamard transformation by introducing the so-called bit-and binary transformations for different orders N is also discussed and illustrated on examples $N = 3, 5, 6$, and 7. In conclusion, the concept of the mixed Fourier transform is given.

In the first part of Chapter 5, the decomposition of the one-dimensional signal by the so-called section basis signals is described. The second part is devoted to the new forms of two-dimensional signal, or image representation. Namely, the 2-D paired representation of the discrete image, which is the 2-D frequency and 1-D time representation, is given. This representation allows for solving the problem of image reconstruction by projections in the discrete model, and defining and processing the image along specific directions. Based on the paired representation, the new concept of the resolution map with all periodic structures of the image is described and used for image enhancement. We then discuss the application of the paired splitting-signals for image enhancement by the fast method of α-rooting. In the last chapter, we present our vision of the problem of signal multiresolution, which is based on the Fourier analysis, namely, we discuss the concept of the Fourier transform wavelet. The representation of the Fourier transform is described by the cosine- and sine-like wavelets. For that, the concepts of A- and B-wavelet transforms are considered. The well-known Fourier integral formula, which leads to frequency-time analysis of the signal, is also considered. And finally, we briefly present the powerful concept of the discrete signal-induced heap transformations with the Haar transform path. Such transformations, which we call the Givens-Haar transformations, are unitary, fast, and can be constructed for any order. The decomposition of the signal and its reconstruction by the Givens-Haar transforms are performed by basic not planar waves that have in many cases complex forms of movement and interaction.

We appreciate all who assisted in preparation of the book. We are grateful to the reviewers for their suggestions and recommendations. We also thank the staff and faculty of the Department of Electrical and Computer Engineering at University of Texas at San Antonio, especially Dr. GVS Raju, who partially support this research through the National Science Foundation under Grant 0551501. Finally, we express our gratitude to our families for their support.

Artyom and Merughan M. Grigoryans, December 24, 2008

1

Discrete Fourier Transform

Since the introduction of Cooley-Tukey fast Fourier transform (FFT) [1], the Fourier transform has been widely used in different areas of signal and image processing, communication systems, data compression, pattern recognition and image reconstruction, interpolation, linear filtering, and spectral analysis [2]-[6]. The Fourier transform determines all frequencies in the function (signal), and transfers the data defined on the real space into the complex, while simplifying the realization of the operation of linear convolution. We start with the definition and properties of the discrete Fourier transformation in the one-dimensional case, and then we will try to reveal the mathematical structure of this transformation for better understanding the transformation and using it in practical applications. The splitting of the transform is based on the paired representation of the signals, in a form of sets of short signals which can be analyzed and processed separately. The paired representation of signals is referred to as a time-frequency representation; however, the paired transform is not the wavelet transform, different types of which were developed after the Haar transform. It is interesting to note that the matrices of the paired and Haar transformations are equal up to a permutation of rows and columns. To show that, we describe the complete set of the one-dimensional paired functions and, then, analyze the relation between the paired and Haar transformations.

1.1   Properties of the discrete Fourier transform

Let f_n be a finite sequence or discrete-time signal of length $N > 1$. The N-point discrete Fourier transform (DFT) of the signal f_n is defined by

$$F_p = (\mathcal{F}_N \circ f_n)_p = \sum_{n=0}^{N-1} f_n W^{np}, \qquad p = 0 : (N-1), \qquad (1.1)$$

where $W = W_N = \exp(-2\pi j/N)$ and the notation $p = 0 : (N-1)$ denotes integer numbers that run from 0 to $(N-1)$.

This transform can be considered as the discrete-time Fourier transform

$$F(e^{j\omega}) = \sum_{n=0}^{N-1} f_n e^{-jwn}$$

defined only at N points which are placed uniformly on the unit circle. The corresponding frequency-points are $\omega = \omega_p = \frac{2\pi}{N}p$, $p = 0 : (N-1)$, and at these points $F(e^{j\omega_p}) = F_p$. The transform F_p is a periodic sequence with period N, i.e., we consider that $F_p = F_{p \bmod N}$, for any integer p. Thus, the discrete Fourier transformation converts finite discrete signals to discrete periodic signals.

We consider properties of the discrete Fourier transformation

$$\mathcal{F} : f_n \rightarrow F_p = \sum_{n=0}^{N-1} f_n W^{np} \rightarrow \frac{1}{N} \sum_{k=0}^{N-1} F_p W^{-pn} = f_n, \qquad (1.2)$$

where $n = 0 : (N-1)$. The kernel of the transform is periodic, $W^{-p(n+N)} = W^{-pn}$, and the second sum thus defines the periodic sequence $\hat{f}_n$ ($n = 0, \pm 1, \pm 2, \dots$), and f_n is one period of this sequence.

1. *(Linearity)* DFT is a linear transformation, i.e., $\mathcal{F}[f_n + kg_n] = \mathcal{F}[f_n] + k\mathcal{F}[g_n]$, for any two sequences f_n, g_n, and a constant k.

2. *(Duality)*

$$\begin{array}{cc} f_n & F_n \\ \downarrow \mathcal{F} \searrow & \downarrow \mathcal{F} \\ F_p & Nf_{-p} \end{array}$$

where $f_{-p} = f_{N-p}$. Indeed, it follows directly from (1.2) that

$$\sum_{n=0}^{N-1} F_n W^{np} = \sum_{n=0}^{N-1} F_n W^{-n(-p)} = Nf_{-p}, \quad p = 0 : (N-1).$$

3. *(Time reversal)*

$$\begin{array}{cc} f_n & \rightarrow f_{-n} \, (f_{N-n}) \\ \downarrow \mathcal{F} & \downarrow \mathcal{F} \\ F_p & \rightarrow F_{-p} \, (F_{N-p}) \end{array} \qquad (1.3)$$

Indeed, the following calculations hold:

$$\sum_{n=0}^{N-1} f_{N-n} W^{np} = \sum_{n=0}^{N-1} f_{N-n} W^{-(N-n)p} = \sum_{n=0}^{N-1} f_{N-n} W^{(N-n)(-p)}$$

$$= \sum_{n=1}^{N} f_n W^{n(-p)} = \sum_{n=0}^{N-1} f_n W^{n(-p)} = F_{-p}$$

since we consider $f_N = f_0$, and $W^{N(-p)} = 1$ for any integer p.

If the discrete-time signal is even, $f_n = f_{-n}$, then $F_p = F_{-p}$. If the discrete-time signal is odd, $f_n = -f_{-n}$, then $F_p = -F_{-p}$. In both the cases, it suffices to calculate $N/2+1$ (or $(N+1)/2$) first values of F_p and the rest $N/2-1$ (or $(N-1)/2$) of values to calculate by conjugation, if N is even (or odd). The amount of computation necessary to determine the DFT can be thus halved.

4. *(Conjugate)*

$$\mathcal{F} : \bar{f}_n \rightarrow \overline{F_{-p}} = \overline{F_{N-p}}.$$

Indeed

$$\sum_{n=0}^{N-1} \bar{f}_n W^{np} = \overline{\sum_{n=0}^{N-1} f_n W^{-np}} = \overline{\sum_{n=0}^{N-1} f_n W^{n(-p)}} = \overline{F_{-p}}.$$

If the signal is real, then $f_n = \bar{f}_n \rightarrow F_p = \overline{F_{-p}}$. The amplitude of the Fourier transform does not change, but phase changes its sign

$$Arg\, F_{-p} = -Arg\, \overline{F_{-p}} = -Arg\, F_p.$$

5. *(Time shift)*

$$
\begin{array}{ccc}
f_n & \rightarrow & g_n = f_{n-n_0} \\
\downarrow \mathcal{F} & & \downarrow \mathcal{F} \\
F_p & \rightarrow & G_p = W^{pn_0} F_p
\end{array}
$$

where n_0 is an integer number. This important property holds because of the periodicity of the extended sequence and the kernel of the transform,

$$\sum_{n=0}^{N-1} g_n W^{np} = \sum_{n=0}^{N-1} f_{n-n_0} W^{[n-n_0]p+n_0p} = \left[\sum_{n=0}^{N-1} f_{n-n_0} W^{[n-n_0]p}\right] W^{n_0p}$$

$$= \left[\sum_{n=-n_0}^{N-1-n_0} f_n W^{np}\right] W^{n_0p} = \left[\sum_{n=0}^{N-1} f_n W^{np}\right] W^{n_0p} = F_p W^{n_0p}.$$

By shifting the signal, the amplitude F_p of the spectrum does not change, but the number $\vartheta(p) = (2\pi n_0/N)p$ is added to the phase $\arg(F_p)$ at frequency-point p. In other words, $Arg\, G_p = Arg\, F_p - \vartheta(p)$, $p = 0 : (N-1)$, where the function $\vartheta(p)$ is linear with respect to p.

Example 1.1
Let N be an even number greater than 10, and let f_n be the following periodic sequence with seven unit pulses placed in one period $n = -N/2 : N/2 - 1$ by

$$f_n = \sum_{m=-3}^{3} \delta_N[n-m] = \begin{cases} 1, & n = -3 : 3, \\ 0, & 3 < |n| < N/2, \end{cases}$$

where $\delta_N[n] = 1$ if n is an integer multiple of N, and 0 otherwise. Then

$$F_p = \sum_{n=-N/2}^{N/2-1} f_n W^{np} = \sum_{n=-3}^{3} W^p = \begin{cases} 7, & p = 0, \\ \dfrac{\sin\left(\frac{7\pi p}{N}\right)}{\sin\left(\frac{\pi p}{N}\right)}, & p = 1 : (N-1). \end{cases}$$

As an example, Figure 1.1 shows the signal f_n of length $N = 1024$ in the time interval $[-5, 5]$ in part a, along with the amplitude and phase of the 1024-point DFT of the signal in parts b and c, respectively.   ⧠

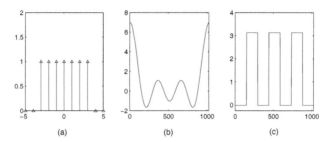

(a)   (b)   (c)

FIGURE 1.1
(a) Signal, (b) the 1024-point DFT, and (c) the phase of the DFT.

6. *(Shift in frequency domain)*

$$\begin{array}{ccc} F_p & \leftarrow & f_n \\ \downarrow & & \downarrow \\ F_{p-p_0} & \to & W^{-p_0 n} f_n \end{array}$$

For instance, the shift by $p_0 = N/2$, when N is even, leads to the change of the sign of every second component of the signal: $f_n \to W^{-N/2n} f_n = W_2^{-n} f_n = (-1)^n f_n$.

7. *(Circular convolution)*

$$\begin{array}{ccc} f_n, \ y_n & \to & f_n \otimes y_n \\ \downarrow \mathcal{F} \downarrow \mathcal{F} & & \downarrow \mathcal{F} \\ F_p, \ Y_p & \to & F_p \cdot Y_p \end{array}$$

where *the circular,* or *periodic convolution* of length N is defined by

$$f_n \otimes y_n = (f \otimes y)_n = \sum_{m=0}^{N-1} f_m y_{(n-m) \bmod N}, \quad n = 0 : (N-1).$$

To demonstrate the importance of this property, we consider a random noisy signal f_n of length $N = 512$ shown in Figure 1.2 in part a, which has been obtained from the original signal o_n convoluted with the window h_n shown in b, plus a random noise has been added. The amplitude of the DFT of the noisy signal is shown in c, and the filtered signal $f_n \otimes y_n$ in d, which has been defined first in the frequency domain. Namely, the noisy signal has been

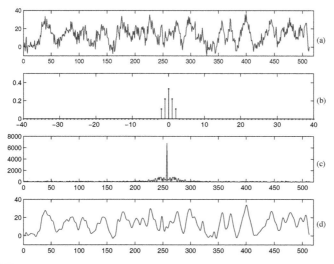

FIGURE 1.2

(a) Noisy signal, (b) the impulse characteristic (the window $[0, 1, 2, \underline{3}, 2, 1, 0]/9$), (c) the 512-point DFT of the signal in the absolute scale (shifted to the center), and (d) the filtered signal.

convoluted with the filter, or sequence y_n whose response function is defined by

$$Y_p = \frac{\bar{H}_p}{|H_p|^2 + \phi_{N/O}(p)}, \quad p = 0 : (N-1), \tag{1.4}$$

where $\phi_{N/O}(p)$ denotes the ratio noise-signal. This process is called the *optimal filtration* of the signal. This linear filter depends on both the original signal and degradation. The characteristics of the filter are shown in Figure 1.3.

8. *(Parseval's equality)*

The energy of a discrete-time signal f can be expressed in the time and frequency domains as

$$E^2(f) = \sum_{n=0}^{N-1} |f_n|^2 = \frac{1}{N} \sum_{p=0}^{N-1} |F_p|^2. \tag{1.5}$$

Such a nonsymmetric form of the equation arises because of luck of the normalized coefficient $1/\sqrt{N}$ in the definition of the DFT in (1.1). To derive this equation, we consider the cyclic convolution, or autocorrelation of the signal, which corresponds to $|F_p|^2$ in the frequency domain, i.e.,

$$R_{f,f}(k) = \sum_{n=0}^{N-1} f_n f_{n-k \bmod N} = \frac{1}{N} \sum_{p=0}^{N-1} |F_p|^2 W^{-kp}, \quad k = 0 : (N-1),$$

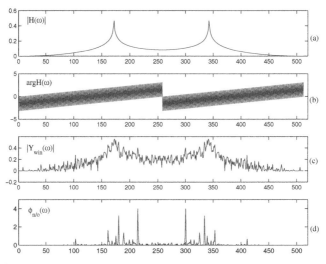

FIGURE 1.3
(a) The amplitude (in the logarithmic scale) and (b) phase of the DFT of h_n, (c) the response function of the optimal filter, and (d) the noise-signal ratio.

which leads to (1.5), when $k = 0$.

The distance between discrete-time signals f_n and g_n of the same length N equals thus to the distance between their Fourier transforms F_p and G_p, i.e.,

$$d_2(f,g) = \sqrt{\sum_{n=0}^{N-1} |f_n - g_n|^2} = d_2(F,G) = \sqrt{\frac{1}{N} \sum_{p=0}^{N-1} |F_p - G_p|^2}.$$

The response function Y_p of the optimal filter defined in (1.4) is derived from the condition of minimization of the square-root error of approximation,

$$\min_{\hat{O}=YF} d_2^2(O,\hat{O}) = < \frac{1}{N} \sum_{p=0}^{N-1} |O_p - Y_p F_p|^2 >,$$

which guarantees the minimum of the distance $d_2(o, f \otimes y)$ in the time domain. Here $< \cdot >$ denotes an expected value.

1.2   Fourier transform splitting

In this section, we describe a splitting of the discrete Fourier transform (DFT) by sections, that leads to the concept of wavelet-like unitary transform, or the so-called *paired transform* [8]. This transform is considered as a core part of the mathematical structure of the discrete Fourier transform, which defines the frequency-time representation of the signal and allows for minimizing not only the computational cost of the fast Fourier transform, but other transforms as well [9]. The paired transform represents the signal as a unique set of separate short and independent signals that can be processed separately when solving different problems of signal processing. The splitting-signals have different lengths and carry the spectral information of the represented signal in disjoint subsets of frequency-points. The paired transform has a fast algorithm and leads to an effective decomposition of the signal. We consider the concept of the paired representation of the signal with respect to the Fourier transform.

Let f_n be a finite sequence or discrete-time signal of length N, where N is a power of two, $N = 2^r$, $r \geq 1$. The N-point DFT of the signal f_n

$$F_p = (\mathcal{F}_N \circ f_n)_p = \sum_{n=0}^{N-1} f_n W^{np}, \qquad p = 0 : (N-1), \qquad (1.6)$$

can be divided by subsets of its components, which are images of short 1-D signals describing the original signal in a new representation. In the paired representation, the signal f_n is transformed into a set of $(r+1)$ short signals

$$f_n \xrightarrow{\chi'} \begin{cases} \mathbf{f}'_1 = \{f'_{1,t}; \, t = 0 : (N/2 - 1)\} \\ \mathbf{f}'_2 = \{f'_{2,2t}; \, t = 0 : (N/4 - 1)\} \\ \mathbf{f}'_4 = \{f'_{4,4t}; \, t = 0 : (N/8 - 1)\} \\ \dots \qquad \dots \qquad \dots \\ \mathbf{f}'_{N/2} = \{f'_{N/2,0}\} \\ \mathbf{f}'_0 = \{f'_{0,0}\}. \end{cases} \qquad (1.7)$$

Components of these signals are defined by

$$f'_{p,t} = \sum_{np = t \bmod N} f_n - \sum_{np = t + N/2 \bmod N} f_n \qquad (1.8)$$

where $t = 0 : (N/2 - 1)$. The last one-component signal $\{f'_{0,0}\}$ is the power of the signal f_n. The sum of lengths of these short signals equals N and they together represent uniquely the signal f_n.

The N-point DFT is split by disjoint subsets of frequency-points as follows:

$$F_p \rightarrow \begin{cases} \{F_{2k+1}; \, k = 0 : (N/2 - 1)\} \\ \{F_{(2k+1)2 \bmod N}; \, k = 0 : (N/4 - 1)\} \\ \{F_{(2k+1)4 \bmod N}; \, k = 0 : (N/8 - 1)\} \\ \dots \qquad \dots \qquad \dots \\ \{F_{N/2}\} \\ \{F_0\}. \end{cases} \qquad (1.9)$$

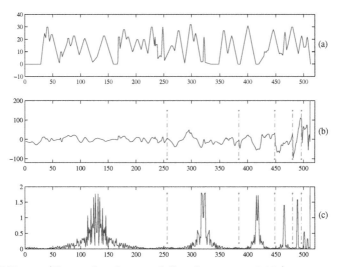

FIGURE 1.4 (See color insert following page 242.)
(a) The signal of length 512, (b) the paired transform of the signal, and (c)
the splitting of the DFT of the signal (shown in absolute scale and cyclicly
shifted to the centers). (The last value of the transform has been truncated.)
These short DFTs together compose the 512-point DFT of the signal.

As an example, Figure 1.4 shows a signal of length 512 in part a, along
with the 512-point paired transform composed by ten splitting-signals (the
first five of which are separated by vertical dot lines) in b, and ten short
DFTs that split the 512-point DFT of the signal in c. These DFTs are of
lengths $256, 128, 64, 32, 16, 8, 4, 2, 1, 1$.

The short signals $\mathbf{f}'_{2^k}$ representing the signal f_n in (1.7) are called *splitting-
signals*. The representation of the signal f_n in the form of $(r+1)$ splitting-
signals $\{\{\mathbf{f}'_{2^k}; k = 0 : (r-1)\}, \mathbf{f}'_0\}$ is called the *paired representation* of the
signal f_n, and χ' is the *paired* transformation [8]. Each splitting-signal defines
components of the Fourier transform of the signal f_n at frequency-points of
the corresponding subset

$$T'_p = \{(2k+1)p; \ k = 0 : (L_p - 1)\}, \tag{1.10}$$

where p is considered from the set $J' = \{1, 2, 4, 8, \ldots, N/2, 0\}$, and $L_p = N/(2p)$, if $p \neq 0$, and $L_0 = 1$. Therefore we denote these signals by $f_{T'_p}$. It
should be noted that the set J' of selected frequencies p can be chosen in
different ways. For instance, we can take $p = 3$ instead of $p = 1$ and consider
$J' = \{3, 2, 4, 8, \ldots, N/2, 0\}$ and $L_3 = N/2$. However, the sets T'_3 and T'_1 are
equal up to a permutation P; therefore, components of the splitting-signal $\mathbf{f}'_3$
can be expressed by $\mathbf{f}'_1$, as $f'_{3,t} = \pm f'_{1,P(t)}$, $t = 0 : (N/2 - 1)$. The set J' is
defined from the condition of partitioning of the set of all frequencies in one
period by subsets T'_p. The cardinality of such a partition equals $(\log_2 N + 1)$

and the chosen set J' with frequencies being powers of two is very convenient for our further calculations.

The following property holds for the paired representation:

$$F_{(2k+1)p \bmod N} = \sum_{t=0}^{L_p-1} (f'_{p,pt} W^t_{2L_p}) W^{kt}_{L_p}, \qquad k = 0 : (L_p - 1), \qquad (1.11)$$

for $p \in J'$. In particular, when $p = 1$, the $N/2$-point DFT over the first splitting-signal $f_{T'_1}$ modified by the vector of twiddle factors $\{1, W, W^2, W^3, ..., W^{N/2-1}\}$,

$$f_{T'_1} = \{f'_{1,t}; t = 0 : (N/2 - 1)\} \rightarrow \{f'_{1,t} W^t; t = 0 : (N/2 - 1)\},$$

coincides with the N-point DFT over signal f_n at frequency-points of the subset $T'_1 = \{1, 3, 5, 7, ..., N-1\}$. When $p = 2$, the $N/4$-point DFT over the second splitting-signal $f_{T'_2}$ modified by twiddle factors $\{1, W^2, W^4, ..., W^{N/2-2}\}$,

$$f_{T'_2} = \{f'_{2,2t}; t = 0 : (N/4 - 1)\} \rightarrow \{f'_{2,2t} W^t_{N/2}; t = 0 : (N/4 - 1)\},$$

coincides with the N-point DFT over signal f_n at frequency-points of the subset $T'_2 = \{2, 6, 10, ..., N - 2\}$, and so on.

Example 1.2

Let $N = 8$, and let $\{f_n\}$ be the following signal $\{1, 4, 2, 3, 5, 7, 6, 8\}$. The set of frequency-points $X_8 = \{0, 1, ..., 7\}$, which we call also *the period*, is covered by the partition

$$\sigma' = \left\{ \begin{array}{l} T'_1 = \{1, 3, 5, 7\} \\ T'_2 = \{2, 6\} \\ T'_4 = \{4\} \\ T'_0 = \{0\} \end{array} \right\}.$$

The splitting-signals $f_{T'_1}$, $f_{T'_2}$, $f_{T'_4}$, and $f_{T'_0}$ are defined by the 8-point paired transformation χ'_8, and their calculation can be written in matrix form as

$$[\chi'_8]\mathbf{f} = \begin{bmatrix} 1 & 0 & 0 & 0 & -1 & 0 & 0 & 0 \\ 0 & 1 & 0 & 0 & 0 & -1 & 0 & 0 \\ 0 & 0 & 1 & 0 & 0 & 0 & -1 & 0 \\ 0 & 0 & 0 & 1 & 0 & 0 & 0 & -1 \\ 1 & 0 & -1 & 0 & 1 & 0 & -1 & 0 \\ 0 & 1 & 0 & -1 & 0 & 1 & 0 & -1 \\ 1 & -1 & 1 & -1 & 1 & -1 & 1 & -1 \\ 1 & 1 & 1 & 1 & 1 & 1 & 1 & 1 \end{bmatrix} \begin{bmatrix} 1 \\ 4 \\ 2 \\ 3 \\ 5 \\ 7 \\ 6 \\ 8 \end{bmatrix} = \begin{bmatrix} \left. \begin{array}{r} -4 \\ -3 \\ -4 \\ -5 \end{array} \right\} = f_{T'_1} \\ \left. \begin{array}{r} -2 \\ 0 \end{array} \right\} = f_{T'_2} \\ \{-8\} = f_{T'_4} \\ \{36\} = f_{T'_0} \end{bmatrix}.$$

Four components of the Fourier transform F_1, F_3, F_5, and F_7 are defined by the splitting-signal $f_{T'_1} = \{-4, -3, -4, -5\}$. These four components together are referred to as *one section* of the DFT, in a sense that this spectral information of the original signal is not carried by the other three splitting-signals. In a similar way, the splitting-signal $f_{T'_2} = \{-2, 0\}$ defines two components F_2 and F_6 which together compose another section of the eight-point DFT. The components F_4 and F_0 are also considered as two last sections of the DFT.

The four-point DFT of the modified splitting-signal $f_{T'_1}$ can be split by the four-point paired transformation χ'_4 in a similar way. The calculation of the eight-point DFT of f_n can thus be written in matrix form as:

$$
\begin{bmatrix} F_7 \\ F_3 \\ F_5 \\ F_1 \\ F_6 \\ F_2 \\ F_4 \\ F_0 \end{bmatrix}
=
\begin{bmatrix} \begin{array}{cc} [\mathcal{F}_2] & \\ & 1 \\ & \ \ 1 \end{array} \end{bmatrix}
\operatorname{diag}\left\{ \begin{array}{c} 1 \\ -j \\ 1 \\ 1 \end{array} \right\} [\chi'_4]
\begin{bmatrix} \begin{array}{cc} [\mathcal{F}_2] & \\ & 1 \\ & \ 1 \end{array} \end{bmatrix}
\operatorname{diag}\left\{ \begin{array}{c} 1 \\ W \\ -j \\ W^3 \\ 1 \\ -j \\ 1 \\ 1 \end{array} \right\} [\chi'_8]
\begin{bmatrix} 1 \\ 4 \\ 2 \\ 3 \\ 5 \\ 7 \\ 6 \\ 8 \end{bmatrix}
$$

where

$$
[\chi'_4] = \begin{bmatrix} 1 & 0 & -1 & 0 \\ 0 & 1 & 0 & -1 \\ 1 & -1 & 1 & -1 \\ 1 & 1 & 1 & 1 \end{bmatrix}, \qquad [\mathcal{F}_2] = [\chi'_2] = \begin{bmatrix} 1 & -1 \\ 1 & 1 \end{bmatrix},
$$

and the twiddle factors $W = \exp(-2\pi j/8) = 0.7071(1 - j)$ and $W^3 = -0.7071(1 + j)$. □

Example 1.3
Let f_n be a signal of length $N = 16$. We consider the paired transformation, $\chi' = \chi'_{16}$, of the signal into five splitting-signals

$$
f \to \{f_{T'_1}, f_{T'_2}, f_{T'_4}, f_{T'_8}, f_{T'_0}\} \tag{1.12}
$$

whose DFTs define the transform of f_n at frequency-points of corresponding subsets T'_p that completely fill the period $X = \{0, 1, 2, \ldots, 15\}$. These subsets are $T'_1 = \{1, 3, 5, 7, 9, 11, 13, 15\}$, $T'_2 = \{2, 6, 10, 14\}$, $T'_4 = \{4, 12\}$, $T'_8 = \{8\}$, $T'_0 = \{0\}$. The corresponding splitting-signals of lengths $8, 4, 2, 1$, and 1 are defined by

$$
\begin{aligned}
f_{T'_1} &= \{f_0 - f_8, f_1 - f_9, f_2 - f_{10}, f_3 - f_{11}, f_4 - f_{12}, f_5 - f_{13}, f_6 - f_{14}, f_7 - f_{15}\} \\
f_{T'_2} &= \{f_0 - f_4 + f_8 - f_{12}, f_1 - f_5 + f_9 - f_{13}, f_2 - f_6 + f_{10} - f_{14}, \\
&\qquad f_3 - f_7 + f_{11} - f_{15}\} \\
f_{T'_4} &= \{f_0 - f_2 + f_4 - f_6 + f_8 - f_{10} + f_{12} - f_{14}, \\
&\qquad f_1 - f_3 + f_5 - f_7 + f_9 - f_{11} + f_{13} - f_{15}\} \\
f_{T'_8} &= \{f_0 - f_1 + f_2 - f_3 + \cdots - f_{13} + f_{14} - f_{15}\} \\
f_{T'_0} &= \{f_0 + f_1 + f_2 + f_3 + \cdots + f_{13} + f_{14} + f_{15}\}.
\end{aligned}
$$

The 16-point DFT of f_n is split by five transforms of orders $8, 4, 2, 1$, and 1, i.e., $\{\mathcal{F}_8, \mathcal{F}_4, \mathcal{F}_2, \mathcal{F}_1, \mathcal{F}_1\}$. Components of splitting-signals of the paired representation in (1.12) are multiplied by twiddle factors $W_{16}^t = \exp(-j2\pi t/16)$ and, then, the DFTs over the splitting-signals are calculated. For instance, at eight frequency-points of the subset T_1', the DFT is calculated by

$$
\begin{bmatrix} F_1 \\ F_3 \\ F_5 \\ F_7 \\ F_9 \\ F_{11} \\ F_{13} \\ F_{15} \end{bmatrix} = [\mathcal{F}_8] \begin{bmatrix} 1 & & & & & & & \\ & W_{16} & & & & & & \\ & & W_{16}^2 & & & & & \\ & & & W_{16}^3 & & & & \\ & & & & -j & & & \\ & & & & & W_{16}^5 & & \\ & & & & & & W_{16}^6 & \\ & & & & & & & W_{16}^7 \end{bmatrix} \left[f_{T_1'} \right]^T,
$$

where $[\mathcal{F}_8]$ is the matrix (8×8) of the eight-point DFT,

$$
[\mathcal{F}_8] = \begin{bmatrix} 1 & 1 & 1 & 1 & 1 & 1 & 1 & 1 \\ 1 & W_8^1 & W_8^2 & W_8^3 & W_8^4 & W_8^5 & W_8^6 & W_8^7 \\ 1 & W_8^2 & W_8^4 & W_8^6 & 1 & W_8^2 & W_8^4 & W_8^6 \\ 1 & W_8^3 & W_8^6 & W_8^1 & W_8^4 & W_8^7 & W_8^2 & W_8^5 \\ 1 & W_8^4 & 1 & W_8^4 & 1 & W_8^4 & 1 & W_8^4 \\ 1 & W_8^5 & W_8^2 & W_8^7 & W_8^4 & W_8^1 & W_8^6 & W_8^3 \\ 1 & W_8^6 & W_8^4 & W_8^2 & 1 & W_8^6 & W_8^4 & W_8^2 \\ 1 & W_8^7 & W_8^6 & W_8^5 & W_8^4 & W_8^3 & W_8^2 & W_8^1 \end{bmatrix}
$$

which can be decomposed by the paired transforms as described in Example 1.2. At frequency-points of other two subsets T_2' and T_4', the DFT is calculated as follows:

$$
\begin{bmatrix} F_2 \\ F_6 \\ F_{10} \\ F_{14} \end{bmatrix} = \begin{bmatrix} 1 & 1 & 1 & 1 \\ 1 & -j & -1 & j \\ 1 & -1 & 1 & -1 \\ 1 & j & -1 & -j \end{bmatrix} \begin{bmatrix} 1 & & & \\ & 0.7071(1-j) & & \\ & & -j & \\ & & & -0.7071(1+j) \end{bmatrix} \left[f_{T_2'} \right]^T
$$

$$
\begin{bmatrix} F_4 \\ F_{12} \end{bmatrix} = \begin{bmatrix} 1 & 1 \\ 1 & -1 \end{bmatrix} \begin{bmatrix} 1 & \\ & -j \end{bmatrix} \left[f_{T_4'} \right]^T.
$$

▯

In the general $N = 2^r$ case, the totality of subsets

$$
\sigma' = \{T_1', T_2', T_4', \ldots, T_{N/2}', T_0'\}
$$

is the partition of $X_N = \{0, 1, 2, \ldots, N-1\}$. The N-point discrete Fourier transformation, $\mathcal{F}_N$, is thus revealed by partition σ' by a set of short transformations, and we write this fact as

$$
\mathcal{F}_N \sim \{\mathcal{F}_{N/2}, \mathcal{F}_{N/4}, \mathcal{F}_{N/8}, \ldots, \mathcal{F}_2, \mathcal{F}_1, 1\}.
$$

In matrix form, the decomposition by the paired transformation χ'_{2^r} can be written as

$$[\mathcal{F}_{2^r}] = \left[\left(\bigoplus_{k=0}^{r-1} [\mathcal{F}_{2^{r-k-1}}] \right) \oplus 1 \right] D[\chi_{2^r}], \qquad (1.13)$$

where $\oplus$ denotes the operation of the Kronecker sum of matrices and the diagonal matrix D equals

$$\text{diag}\{1, W, W^2, ..., W^{2^r-1}, 1, W^2, W^4, ..., W^{2^r-2}, 1, W^4, W^8, ..., W^{2^r-4}, 1, ..., 1\}.$$

Each short transformation in this splitting can be split similarly by the $N/2^k$-point paired transform, where $1 \le k \le r-2$, into a set of shorter transformations. The full splitting of the Fourier transform by paired transforms leads to the known paired algorithm of the DFT, which requires $2^{r-1}(r-3)+2$ operations of multiplications [13, 8].

1.3   Fast Fourier transform

In this section, we continue the presentation of effective calculation of the fast Fourier transform, which is based on the simplification of the signal-flow graph of calculation of the transform by the paired transform. The paired transform splits the DFT into a minimum set of short transforms, and the algorithm of calculation of the DFT by the paired transform uses a minimum number of multiplications by twiddle factors. The question arises how to define the exact minimum number of real multiplications by maximum simplifying of the flow graph of the algorithm. This question also applies to many other algorithms of the DFT. Instead of finding new effective formulas for calculation of transform coefficients, we will work directly on the flow graph of the transform. In many cases of order N of transform, the simplification of the signal-flow graph can be done easily. We here consider in detail the $N = 8$ and 16 cases.

We refer to the paired algorithm of the DFT, but we believe that signal-flow graphs of other fast algorithms, such as the fractional DFT [15], split-radix, vector split-radix, mixed radix [16, 17], can be considered and modified in a similar way. The advantage of using the paired transform is in the fact that this transform reveals completely the mathematical structure of many other unitary transforms, such as cosine, Hartley, and Hadamard transforms, and requires a minimum number of operations [13, 10, 11, 9]. Therefore the method described here for the DFT can be applied for fast calculation of these transforms, too. As an example, Figure 1.5 shows the diagram of splitting a 16-point discrete unitary transform (DUT) by the paired transform χ'_{16}. The calculation of the 16-point DUT is reduced to calculation of the 8-, 4-, 2-, and 1-point DUTs. When weighted coefficients w_k for the output of the paired transform are equal to $\exp(-j2\pi k/N)$, $k = 1 : 7$, and DUT is the discrete Fourier transform (DFT), then the diagram describes the algorithm of calculation of the 16-point DFT. When all coefficients $w_k \equiv 1$ and DUT is considered to be the discrete Hadamard transform (DHdT), then we obtain the diagram of calculation of the 16-point DHdT.

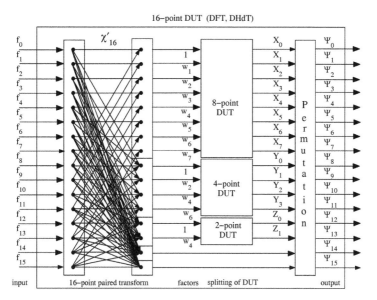

FIGURE 1.5
Diagram of calculation of the 16-point discrete unitary transform (DUT) by the paired transforms and 8-, 4-, and 2-point DUTs. The permutation is calculated by $(2k + 1) \mod 16 \leftarrow k$, $k = 0 : 15$.

The paired transform is fast and requires $(2N - 2)$ operations of addition/subtraction. The arithmetical complexity of the paired algorithm for the N-point DFT, when N is a power of 2, is calculated by $\alpha(N) = N/2(r+9) - r^2 - 3r - 6$ and $m(N) = N/2(r - 3) + 2$ operations, where functions $\alpha(N)$ and $m(N)$ stand respectively for number of additions and multiplications by nontrivial twiddle factors. The detail description of the paired transform-based algorithms for calculation of the discrete Fourier and Hadamard transforms, as well as examples of MATLAB-based programs for calculation of these and paired transforms, can be found in [13, 9]. New versions of these codes are given in §1.4.

1.3.1 Unitary paired transform

We here analyze the concept of paired functions that define the unitary paired transformation. The complete set of paired functions are frequency-time type wavelets [8]. The system of paired functions is numbered by two parameters, namely, one parameter for the frequency and one parameter for the time. The change in time determines the series of functions, and the total number of pairs numbering the system of functions, if such is complete, has to be equal to N. Here we consider the case only when N is a power of two, although the paired functions and their complete sets can be constructed for the general $N = L^r$ case, when $L \geq 2$ and $r > 1$.

The splitting of the signal is performed by the N-point discrete paired transformation χ'_N whose basis functions $\chi'_{p,t}(n)$ are defined by

$$\chi'_{p,t}(n) = \begin{cases} 1, & \text{if } np = t \bmod N; \\ -1, & \text{if } np = t + N/2 \bmod N; \\ 0, & \text{otherwise,} \end{cases} \qquad n = 0 : (N-1), \qquad (1.14)$$

if $p > 0$, and $\chi'_{0,0}(n) \equiv 1$.

In the paired representation, the sequence f_n is considered as a set of $(r+1)$ splitting-signals whose components are defined by

$$f'_{p,t} = \chi'_{p,t} \circ f = \left[\sum_{np=t \bmod N} f_n \right] - \left[\sum_{np=(t+N/2) \bmod N} f_n \right]. \qquad (1.15)$$

The complete system of paired functions is composed as

$$\left\{ \{\chi'_{2^k, 2^k t};\ t = 0 : (2^{r-k-1} - 1),\ k = 0 : (r-1)\}, 1 \right\} \qquad (1.16)$$

which defines the unitary paired transformation χ'. It should be noted that the basic paired functions can be defined by extreme and zero values of certain cosine waves, when they run through the interval $[0, N-1]$ with different frequencies $\omega_k = (2\pi/N)2^k$, where $k = 0 : (r-1)$. Indeed, we can write that

$$\chi'_{2^k, 2^k t}(n) = Q\left[\cos\left(\frac{2\pi(n-t)}{2^{r-k}}\right)\right], \qquad (\chi'_{0,0} \equiv 1), \qquad (1.17)$$

where $t = 0 : (2^{r-k-1} - 1)$, $k = 0 : (r-1)$, and $Q[x]$ denotes the following quantization function of the interval $[-1, 1] : Q[x] = x$, if $|x| = 1$, and $Q[x] = 0$, otherwise.

As an example, Figure 1.6 illustrates the process of composition of these functions from the corresponding cosine waves defined in the interval $[0, 7]$. For the $N = 16$ case, Figure 1.7 shows the system of the sixteen cosine waves defined in the interval $[0, 15]$. The first series consists of eight shifted versions of the cosine functions with frequency $\omega_0 = \pi/8$. The next series consists of four shifted versions of the cosine function with frequency $2\omega_0$. There are also

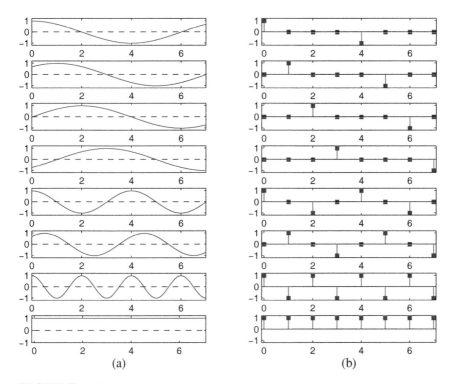

FIGURE 1.6

(a) Cosine waves and (b) discrete paired functions for $N = 8$.

two shifted cosine waves with frequencies $4\omega_0$, one cosine function with frequency $8\omega_0$, and one constant function. Extremum values of all these sixteen waves define exactly the matrix of the 16-point discrete paired transformation. The image-matrix of the 16-point paired transformation is also shown in this figure. Image elements of white, gray, and dark intensities correspond respectively to vales $0, -1$, and 1 of coefficients of the matrix.

One can notice that the last eight waves of the system represent the complete system of waves defined for the $N = 8$ case (shown in Figure 1.6). Namely, they coincide in the time interval $[0, 7]$ and then periodically extend in the rest of the interval. The first eight waves of the system $N = 16$ carry one pike, or impulse, which is moving from the left to right, and at point $t = 8$ it changes the sign and returns back to the first wave. The complete system of waves defined for the $N = 8$ is composed from the system of waves defined for the $N = 4$ and then for $N = 2$ cases in a similar way. In matrix form, the described composition of the complete system of functions from its

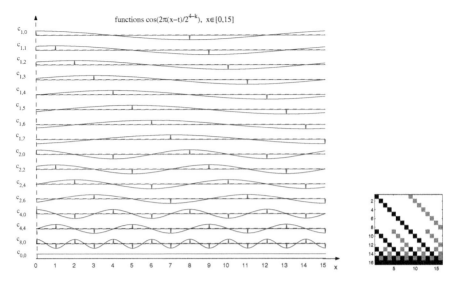

FIGURE 1.7
Sixteen cosine waves that define the complete system of paired functions of the 16-point discrete paired transformation (with matrix shown on the right).

subsystems can be written as

$$[\chi'_{16}] = \begin{bmatrix} I_8 & -I_8 \\ [\chi'_8] & [\chi'_8] \end{bmatrix} = \begin{bmatrix} I_8 & & -I_8 \\ I_4 & -I_4 & I_4 & -I_4 \\ [\chi'_4] & [\chi'_4] & [\chi'_4] & [\chi'_4] \end{bmatrix} =$$

$$= \begin{bmatrix} I_8 & & & & -I_8 \\ I_4 & & -I_4 & I_4 & & -I_4 \\ I_2 & -I_2 & I_2 & -I_2 & I_2 & -I_2 & I_2 & -I_2 \\ [\chi'_2] & [\chi'_2] & [\chi'_2] & [\chi'_2] & [\chi'_2] & [\chi'_2] & [\chi'_2] & [\chi'_2] \end{bmatrix},$$

where I_M denotes the identity matrix $(M \times M)$. In the general $N = 2^r$ case, where $r > 1$, the system of complete paired functions is generated sequentially from the systems of small orders $N/2, N/4, ...,$ which can be expressed by

$$[\chi'_N] = \begin{bmatrix} I_{N/2} & -I_{N/2} \\ [\chi'_{N/2}] & [\chi'_{N/2}] \end{bmatrix} = \begin{bmatrix} I_{N/2} & & -I_{N/2} \\ I_{N/4} & -I_{N/4} & I_{N/4} & -I_{N/4} \\ [\chi'_{N/4}] & [\chi'_{N/4}] & [\chi'_{N/4}] & [\chi'_{N/4}] \end{bmatrix} = \dots.$$

Example 1.4
We consider the $N = 4$ case. The set $X = \{0, 1, 2, 3\}$ is covered by the partition $\sigma' = (T'_1, T'_2, T'_0)$ with subsets $T'_1 = \{1, 3\}$, $T'_2 = \{2\}$, and $T'_0 = \{0\}$. For the generators $p = 1, 2,$ and 0 of the subsets T'_p, the following matrix

(4×3) with values of $t = (np) \bmod 4$, when $n = 0 : 3$, is composed:

$$\|t\|_{n=0:3,p=1,2,0} = \begin{Vmatrix} 0 & 1 & 2 & 3 \\ 0 & 2 & 0 & 2 \\ 0 & 0 & 0 & 0 \end{Vmatrix}.$$

A length-4 sequence f_n can thus be represented as three short sequences $f_{T_1'} = \{f_{1,0}', \ f_{1,1}'\} = \{f_0 - f_2, f_1 - f_3\}$, $f_{T_2'} = \{f_{2,0}'\} = \{f_0 - f_1 + f_2 - f_3\}$, and $f_{T_0'} = \{f_{0,0}'\} = \{f_0 + f_1 + f_2 + f_3\}$. This representation is performed by the paired transformation χ_4' with the matrix

$$[\chi_4'] = \begin{bmatrix} [\chi_{1,0}'] \\ [\chi_{1,1}'] \\ [\chi_{2,0}'] \\ [\chi_{0,0}'] \end{bmatrix} = \begin{bmatrix} 1 & 0 & -1 & 0 \\ 0 & 1 & 0 & -1 \\ 1 & -1 & 1 & -1 \\ 1 & 1 & 1 & 1 \end{bmatrix}.$$

The decomposition of the 4-point DFT by the paired transform into the 2- and 1-point DFTs can be written in matrix form as

$$[\mathcal{F}_4] = ([\mathcal{F}_2] \oplus 1 \oplus 1) \operatorname{diag}\{1, -j, 1, 1\}[\chi_4'].$$

▯

We now describe the paired method of fast calculation of the eight-point DFT and analyze its signal-flow graph, which will be used for effective calculation of the reversible integer DFT. For that, we start with the $N = 8$ case and generalize the result of Example 1.2.

1.3.2 Fast 8-point DFT

Let f_n be a sequence (signal) of length 8. The set $X = \{0, 1, ..., 7\}$ is covered by the following partition of subsets $\sigma' = (T_1', T_2', T_4', T_0')$, where subsets $T_1' = \{1, 3, 5, 7\}$, $T_2' = \{2, 6\}$, $T_4' = \{4\}$, and $T_0' = \{0\}$. The partition of X by subsets T_p' is unique. For the generators of these subsets T_p', the following 8×4 matrix is composed:

$$\|t = (np) \bmod 8\|_{\substack{n = 0 : 7 \\ p = 1, 2, 4, 0}} = \begin{Vmatrix} 0 & 1 & 2 & 3 & 4 & 5 & 6 & 7 \\ 0 & 2 & 4 & 6 & 0 & 2 & 4 & 6 \\ 0 & 4 & 0 & 4 & 0 & 4 & 0 & 4 \\ 0 & 0 & 0 & 0 & 0 & 0 & 0 & 0 \end{Vmatrix}.$$

Elements of this matrix show a way of calculating the matrix of the eight-point discrete paired transform. Indeed, it follows from this matrix that the signal f_n is represented in paired form as the following four splitting-signals:

$$f_{T_1'} = \{f_{1,0}', \ f_{1,1}', \ f_{1,2}', \ f_{1,3}'\} = \{f_0 - f_4, f_1 - f_5, f_2 - f_6, f_3 - f_7\}$$
$$f_{T_2'} = \{f_{2,0}', f_{2,2}'\} = \{f_0 - f_2 + f_4 - f_6, f_1 - f_3 + f_5 - f_7\}$$
$$f_{T_4'} = \{f_{4,0}'\} = \{f_0 - f_1 + f_2 - f_3 + f_4 - f_5 + f_6 - f_7\}$$
$$f_{T_0'} = \{f_{0,0}'\} = \{f_0 + f_1 + f_2 + f_3 + f_4 + f_5 + f_6 + f_7\}.$$

All components of these four signals are calculated by the paired transformation χ'_8 with the matrix

$$
[\chi'_8] =
\begin{bmatrix}
[\chi'_{1,0}] \\
[\chi'_{1,1}] \\
[\chi'_{1,2}] \\
[\chi'_{1,3}] \\
[\chi'_{2,0}] \\
[\chi'_{2,2}] \\
[\chi'_{4,0}] \\
[\chi'_{0,0}]
\end{bmatrix}
=
\begin{bmatrix}
1 & 0 & 0 & 0 & -1 & 0 & 0 & 0 \\
0 & 1 & 0 & 0 & 0 & -1 & 0 & 0 \\
0 & 0 & 1 & 0 & 0 & 0 & -1 & 0 \\
0 & 0 & 0 & 1 & 0 & 0 & 0 & -1 \\
1 & 0 & -1 & 0 & 1 & 0 & -1 & 0 \\
0 & 1 & 0 & -1 & 0 & 1 & 0 & -1 \\
1 & -1 & 1 & -1 & 1 & -1 & 1 & -1 \\
1 & 1 & 1 & 1 & 1 & 1 & 1 & 1
\end{bmatrix}.
$$

Each splitting-signal carries the spectral information at frequency-points of the corresponding subset T'_p. Thus, for $p = 1$

$$
\sum_{n=0}^{3} \left(f'_{1,t} W_8^t \right) W_4^{kt} = F_{2k+1}, \qquad k = 0, 1, 2, 3,
$$

for $p = 2$

$$
f'_{2,0} + \left(f'_{2,2} W_4 \right)(-1)^k = F_{4k+2}, \qquad k = 0, 1,
$$

and for $p = 4$ and 0, we obtain respectively $f'_{4,0} = F_4$ and $f'_{0,0} = F_0$.

The decomposition of the 8-point DFT by the paired transform can thus be written in matrix form as

$$
[\mathcal{F}_8] = ([\mathcal{F}_4] \oplus [\mathcal{F}_2] \oplus 1 \oplus 1) D_8 [\chi'_8],
$$

where D_8 is the diagonal matrix with coefficients $\{1, W^1, -j, W^3, 1, -j, 1, 1\}$ and $W = \exp(-j\pi/4)$. The signal-flow graph for calculation of the eight-point DFT by the paired transforms is shown in Figure 1.8 and its block-diagram in Figure 1.9. The matrix of the 2-point paired transformation equals $[\chi'_2] = [1 \ -1, 1 \ \ 1]$.

We now consider the signal-flow graph of Figure 1.9 for the case when the input sequence f_n is real. There are two operations of multiplication by non-trivial complex factors $W = (1 - j)a$ and $W^3 = (-1 - j)a$, where $a = \sqrt{2}/2$, to be used over the output of the 8-point paired transform. Three trivial multiplications by $-j$ are also used in calculation. All inputs and outputs of the paired transforms χ'_8, χ'_4, and χ'_2 as well as multiplications by complex factors can be split into two parts as shown in Figure 1.10. The new signal-flow graph can then be simplified. One can see that calculations for real R_p and imaginary I_p parts of transform coefficients F_p have been separated in a symmetric way. This figure shows also values of all inputs and outputs for each transform in the block-scheme, when the sequence $\{f_n\}$ equals $\{1, 2, 4, 4, 3, 7, 5, 8\}$.

Using the property of complex conjugacy of Fourier coefficients for a real input, $F_{N-p} = \bar{F}_p$ for $p = 1 : (N/2 - 1)$, we obtain $R_{N-p} = R_p$, $I_{N-p} = -I_p$,

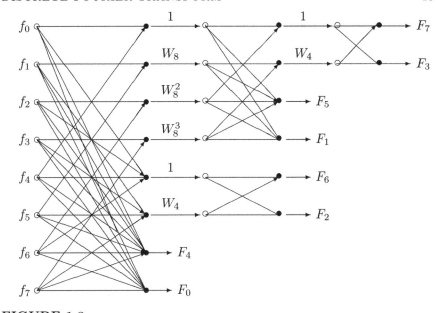

FIGURE 1.8
The signal-flow graph for computing the eight-point DFT.

$p = 1 : (N/2 - 1)$, and $I_N = I_{N/2} = 0$. Therefore, the signal-flow graph of Figure 1.10 can be reduced to the signal-flow graph shown in Figure 1.11.

We denote by $\chi'_{4;in}$ two incomplete four-point paired transforms for which only the last two outputs are calculated. Since one of the inputs for both the incomplete four-point paired transforms is zero, they can be considered as 3-to-2-point transforms with the following matrices

$$[\chi'_{4;in}] = \begin{bmatrix} 1 & -1 & -1 \\ 1 & 1 & 1 \end{bmatrix}, \quad [\chi'_{4;in}] = \begin{bmatrix} -1 & 1 & -1 \\ 1 & 1 & 1 \end{bmatrix}.$$

The calculation of the eight-point DFT by the signal-flow graph of Figure 1.11 requires two real multiplications by factor $a = \sqrt{2}/2$ and $14 + 2 \times 3 = 20$ additions. Indeed, the fast 2^n-point paired transform uses $2^{n+1} - 2$ additions [8]. The eight-point paired transform requires 14 additions, and each of the four-point incomplete paired transforms uses 3 additions.

1.3.3   Fast 16-point DFT

The 16-point paired transform is defined by the partition $\sigma' = (T'_1, T'_2, T'_4, T'_8, T'_0)$ of the fundamental period $X = \{0, 1, 2, 3, \ldots, 15\}$. Let f_n be a sequence of length 16, which is split in the paired representation by five signals

$$f \xrightarrow{\chi'_{16}} \left\{ f'_{T'_1}, f'_{T'_2}, f'_{T'_4}, f'_{T'_8}, f'_{T'_0} \right\}.$$

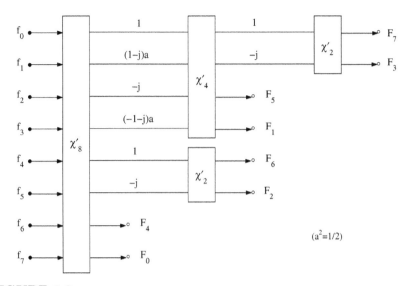

FIGURE 1.9
Block-scheme of calculation of the 8-point DFT by paired transforms.

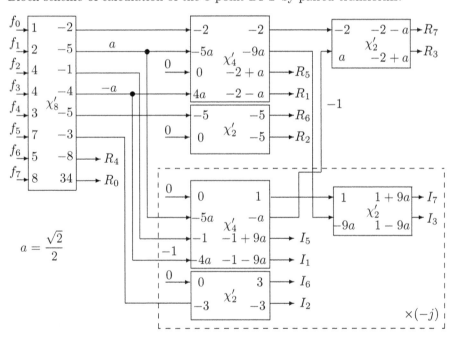

FIGURE 1.10
Block-scheme of calculation of real and imaginary parts of the 8-point DFT.

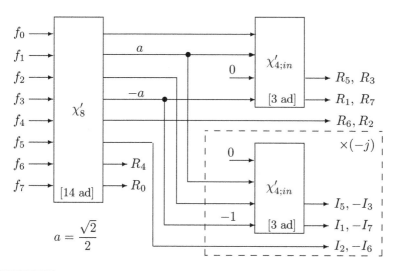

FIGURE 1.11
Simplified block-scheme of calculation of real and imaginary parts of the 8-point DFT.

The signal-flow graph for calculation of the 16-point DFT by paired transforms is given in Figure 1.12. Ten multiplications by non-trivial twiddle factors W_{16}^k, $k = 1, 2, 3, 5, 6, 7$, plus seven trivial multiplications by $-j$ are used in the calculation. For the case when the sequence f_n is real, for calculation of the 16-point DFT, we can use only a part of this signal-flow graph, as shown in Figure 1.13. On the second stage of calculations, three incomplete paired transforms of orders 8, 4, and 2 are used, which reduces the number of operations of addition by $(4 + 2 + 1) + (6 + 2) + 2 = 17$ and multiplication by 2.

Similar to the $N = 8$ case described above, for a real input f_n we can redraw the signal-flow graph of the 16-point DFT by separating the calculation for the real and imaginary parts of Fourier coefficients F_p, $p = 0 : 15$ [12]. After such a separation the signal-flow graph can be simplified, because of the following relations between Fourier coefficients for a real input, $R_{16-p} = R_p$, $I_{16-p} = -I_p$, when $p = 1 : 7$. The simplified signal-flow graph for calculating the 16-point DFT is shown in Figure 1.14.

The calculation of the 16-point DFT by the simplified signal-flow graph requires $m'(16) = 12$ real multiplications by factors $a = \cos(\pi/4)$, $b = \cos(\pi/8)$, and $c = \cos(3\pi/8)$. We denote by $\chi'_{8;in}$ two incomplete 8-point paired transforms with one zero input and for which only the last four outputs are calculated. These transforms over an input $\mathbf{x} = (x_0, x_1, ..., x_7)'$ are described in

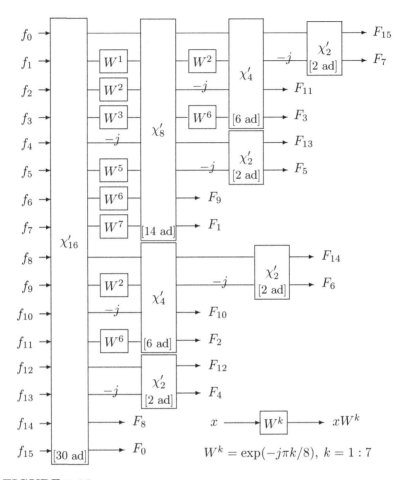

FIGURE 1.12
Block-scheme of calculation of the 16-point DFT by paired transforms.

matrix form as

$$[\chi'_{8;in} \circ \mathbf{x}] = \begin{bmatrix} 1 & 0 & -1 & 0 & 0 & -1 & 0 \\ 0 & 1 & 0 & -1 & 1 & 0 & -1 \\ 1 & -1 & 1 & -1 & -1 & 1 & -1 \\ 1 & 1 & 1 & 1 & 1 & 1 & 1 \end{bmatrix} \begin{bmatrix} x_0 \\ x_1 \\ x_2 \\ x_3 \\ x_5 \\ x_6 \\ x_7 \end{bmatrix}$$

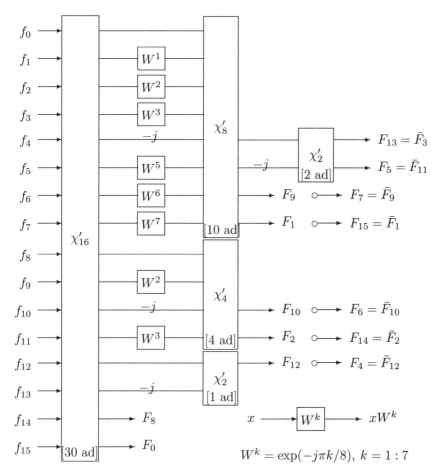

FIGURE 1.13
Block-scheme of the 16-point DFT of the real data.

and

$$[\chi'_{8;in} \circ \mathbf{x}] = \begin{bmatrix} 0 & -1 & 0 & 1 & 0 & -1 & 0 \\ 1 & 0 & -1 & 0 & 1 & 0 & -1 \\ -1 & 1 & -1 & 1 & -1 & 1 & -1 \\ 1 & 1 & 1 & 1 & 1 & 1 & 1 \end{bmatrix} \begin{bmatrix} x_1 \\ x_2 \\ x_3 \\ x_4 \\ x_5 \\ x_6 \\ x_7 \end{bmatrix}.$$

Two incomplete 8-point paired transforms require 9 additions each. Two incomplete 4-point paired transforms $\chi'_{4;in}$ are used for calculation of the last two outputs. The incomplete 4-point paired transforms require 3 additions

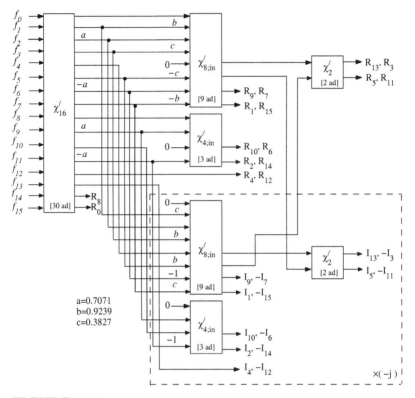

FIGURE 1.14

Block-scheme of calculation of the 16-point DFT of a real input f_n, $n = 0 : 15$.

each. The total number of the required additions is thus calculated as

$$\alpha'(16) = \alpha(\chi'_{16}) + 2\alpha(\chi'_{8;in}) + 2\alpha(\chi'_{4;in}) = 30 + 2 \times 9 + 2 \times 3 + 2 \times 2 = 58.$$

The proposed calculation of the N-point DFT by the simplified flow graph can be used for real and imaginary inputs separately. Therefore the number of operations of multiplication is counted as twice those estimates derived for real inputs. The number of additions is counted as twice those estimates derived for real inputs, plus extra additions are needed to combine the first $(N - 2)$ DFT outputs produced from real and imaginary inputs. For instance, for the 16-point DFT of complex data, the number of additions equals $2(58)+2(14) = 144$. For the 8-point DFT of complex data, the number of additions equals $2(20) + 2(6) = 52$. The same estimates were also reported in [19, 33].

Table 1.1 shows the estimates for numbers of multiplication and addition that have been received in radix-2 algorithms with 1, 2, 3, and 5 butterflies [7],[16]-[18]. The data are given for a complex input f_n. It is assumed for these estimates that the complex multiplication by a non-trivial twiddle factor in radix-2 algorithms is performed with two additions and four multiplications. One can see that the paired algorithm is the best, by operations of multiplication and addition.

TABLE 1.1
Number of multiplications/additions for calculating the 8- and 16-point FFTs by the radix-2 by 1,2,3,5 butterflies and paired algorithm.

N	$m_{2\mid1}$	$a_{2\mid1}$	$m_{2\mid2}$	$a_{2\mid2}$	$m_{2\mid3}$	$a_{2\mid3}$	$m_{2\mid5}$	$a_{2\mid5}$	m_p'	α_p'
8	48	72	20	58	8	52	4	52	4	52
16	128	192	68	162	40	148	28	148	24	144

Since the real and imaginary parts of the Fourier transform are calculated separately when using the simplified block-diagram of Figure 1.14, this block-diagram can be used for the calculation of the 16-point discrete Hartley transform (DHT)

$$H_p = \sum_{n=0}^{15} f_n[\cos\left(\frac{2\pi np}{N}\right) + \sin\left(\frac{2\pi np}{N}\right)] = \text{Real}[F_p] - \text{Imag}[F_p], \ p = 0:15.$$

For that, we need to remove the multiplication by $(-j)$ of imaginary parts. The arithmetical complexity of this algorithm applied for both DFT and DHT is the same. In the same way, the simplified block-diagram of Figure 1.11 can be used for calculating the eight-point DHT.

1.4 Codes for the paired FFT

Below are simple examples of MATLAB-based codes for computing the discrete Fourier transform by the paired transform. The code can be optimized, but it is given in a simple and recursive form for better understanding. The paired transform is also written recursively.

```
% ------------------------------------------------------------------
%  demo_pfft.m file of programs (library of codes of Grigoryans)
%  List of codes for processing signals of length N=2^r, r>1:
%      1-D fast direct paired transform    -  'fastpaired_1d.m'
%      1-D paired fast Fourier transform   -  'paired_1dfft.m'
%      permutation of the output           -  'fastpermut.m'
%
%  1. Read and then plot the input signal of length N=512
     fid=fopen('Boli.sig','rb');
```

```
      o=fread(fid,'float');  fclose(fid); clear fid; o=o';
      N=length(o)-1;
      signal_test=o(1:N);
      figure;
      subplot(2,2,1); plot(signal_test);
      axis([-5,N+8,-5,35]);
      h_a=xlabel('(a)');
      set(h_a,'FontName','Times','FontSize',12);
%     2. Calculate the paired transform of the input signal
      paired_transform=fastpaired_1d(signal_test);
      subplot(2,2,2); plot(paired_transform,'r');
      axis([-5,N+8,-100,400]);
      h_b=xlabel('(b)');
      set(h_b,'FontName','Times','FontSize',12);
%     3. Calculate the Fourier transform of the input signal
      MatlFFT = fft(signal_test,N);   % MATLAB code is used
      PairFFT = paired_1dfft(signal_test);
      % 8358.9 at zero, i.e. PairFFT(N)=8358.9
      subplot(2,2,3);
      plot(abs(PairFFT),'Color',[0 .5 1]);
      axis([-5,N+8,-100,5E3]);
      h_c=xlabel('(c)');
      set(h_c,'FontName','Times','FontSize',12);
%     4. Reordering and shift of the DFT (if needed)
      RR1=permut(N);
      PPairFFT=zeros(1,N);
      PPairFFT=[PairFFT(N) PairFFT(RR1(1:N-1))];
      % PPairFFT=fastpermut(PairFFT);  % not working ???
      subplot(2,2,4);
      plot([fftshift(abs(PPairFFT(:))) fftshift(abs(MatlFFT(:)))]);
      axis([-5,N+8,-100,5E3]);
      h_d=xlabel('(d)');
      set(h_d,'FontName','Times','FontSize',12);
%     print -dpsc demo_pairedfft.ps
% ----------------------------------------------------------------
%     call: paired_1dfft.m
      function h=paired_1dfft(x)
      N=length(x);
%     1. Calculation of the exponential coefficients. The calculations
%     can be performed outside the body of this function. For instance,
%     all coefficients wn can be considered as static variables in the code.
%     Note, if all wn=1, then this code results in the Hadamard transform.
      m=bitshift(N,-1);
      t=0:m-1;
      wn=exp(-(pi*j/m)*t);
      w=[];
      while m>1
         w=[w wn];
         wn=wn(1:2:end);
```

```
         m=bitshift(m,-1);
     end
     w=[w 1 1];
%    2. Calculation of the short DFTs of the modified splitting signals
     if N==1
         h=x;
     else
         z=fastpaired_1d(x);
         z=z.*w;
         nn=1;
         nk0=bitshift(N,-1);
         nk=nk0;
         p=[];
         while nk0>1
             t=nn:nk;
             y=z(t);
             nk0=bitshift(nk0,-1);
             p=[p paired_1dfft(y)];
             nn=nk+1;
             nk=nk+nk0;
         end
         h=[p -z(end-1) z(end)];
     end;

%    call: fastpaired_1d.m
     function y=fastpaired_1d(x)
     N=length(x);
     if N==1
          y=x;
     else
         N2=bitshift(N,-1);
         x1=x(1:N2);
         x2=x(N2+1:N);
         y1=x1+x2;
         y2=x1-x2;
         y=[y2 fastpaired_1d(y1)];
     end;

%    call: fastpermut.m
%    Code is written by UTSA MS student, Elias Gonzales (class EE 6363)
%    to substitute the complex and non elegant code that was used in [9,13]
     function output=fastpermut(input)
     N=length(input);
     m=log2(N);
     output=input;
     for a=1:N
         output(mod(binvec2dec(fliplr(dec2binvec(a-1,m)))+1,N)+1)=input(a);
     end
%    -------------------------------------------------------------------------
```

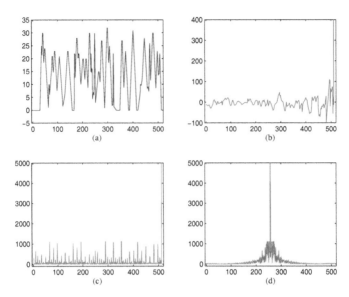

FIGURE 1.15
(a) The signal, (b) the paired transform of the signal, (c) 512-point paired
DFT (in absolute scale), and (d) the DFT with the permutation and shifted
to the middle.

Figure 1.15 shows the result of the above program "demo_pfft.m". The
original signal of length 512 is shown in part a, along with the paired transform
in b, the amplitude of the 512-point DFT of the signal in part c, and the
shifted spectrums of the DFT calculated by this program together with the
DFT calculated by using MATLAB code "fft" in d, for comparison.

1.5   Paired and Haar transforms

In this section, the Haar and paired transformations are analyzed and relations between them are described. The discrete Haar transformation [21] is
considered as the particular case of the paired transformations, namely, the
2-paired transformation, as well as a threshold version of a cosine transformation. Fast algorithms for calculating the 16-, 8-, and 4-point Haar transforms
by paired transforms are described in detail.

The Haar transformation is used in speech processing and communication.
This transform is the first fast transform developed after the Fourier transformation. The Haar transformation used to be considered a transformation
that not only differs much from the Fourier transformation, but that does

not relate to the Fourier transform. Indeed, the basic functions of the Fourier transformation are continuous trigonometric cosine and sine functions, and the performance of the transformation of large order requires multiplications by many irrational numbers. Nonnormalized basic functions of the Haar transformation take values of ± 1 and 0, and the computation of the transform requires no operations of multiplication. The values of basic functions of the discrete Haar transformation (DHT) are zero at many points, the matrix of the transformation is sparse, and that makes the transformation very simple in calculation.

1.5.1 Haar functions

The complete system of the Haar transformation is composed by series of shifted functions. Inside each series, the functions represent themselves the joint, identical, but different by sign impulses running on the unit interval $[0, 1)$ by a discrete interval of time. The functions are the precise and shifted copies of each other. The resemblance property of the functions makes the Haar system of functions popular in wavelets [22]-[26]. The Haar system of functions is the first system for which the concepts of frequency and time have been connected together into one parameter (number) of functions.

The Haar basic functions, $h_m(t)$, are defined on the interval $[0, 1)$. For $N = 2^r$, the basic function with number $m = 0 : (N - 1)$ is defined as follows:

$$h_0(t) = 1$$

$$h_m(t) = \begin{cases} 2^{l/2}, \text{ if } \dfrac{k-1}{2^l} \leq t < \dfrac{k-0.5}{2^l}, \\ -2^{l/2}, \text{ if } \dfrac{k-0.5}{2^l} \leq t < \dfrac{k}{2^l}, \\ 0, \text{ for all other } t \in [0, 1), \end{cases}$$

where $m = 2^l + k - 1$, $0 \leq l \leq r - 1$, and $0 \leq k \leq 2^l$. Figure 1.16 shows the basic functions of the 2-, 4,- and 8-point discrete Haar transformations. One can note that the basic functions of the $N/2$-point Haar transformation are the first $N/2$ basic functions of the N-point Haar transformation.

The basic functions, $h_m(n)$, of the discrete N-point Haar transformation are defined as the sampled version of the Haar functions, namely, $h_m(n) = h_m(nT)$, $n = 0 : (N - 1)$, $T = 1/N$.

Example 1.5

Let $N = 8, 4$, and 2, then the following matrices correspond to the N-point

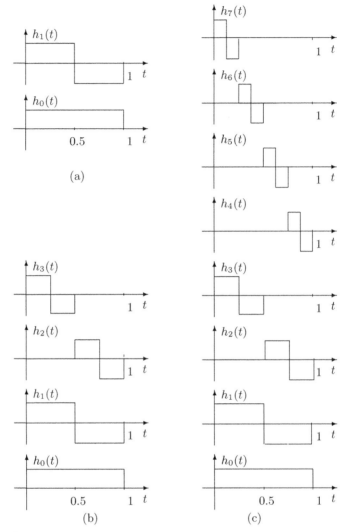

FIGURE 1.16
Nonnormalized basic functions of the 2-, 4-, and 8-point Haar transformations.

Haar transformations (up to the normalized coefficients):

$$[H_8] = \begin{bmatrix} [h_0] \\ [h_1] \\ [h_2] \\ [h_3] \\ [h_4] \\ [h_5] \\ [h_6] \\ [h_7] \end{bmatrix} = \begin{bmatrix} 1 & 1 & 1 & 1 & 1 & 1 & 1 & 1 \\ 1 & 1 & 1 & 1 & -1 & -1 & -1 & -1 \\ \sqrt{2} & \sqrt{2} & -\sqrt{2} & -\sqrt{2} & 0 & 0 & 0 & 0 \\ 0 & 0 & 0 & 0 & \sqrt{2} & \sqrt{2} & -\sqrt{2} & -\sqrt{2} \\ 2 & -2 & 0 & 0 & 0 & 0 & 0 & 0 \\ 0 & 0 & 2 & -2 & 0 & 0 & 0 & 0 \\ 0 & 0 & 0 & 0 & 2 & -2 & 0 & 0 \\ 0 & 0 & 0 & 0 & 0 & 0 & 2 & -2 \end{bmatrix}$$

$$[H_4] = \begin{bmatrix} [h_0] \\ [h_1] \\ [h_2] \\ [h_3] \end{bmatrix} = \begin{bmatrix} 1 & 1 & 1 & 1 \\ 1 & 1 & -1 & -1 \\ \sqrt{2} & -\sqrt{2} & 0 & 0 \\ 0 & 0 & \sqrt{2} & -\sqrt{2} \end{bmatrix}, \quad [H_2] = \begin{bmatrix} [h_0] \\ [h_1] \end{bmatrix} = \begin{bmatrix} 1 & -1 \\ 1 & 1 \end{bmatrix}.$$

▯

We consider the decomposition of the signal by the Haar transformation, for the $N = 8$ case, when the signal is $\mathbf{f} = (1, 3, 2, 6, 7, 5, 4, 2)'$. On each stage of the decomposition, the averaging and differencing process is performed as follows. *Step 1:*

$$\left(\frac{1+3}{2}, \frac{2+6}{2}, \frac{7+5}{2}, \frac{4+2}{2}, \frac{1-3}{2}, \frac{2-6}{2}, \frac{7-5}{2}, \frac{4-2}{2} \right)' = (2, 4, 6, 3, -1, -2, 1, 1)'.$$

Step 2: Calculation over the first part of the data:

$$\left(\frac{2+4}{2}, \frac{6+3}{2}, \frac{2-4}{2}, \frac{6-3}{2}, -1, -2, 1, 1 \right)' = (3, 4.5, -1, 1.5, -1, -2, 1, 1)'.$$

Step 3: Calculation over the first quarter of the data:

$$\left(\frac{3+4.5}{2}, \frac{3-4.5}{2}, -1, 1.5, -1, -2, 1, 1 \right)' = (3.75, -0.75, -1, 1.5, -1, -2, 1, 1)'.$$

The obtained signal $(3.75, -0.75, -1, 1.5, -1, -2, 1, 1)'$ is the Haar transform of $\mathbf{f}$.

In matrix form, the above described calculation of the Haar transform is described by three sparse matrices which are calculated as follows.
Step 1: (The first matrix of decomposition, T_1)

$$\begin{bmatrix} \frac{1}{2} & \frac{1}{2} & & & & & & \\ & & \frac{1}{2} & \frac{1}{2} & & & & \\ & & & & \frac{1}{2} & \frac{1}{2} & & \\ & & & & & & \frac{1}{2} & \frac{1}{2} \\ \frac{1}{2} & -\frac{1}{2} & & & & & & \\ & & \frac{1}{2} & -\frac{1}{2} & & & & \\ & & & & \frac{1}{2} & -\frac{1}{2} & & \\ & & & & & & \frac{1}{2} & -\frac{1}{2} \end{bmatrix} \begin{bmatrix} 1 \\ 3 \\ 2 \\ 6 \\ 7 \\ 5 \\ 4 \\ 2 \end{bmatrix} = \begin{bmatrix} 2 \\ 4 \\ 6 \\ 3 \\ -1 \\ -2 \\ 1 \\ 1 \end{bmatrix}.$$

Step 2: (The second matrix of decomposition, T_2)

$$\begin{bmatrix} \frac{1}{2} & \frac{1}{2} & & & & & & \\ & & \frac{1}{2} & \frac{1}{2} & & & & \\ \frac{1}{2} & -\frac{1}{2} & & & & & & \\ & & \frac{1}{2} & -\frac{1}{2} & & & & \\ & & & & 1 & & & \\ & & & & & 1 & & \\ & & & & & & 1 & \\ & & & & & & & 1 \end{bmatrix} \begin{bmatrix} 2 \\ 4 \\ 6 \\ 3 \\ -1 \\ -2 \\ 1 \\ 1 \end{bmatrix} = \begin{bmatrix} 3 \\ 4.5 \\ -1 \\ 1.5 \\ -1 \\ -2 \\ 1 \\ 1 \end{bmatrix}.$$

Step 3: (The third matrix of decomposition, T_3)

$$
\begin{bmatrix}
\frac{1}{2} & \frac{1}{2} & & & & & & \\
\frac{1}{2} & -\frac{1}{2} & & & & & & \\
& & 1 & & & & & \\
& & & 1 & & & & \\
& & & & 1 & & & \\
& & & & & 1 & & \\
& & & & & & 1 & \\
& & & & & & & 1
\end{bmatrix}
\begin{bmatrix}
3 \\ 4.5 \\ -1 \\ 1.5 \\ -1 \\ -2 \\ 1 \\ 1
\end{bmatrix}
=
\begin{bmatrix}
3.75 \\ 0.75 \\ -1 \\ 1.5 \\ -1 \\ -2 \\ 1 \\ 1
\end{bmatrix}.
$$

The matrix of the eight-point Haar transformation equals the product of the obtained three matrices,

$$
H_8 = T_3 T_2 T_1 =
\begin{bmatrix}
\frac{1}{8} & \frac{1}{8} & \frac{1}{8} & \frac{1}{8} & \frac{1}{8} & \frac{1}{8} & \frac{1}{8} & \frac{1}{8} \\
\frac{1}{8} & \frac{1}{8} & \frac{1}{8} & \frac{1}{8} & -\frac{1}{8} & -\frac{1}{8} & -\frac{1}{8} & -\frac{1}{8} \\
\frac{1}{4} & \frac{1}{4} & -\frac{1}{4} & -\frac{1}{4} & & & & \\
& & & & \frac{1}{4} & \frac{1}{4} & -\frac{1}{4} & -\frac{1}{4} \\
\frac{1}{2} & -\frac{1}{2} & & & & & & \\
& & \frac{1}{2} & -\frac{1}{2} & & & & \\
& & & & \frac{1}{2} & -\frac{1}{2} & & \\
& & & & & & \frac{1}{2} & -\frac{1}{2}
\end{bmatrix}.
\tag{1.18}
$$

This matrix in normalized form equals

$$
H_8 = \frac{1}{\sqrt{8}}
\begin{bmatrix}
1 & 1 & 1 & 1 & 1 & 1 & 1 & 1 \\
1 & 1 & 1 & 1 & -1 & -1 & -1 & -1 \\
\sqrt{2} & \sqrt{2} & -\sqrt{2} & -\sqrt{2} & & & & \\
& & & & \sqrt{2} & \sqrt{2} & -\sqrt{2} & -\sqrt{2} \\
2 & -2 & & & & & & \\
& & 2 & -2 & & & & \\
& & & & 2 & -2 & & \\
& & & & & & 2 & -2
\end{bmatrix}.
$$

Thus the decomposition of the eight-point signal $\mathbf{f}$, when using the Haar coefficients, can also be written in the following form (up to the normalized coefficient): $H_8 \mathbf{f} = ((T_3 \oplus I_6)(T_2 \oplus I_4) T_1) \mathbf{f}$.

In the general $N = 2^r$ case, when $r > 2$, in the first stage of calculation of the Haar transform, the following two signals $\mathbf{l}_1$ and $\mathbf{h}_1$ of length $N/2$ each are formed from the signal $\mathbf{f}$ of length N :

$$
\mathbf{l}_1 = \left\{ l_0^1, l_1^1, ..., l_n^1, ..., l_{N/2-2}^1, l_{N/2-1}^1 \right\} = \left\{ \frac{1}{2}(f_0 + f_1), \frac{1}{2}(f_2 + f_3), ..., \right.
$$

$$
\left. \frac{1}{2}(f_{2n} + f_{2n+1}), ..., \frac{1}{2}(f_{N-4} + f_{N-3}), \frac{1}{2}(f_{N-2} + f_{N-1}) \right\}.
$$

$$
\mathbf{h}_1 = \left\{ h_0^1, h_1^1, ..., h_n^1, ..., h_{N/2-2}^1, h_{N/2-1}^1 \right\} = \left\{ \frac{1}{2}(f_0 - f_1), \frac{1}{2}(f_2 - f_3), ..., \right.
$$

$$\left. \frac{1}{2}(f_{2n} - f_{2n+1}), ..., \frac{1}{2}(f_{N-4} - f_{N-3}), \frac{1}{2}(f_{N-2} - f_{N-1}) \right\}$$

The second signal $\mathbf{h}_1$ defines the differences of the signal at point pairs $2n$ and $2n + 1$, for $n = 0 : (N/2 - 1)$. This signal is half of the Haar transform. The first signal $\mathbf{l}_1$ defines the averages of the signals at those pairs; therefore $l_n^1 + h_n^1 = f_{2n}$, and $l_n^1 - h_n^1 = f_{2n+1}$, $n = 0 : (N/2 - 1)$.

The same process then is applied to $\mathbf{l}_1$, and the other two short signals $\mathbf{h}_2$ and $\mathbf{l}_2$ of length $N/4$ each that are calculated,

$$\mathbf{l}_2 = \left\{ l_0^2, l_1^2, ..., l_n^2, ..., l_{N/4-2}^1, l_{N/4-1}^1 \right\}$$

$$= \left\{ \frac{1}{2}(l_0^1 + l_1^1), \frac{1}{2}(l_2^1 + l_3^1), ..., \frac{1}{2}(l_{2n}^1 + l_{2n+1}^1), ..., \right.$$

$$\left. \frac{1}{2}(l_{N/4-4}^1 + l_{N/4-3}^1), \frac{1}{2}(l_{N/4-2}^1 + l_{N/4-1}^1) \right\}$$

$$\mathbf{h}_2 = \left\{ h_0^2, h_1^2, ..., h_n^2, ..., h_{N/4-2}^2, h_{N/4-1}^2 \right\}$$

$$= \left\{ \frac{1}{2}(l_0^1 - l_1^1), \frac{1}{2}(l_2^1 - l_3^1), ..., \frac{1}{2}(l_{2n}^1 - l_{2n+1}^1), ..., \right.$$

$$\left. \frac{1}{2}(l_{N/4-4}^1 - l_{N/4-3}^1), \frac{1}{2}(l_{N/4-2}^1 - l_{N/4-1}^1) \right\}.$$

Continuing the process, we obtain the Haar transform $\bar{H}\mathbf{f}$ of the signal $\mathbf{f}$, i.e., the sequence of short signals that compose the signal decomposition: $\bar{H}\mathbf{f} = \{\mathbf{l}_r, \mathbf{h}_r\mathbf{h}_{r-1}, ..., \mathbf{h}_3, \mathbf{h}_2, \mathbf{h}_1\}$. We denote this decomposition by $\bar{H}\mathbf{f}$, not by $H\mathbf{f}$, since the transform was considered without the normalized coefficient. In the above considered case $N = 8$ and $\mathbf{f} = (1, 3, 2, 6, 7, 5, 4, 2)'$, we obtain

$$\bar{H}\mathbf{f} = \{\mathbf{l}_3, \mathbf{h}_3, \mathbf{h}_2, \mathbf{h}_1\} = \left\{ \underbrace{3.75}, \underbrace{-0.75}, \underbrace{-1, 1.5}, \underbrace{-1, -2, 1, 1} \right\}.$$

1.5.2 Codes for the Haar transform

The examples of MATLAB-based codes for computing the direct and inverse discrete Haar transforms are given below.

```
% --------------------------------------------------------------
%   mhaar.m file for MATLAB 7 (library of codes of Grigoryans)
%   The direct and inverse Haar transforms (non normalized)
%   The input signal is of length N which equals a power of 2.
%       mhaar.m     - direct discrete Haar transform
%       minvhaar.m  - inverse discrete Haar transform
%       mmathaar.m  - matrix NxN of the Haar transform

    function y=mhaar(x)
        N=length(x);
```

```
          a=1/2;
          if N==1
          y=x;
       else
          x1=x(1:2:N);
          x2=x(2:2:N);
          y1=x1+x2;
          y2=x1-x2;
          y1=y1*a;
          y2=y2*a;
          y=[m_haar(y1) y2];
       end;

    function y=minvhaar(x)
          N=length(x); r=log2(N);
          y=x;
          a=1;  m=1;
          for k=1:r
              m2=bitshift(m,1);
              z=a*y(1:m2);
              z1=z(1:m);
              z2=z(m+1:m2);
              y(1:2:m2)=z1+z2;
              y(2:2:m2)=z1-z2;
              m=m2;
          end;

    function T=m_mat_haar(N)
          T=zeros(N);
          for i1=1:N
              y=zeros(1,N);
              y(i1)=1;
              a=m_haar(y);
              T(:,i1)=a(:);
          end;
% --------------------------------------------------------------
```

1.5.3   Comparison with the paired transform

For the $2N$-point Haar transformation, the following recursive formula holds:

$$[H_{2N}] = \begin{bmatrix} [H_N] \otimes [1 \quad 1] \\ \sqrt{N} I_N \otimes [1 \quad -1] \end{bmatrix} \quad [H_2] = \begin{bmatrix} 1 & 1 \\ 1 & -1 \end{bmatrix}$$

where I_N is the identity matrix $(N \times N)$, where $N \geq 2$, and $\otimes$ is the right-hand Kronecker product of matrices. If we consider the matrix of the Haar transformation in the nonnormalized form, the recursive formula will look like

$$[H_{2N}] = \begin{bmatrix} [H_N] \otimes [1 \quad 1] \\ I_N \otimes [1 \quad -1] \end{bmatrix} \quad [H_2] = \begin{bmatrix} 1 & 1 \\ 1 & -1 \end{bmatrix}.$$

For the considered $N = 2^r$ case, the complete system of paired functions χ' is defined as

$$\{\{\chi'_{2^n, 2^n t}; t = 0 : (2^{r-n-1} - 1), n = 0 : (r - 1)\}, 1\}.$$

Figure 1.17 shows the basic functions of the $2, 4$, and 8-point discrete paired transforms. The matrix of the paired transformation has also a recursive formula, when constructing the matrix $(2N \times 2N)$ from the matrix $(N \times N)$,

$$[\chi'_{2N}] = \begin{bmatrix} [1 \quad -1] \otimes I_N \\ [1 \quad 1] \otimes [\chi'_N] \end{bmatrix}, \quad [\chi'_2] = \begin{bmatrix} 1 & -1 \\ 1 & 1 \end{bmatrix}.$$

One can note that the rows of the basic matrices (2×2) for these two transformations are permuted, and the operation of the Kronecker product is also used in the opposite direction. The matrices of the paired and Haar transformations can be transformed to each other after some permutations of rows and columns [20]. We now illustrate how to change the matrix of the paired transformation, in order to obtain the matrix of the Haar transformation.

Example 1.6
Let $N = 8$, and let $[H_8]$ be the Haar matrix (8×8). Then, we perform the following permutation of the columns in the matrix:

$$(1) \rightarrow (1), \ (2) \rightarrow (5), \ (3) \rightarrow (3), \ (4) \rightarrow (7)$$
$$(5) \rightarrow (4), \ (6) \rightarrow (8), \ (7) \rightarrow (2), \ (8) \rightarrow (6)$$

that can be written as the permutation $P_c : (2, 5, 4, 7)(6, 8)$. As a result, we obtain the following matrix

$$[H_{8;c}] = \begin{bmatrix} [h'_0] \\ [h'_1] \\ [h'_2] \\ [h'_3] \\ [h'_4] \\ [h'_5] \\ [h'_6] \\ [h'_7] \end{bmatrix} = \begin{bmatrix} 1 & 1 & 1 & 1 & 1 & 1 & 1 & 1 \\ 1 & -1 & 1 & -1 & 1 & -1 & 1 & -1 \\ \sqrt{2} & 0 & -\sqrt{2} & 0 & \sqrt{2} & 0 & -\sqrt{2} & 0 \\ 0 & -\sqrt{2} & 0 & \sqrt{2} & 0 & -\sqrt{2} & 0 & \sqrt{2} \\ 2 & 0 & 0 & 0 & -2 & 0 & 0 & 0 \\ 0 & 0 & 2 & 0 & 0 & 0 & -2 & 0 \\ 0 & 0 & 0 & 2 & 0 & 0 & 0 & -2 \\ 0 & 2 & 0 & 0 & 0 & -2 & 0 & 0 \end{bmatrix}.$$

We next change the order of rows as

$$(1) \rightarrow (8), \ (2) \rightarrow (7), \ (3) \rightarrow (5), \ (4) \rightarrow (6)$$
$$(5) \rightarrow (1), \ (6) \rightarrow (3), \ (7) \rightarrow (4), \ (8) \rightarrow (2).$$

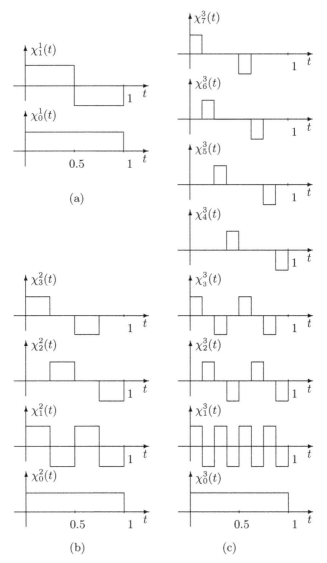

FIGURE 1.17
The basic functions of the 2-, 4-, and 8-point paired transformations.

In other words, we use the following permutation of rows:

$$P_r = \begin{pmatrix} 1\ 2\ 3\ 4\ 5\ 6\ 7\ 8 \\ 8\ 7\ 5\ 6\ 1\ 3\ 4\ 2 \end{pmatrix}. \tag{1.19}$$

In the general case $N \geq 8$, the permutation P_r in (1.19) is the well-known *reverse shuffle* permutation [27]. After performing the permutation by rows, we obtain the following matrix which we denote by $[H_{8;c,r}]$:

$$[H_{8;c,r}] = \begin{bmatrix} [h'_4] \\ [h'_5] \\ [h'_6] \\ [h'_7] \\ [h'_2] \\ [h'_3] \\ [h'_1] \\ [h'_0] \end{bmatrix} = \begin{bmatrix} 2 & 0 & 0 & 0 & -2 & 0 & 0 & 0 \\ 0 & 2 & 0 & 0 & 0 & -2 & 0 & 0 \\ 0 & 0 & 2 & 0 & 0 & 0 & -2 & 0 \\ 0 & 0 & 0 & 2 & 0 & 0 & 0 & -2 \\ \sqrt{2} & 0 & -\sqrt{2} & 0 & \sqrt{2} & 0 & -\sqrt{2} & 0 \\ 0 & -\sqrt{2} & 0 & \sqrt{2} & 0 & -\sqrt{2} & 0 & \sqrt{2} \\ 1 & -1 & 1 & -1 & 1 & -1 & 1 & -1 \\ 1 & 1 & 1 & 1 & 1 & 1 & 1 & 1 \end{bmatrix}.$$

This matrix is the matrix of the 8-point transformation with coefficients of the normalized basic paired functions

$$[H_{8;c,r}] = \text{diag}(2,2,2,2,\sqrt{2},-\sqrt{2},1,1)^T [\chi'_8],$$

where ^{T} is the operation of the transposition of the matrix, and the matrix of the 8-point discrete paired transformation is

$$[\chi'_8] = \begin{bmatrix} 1 & 0 & 0 & 0 & -1 & 0 & 0 & 0 \\ 0 & 1 & 0 & 0 & 0 & -1 & 0 & 0 \\ 1 & 0 & 1 & 0 & 0 & 0 & -1 & 0 \\ 0 & 0 & 0 & 1 & 0 & 0 & 0 & -1 \\ 1 & 0 & -1 & 0 & 1 & 0 & -1 & 0 \\ 0 & 1 & 0 & -1 & 0 & 1 & 0 & -1 \\ 1 & -1 & 1 & -1 & 1 & -1 & 1 & -1 \\ 1 & 1 & 1 & 1 & 1 & 1 & 1 & 1 \end{bmatrix}.$$

As a result, we obtain the calculation of the 8-point discrete Haar transform, which is shown in the signal-flow graph of Figure 1.18. The matrix of the

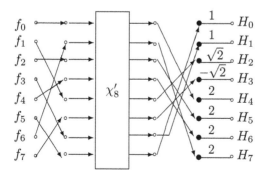

FIGURE 1.18
The signal-flow graph of the 8-point discrete Haar transform by the 8-point discrete paired transform.

transformation is calculated by $[H_8] = D[\chi'_8]T$, where the matrix D with the weighted coefficients and matrix T of the permutation of input are defined as

$$
D = \begin{bmatrix}
0 & 0 & 0 & 0 & 0 & & 0 & 0 & 1 \\
0 & 0 & 0 & 0 & 0 & & 0 & 1 & 0 \\
0 & 0 & 0 & 0 & \sqrt{2} & & 0 & 0 & 0 \\
0 & 0 & 0 & 0 & 0 & -\sqrt{2} & 0 & 0 \\
2 & 0 & 0 & 0 & 0 & & 0 & 0 & 0 \\
0 & 0 & 2 & 0 & 0 & & 0 & 0 & 0 \\
0 & 0 & 0 & 2 & 0 & & 0 & 0 & 0 \\
0 & 2 & 0 & 0 & 0 & & 0 & 0 & 0
\end{bmatrix}, \quad
T = \begin{bmatrix}
1 & 0 & 0 & 0 & 0 & 0 & 0 & 0 \\
0 & 0 & 0 & 0 & 0 & 0 & 1 & 0 \\
0 & 0 & 1 & 0 & 0 & 0 & 0 & 0 \\
0 & 0 & 0 & 0 & 1 & 0 & 0 & 0 \\
0 & 1 & 0 & 0 & 0 & 0 & 0 & 0 \\
0 & 0 & 0 & 0 & 0 & 0 & 0 & 1 \\
0 & 0 & 0 & 1 & 0 & 0 & 0 & 0 \\
0 & 0 & 0 & 0 & 0 & 1 & 0 & 0
\end{bmatrix}.
$$

The calculation requires two operations of multiplication by $\sqrt{2}$ (the multiplication by 2 is considered to be trivial). The fast 2^r-point discrete paired transform uses $2^{r+1} - 2$ operations of addition (subtraction). Thus, the calculation of the 8-point DHT requires $A_8 = (2^4 - 2) = 14$ additions.   ⬜

Example 1.7
In the case $N = 4$, the matrix of the Haar transformation can be written as $[H_4] = D[\widetilde{\chi}'_4]T$, where the diagonal matrix D with coefficients 1 and $\pm\sqrt{2}$, the matrix of the permutation T, and the matrix $[\widetilde{\chi}'_4]$ are defined as follows:

$$
[H_4] = D[\widetilde{\chi}'_4]T = \begin{bmatrix}
1 & 0 & 0 & 0 \\
0 & 1 & 0 & 0 \\
0 & 0 & \sqrt{2} & 0 \\
0 & 0 & 0 & -\sqrt{2}
\end{bmatrix}\begin{bmatrix}
1 & 1 & 1 & 1 \\
1 & -1 & 1 & -1 \\
1 & 0 & -1 & 0 \\
0 & 1 & 0 & -1
\end{bmatrix}\begin{bmatrix}
1 & 0 & 0 & 0 \\
0 & 0 & 0 & 1 \\
0 & 1 & 0 & 0 \\
0 & 0 & 1 & 0
\end{bmatrix}.
$$

$\widetilde{\chi}'_4$ represents the paired transformation whose basic functions are ordered as $\chi'_{0,0}$, $\chi'_{2,0}$, $\chi'_{1,0}$, and $\chi'_{1,1}$. In other words, such permutation Q of the basic paired functions yields the expression $[\widetilde{\chi}'_4] = Q[\chi'_4]$, or

$$
[\widetilde{\chi}'_4] = Q[\chi'_4] = \begin{bmatrix}
1 & 1 & 1 & 1 \\
1 & -1 & 1 & -1 \\
1 & 0 & -1 & 0 \\
0 & 1 & 0 & -1
\end{bmatrix} = \begin{bmatrix}
0 & 0 & 0 & 1 \\
0 & 0 & 1 & 0 \\
1 & 0 & 0 & 0 \\
0 & 1 & 0 & 0
\end{bmatrix}\begin{bmatrix}
1 & 0 & -1 & 0 \\
0 & 1 & 0 & -1 \\
1 & -1 & 1 & -1 \\
1 & 1 & 1 & 1
\end{bmatrix}.
$$

As a result, we obtain the following decomposition of the four-point DHT by the paired transformation: $[H_4] = D[\widetilde{\chi}'_4]T = (DQ)[\chi'_4]T$. The multiplication of matrices D and Q results in the matrix

$$
DQ = \begin{bmatrix}
0 & & 0 & 0 & 1 \\
0 & & 0 & 1 & 0 \\
\sqrt{2} & & 0 & 0 & 0 \\
0 & -\sqrt{2} & 0 & 0
\end{bmatrix}.
$$

Therefore, we obtain the following representation of the four-point DHT:

$$[H_4] = \begin{bmatrix} 0 & 0 & 0 & 1 \\ 0 & 0 & 1 & 0 \\ \sqrt{2} & 0 & 0 & 0 \\ 0 & -\sqrt{2} & 0 & 0 \end{bmatrix} [\chi'_4] \begin{bmatrix} 1 & 0 & 0 & 0 \\ 0 & 0 & 0 & 1 \\ 0 & 1 & 0 & 0 \\ 0 & 0 & 1 & 0 \end{bmatrix}.$$

The calculation of the Haar transform is reduced to the calculation of the paired transform over the reordered input, and then to reordering the output and multiplying them by coefficients of matrix DQ. Figure 1.19 shows the flow-graph of calculating the four-point discrete Haar transform by the paired transform χ'_4. The calculation requires two operations of multiplication by $\sqrt{2}$ and six operations of real or complex addition respectively in the real or complex case. □

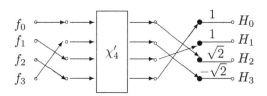

FIGURE 1.19
The signal-flow graph of the four-point DHT by the four-point discrete paired transform.

Example 1.8
Consider the $N = 16$ case. The matrix of the 16-point discrete Haar transform can be decomposed as $H_{16} = D[\chi'_{16}]T$, where D is the matrix of weighted coefficients and T is the matrix of a permutation of the input. The matrix D is composed from the following diagonal matrix (16×16) :

$$\text{diag}\left\{1, 1, \sqrt{2}, \sqrt{2}, 2, 2, 2, 2, 2\sqrt{2}, 2\sqrt{2}, 2\sqrt{2}, 2\sqrt{2}, 2\sqrt{2}, 2\sqrt{2}, 2\sqrt{2}, 2\sqrt{2}\right\},$$

whose rows are rearranged in the order $(16, 15, 13, 14, 9, 11, 10, 12, 1, 5, 3, 7, 2,$ $6, 4, 8)$. The matrix T relates to the permutation $(2, 9)(3, 5)(4, 13)(6, 11)(8, 15)$ $(12, 14)$. The decomposition $D[\chi'_{16}]T$ yields the computation of the 16-point discrete Haar transform by the paired transform, which diagram is given in Figure 1.20. The calculation of the Haar transform requires as many additions as for the paired transform, i.e., $\alpha(16) = 2 \cdot 16 - 2 = 30$. □

The matrix of the Haar transformation up to the permutation of columns and rows is the matrix of the nonnormalized discrete paired transformation.

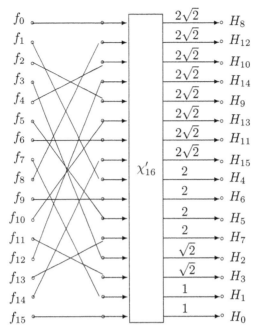

FIGURE 1.20
The signal-flow graph of the 16-point DHT by the 16-point discrete paired transform.

Moreover, the basic functions of the Haar transformation as the basic functions of the paired transformation can be derived from a system of cosine functions of certain frequencies. The paired transformation is fast, requires $2N - 2$ operations of addition and subtraction, and therefore can be used for the fast computing of the Haar transform. The paired transformations split the mathematical structure of the Fourier, Hadamard, and other transformations, being an important part of the transforms, especially in the two- and multi-dimensional cases. For this reason, we may consider the Haar transformation as a transformation that is a compound part of the Fourier and other transformations.

Below are the MATLAB-based codes for calculating the 16-point Haar transform by the paired transform with ten multiplications. The transform is calculated over the signal $\mathbf{x} = \{1, 3, 4, 6, 7, 5, 1, 2, 2, 7, 2, 1, 5, 3, 4, 3\}$. The non-normalized and normalized Haar transforms of this signal equal, respectively,

$$\bar{H}_{16}[\mathbf{x}] = \{56, 2, -1, -3, -6, 9, 6, 1, -2, -2, 2, -1, -5, 1, 2, 1\}$$

and (with precision of one digit after the point)

$$\{56, 2, -1.4, -4.2, -12, 18, 12, 2, -5.7, -5.7, 5.7, -2.8, -14.1, 2.8, 5.7, 2.8\}/4.$$

```
% ------------------------------------------------------------------
%    mahaar.m file for MATLAB 7 (library of codes of Grigoryans)
%    The direct 16-point Haar transform (normalized)
%    The input signal is of length N which equals a power of 2.
%    The functions to be used:
%       matrix_paired.m  - matrix NxN of the paired transform
%       matrix_paired.m -  fast paired transform (see demo_pfft.m)
%
%    1.A Calculation of the Haar transform by the paired transform
       x=[1,3,4,6,7,5,1,2,2,7,2,1,5,3,4,3];
       P=[1,9,5,13,3,11,7,15,2,10,6,14,4,12,8,16];
       x_permuted=x(P);
%    1.B Permutation of the paired transform
       y_paired=fastpaired_1d(x_permuted);
       D=[16,15,13,14,9,11,10,12,1,5,3,7,2,6,4,8];
       y_haar=y_paired(D);
       % 56  2 -1 -3 -6  9  6  1 -2 -2  2 -1 -5  1  2  1
%    1.C Normalization of the outputs (up to the coefficient 4=sqrt(16))
       a=sqrt(2);  b=2*a;
       DD=diag([1,1,a,a,2,2,2,2,b,b,b,b,b,b,b,b]);
       x_haar=y_haar*DD';
       % 56 2 -1.4 -4.2 -12.0 18.0 12.0  2.0 -5.7 ...
       % -5.7  5.7 -2.8 -14.1  2.8  5.7  2.8
%    2. Calculation of the matrix 16x16 of the paired transform
%       and then the matrix 16x16 of the Haar transform
       P16=matrix_paired(16);
       I16=eye(16);
       T16=zeros(16);
       for k=1:16
           T16(k,:)=I16(P(k),:);
       end;
       P2=[16,15,13,14,9,11,10,12,1,5,3,7,2,6,4,8];
       D16=zeros(16);
       for k=1:16
           D16(k,:)=I16(P2(k),:);
       end;
       H16=D16*P16*T16;  % non normalized Haar matrix
       H16n=DD*H16;      % the normalized Haar matrix (can be divided by 4)
%    ------------------------------------------------------------------
       function T=matrix_paired(N)
         T=zeros(N);
         for n=1:N
             y=zeros(1,N);
             y(n)=1;
             m=fastpaired_1d(y);
             T(:,n)=m(:);
         end;
```

Problems

Problem 1.1 Given integer $N > 1$, prove that the system of discrete functions

$$\Phi = \left\{ \varphi_p(n) = \frac{1}{\sqrt{N}} W^{np}; \, p = 0 : (N-1) \right\}, \quad (W = W_N = \exp(-j2\pi/N))$$

is the complete system of orthogonal functions in the N-dimension space of discrete-time signals of length N. In other words, show that

$$(\varphi_p, \varphi_s) = \sum_{n=0}^{N-1} \varphi_p(n) \bar{\varphi}_s(n) = 0, \quad \text{if} \quad p \neq s = 0 : (N-1),$$

and $(\varphi_p, \varphi_p) = 1$.

Problem 1.2 Given integer $N > 1$, prove that the inverse N-point DFT of the signal f_n is calculated by

$$f_n = \frac{1}{\sqrt{N}} \sum_{p=0}^{N-1} F_p \bar{\varphi}_p(n) = \frac{1}{N} \sum_{p=0}^{N-1} F_p W^{-np}, \quad n = 0 : (N-1).$$

Problem 1.3 Consider the discrete-time signal $g_n = 0.01 + 4^{-n}u(n)$ of length $N = 512$. Calculate and plot the DFT magnitude and angle, the real and imaginary parts of this signal.

Problem 1.4 Consider the discrete-time signal f_n of length $N = 512$, which is sampled in the interval $[0, 2\pi]$ from the following signal:

$$f(t) = \begin{cases} 2t\cos(3t), & t \in [1,5], \\ 0, & \text{otherwise.} \end{cases}$$

A. Plot the DFT magnitude and angle (you can use for that the MATLAB commands "*fft*" and "*angle*").

B. Calculate inverse DFT by using the command "*fft*" (not "*ifft*") and plot the real part of the inverse transform.

Problem 1.5 Given a real discrete-time signal f_n of length $N > 1$, compose the new signal g_n as

$$g_n = f_{N-1-n}, \qquad n = 0, 1, 2, ..., (N-1).$$

Express the N-point DFT of the signal g_n by the N-point DFT of the signal f_n.

As an example, take the following signal of length $N = 12$:

$$\{f_n; n = 0 : 12\} = \{1, 2, 5, 3, 1, 4, 7, 3, 1, 2, 7, 3\},$$

and calculate the 12-point DFT of g_n, i.e., G_p, $p = 0 : 11$, by using the 12-point DFT of f_n.

Problem 1.6 Consider a real discrete-time signal f_n of length $N > 1$, for instance,

$$\{f_n; n = 0 : 12\} = \{\underline{1}, 2, 5, 3, 1, 4, 7, 3, 1, 2, 7, 3\}$$

when $N = 12$. Compose the new signal as

$$g_n = \begin{cases} f_0, & n = 0, \\ f_{N-n}, & n = 1, 2, ..., (N-1). \end{cases}$$

Show that the N-point DFT of the signal g_n can be calculated from the N-point DFT of the signal f_n.

Problem 1.7 Consider a random real integer discrete-time signal f_n of length $N = 512$ with values from the interval $[0, 32]$.

A. Calculate and plot DFT magnitudes of the even and odd parts of the signal,

$$e_n = \frac{f_n + f_{N-n}}{2}, \quad o_n = \frac{f_n - f_{N-n}}{2}, \quad n = 0 : (N-1).$$

B. Calculate the DFTs of e_n and o_n by using directly the transform DFT of f_n. Plot and compare the results with the results in A.

Problem 1.8 Given $N = 128$, consider the random signal f_n of length N, which is composed by periodic extension of the signal x_n of length $P = 7$. For instance,

$$f_n = \{\underline{1, 2, 5, 3, 1, 4, 7}, 1, 2, 5, 3, 1, 4, 7, 1, 2, 5, 3, 1, 4, 7, \ldots, \underline{1, 2, 5, 3, 1, 4, 7}, 1, 2\},$$

where the period $\{x_n\} = \{1, 2, 5, 3, 1, 4, 7\}$.

A. Calculate the N-point DFT, F_p, $p = 0 : (N-1)$, of this signal, plot the real and imaginary parts of the DFT, as well as the DFT in polar form, i.e., the DFT magnitude and angle.

B. Figure 1.21 shows the period x_n in part a, along with the DFT magnitude, i.e., $\{|F_p|; p = 0 : (N-1)\}$, in b, and the same spectrum in c, but shifted cyclicly to the center. One can observe a few pikes on the graph of the DFT magnitude, which are located periodicity. Explain this effect and find the locations of the pikes.

C. Verify if a similar effect holds for other signals as well. As an example, consider the signal f_n which is composed by the period $\{x_n\} = \{1, 2, 5, 3, 1, 4, 7, 2, 5\}$ of $P = 9$.

Problem 1.9 Prove analytically that DFTs of two real discrete-time signals f_n and g_n of the same length N can be calculated by using one complex N-point DFT. Demonstrate this fact on the 512-point signals sampled respectively from the functions

$$f(t) = 0.01te^{t/4}, \quad g(t) = 0.1t\cos(4t), \quad t \in [0, 2\pi],$$

with sampling period $T = 2\pi/511$.

Problem 1.10 Model a signal f_n of length $N = 512$ with the random noise of small amplitude, and consider the following window:

$$h_n = \begin{cases} \frac{1}{5}, & n = -2, -1, 0, 1, 2, \\ 0, & n = 3, 4, ..., (N-3). \end{cases}$$

A. Calculate the circular convolution $y_n = f_n \otimes h_n$ of the signal f_n with the window h_n, by using the method of the Fourier transform. Plot the signals x_n, h_n, y_n and

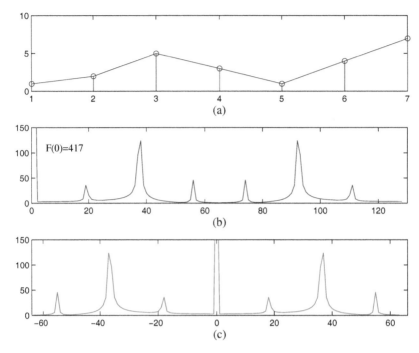

FIGURE 1.21
(a) Seven-point period of the periodic signal, and (b,c) the DFT of this signal in absolute scale in the original time domain and shifted to the center.

the DFT magnitude and phase for each of these signals. As a signal f_n, consider the discrete-time signal stored in the binary file 'Bold_g5.sig' which can be downloaded from http://www.fasttransforms.com.

B. Compare your result with the result obtained when applying the MATLAB command "*cconv(f,h,N)*".

Problem 1.11 Calculate the matrix of the 16-point paired transformation.

Problem 1.12 Calculate the matrix of the 16-point Haar transformation by the paired transformation χ'_{16}.

Problem 1.13 Calculate and plot the first four splitting-signals of the signal f_n of length 512. As an example, take the signal from file "boli.sig" (from the website http://www.fasttransforms.com).

2

Integer Fourier Transform

2.1 Reversible integer Fourier transform

The discrete Fourier transform uses floating-point multiplications which result in noninteger outputs even in the case of most importance in practice, when input data are integer numbers. It is thus desired to develop a reversible transform which approximates the Fourier transform and maps integer data into integer ones. For the Fourier transform, we mention two approaches for defining an integer DFT. The first approach is based on approximation of the transform matrix by a matrix with integer coefficients, and the second one suggests using the lifting scheme with an integer quantizer. The integer Fourier transform with integer entries was introduced in [28] and implementation of this transform requires a large number of the fixed-point multiplications. For instance, for inputs of length eight, this transform requires eight fixed-point multiplications instead of two floating-point multiplications for the conventional FFT. The approach based on substitution of the exponential coefficients (twiddle factors) in the DFT by lifting schemes is described in [29]. Each such substitution may increase the resolution of its input by one bit, and there are different choices in parameterizing the lifting coefficients. The transform is effective, because it can be implemented by using only bit shifts and additions.

We focus here on the new approach of the reversible integer DFT, which is based on the concept of the integer multiplication with control bits. Examples of implementation of this approach will be described in detail for the eight- and sixteen-point DFTs. The concept of integer multiplication with control bits can be used for integer approximation of other transforms, such as the cosine transforms, as well.

2.1.1 Lifting scheme implementation

The method of lifting schemes (LS) is widely used in integer wavelets and perfect reconstruction filter-bank, in lossless coding techniques such as audio and image coding, and DCTs [30]-[32]. The approach is based on substitution of the exponential coefficients (twiddle factors) of the DFT by lifting schemes with different coefficients.

We consider the operation of multiplication of data x by exponential coefficients, $W = e^{-j\phi} = \cos(\phi) - j\sin(\phi)$, where $\phi = \phi_p = (2\pi/N)p$, $p = 1 : (N-1)$. The multiplication Wx, where $x = x_1 + jx_2$, is defined as

$$Wx = (Wx)_1 + j(Wx)_2 = [\cos(\phi)x_1 + \sin(\phi)x_2] + j[-\sin(\phi)x_1 + \cos(\phi)x_2]$$

and in matrix form it can be written as

$$\begin{bmatrix} (Wx)_1 \\ (Wx)_2 \end{bmatrix} = \begin{bmatrix} \cos(\phi) & \sin(\phi) \\ -\sin(\phi) & \cos(\phi) \end{bmatrix} \begin{bmatrix} x_1 \\ x_2 \end{bmatrix} = \begin{bmatrix} c & s \\ -s & c \end{bmatrix} \begin{bmatrix} x_1 \\ x_2 \end{bmatrix}$$

where $c = \cos(\phi)$ and $s = \sin(\phi)$.

Four multiplications that are required to calculate directly Wx can be reduced to three multiplications when using the following well-known lifting scheme:

$$\begin{bmatrix} c & s \\ -s & c \end{bmatrix} \begin{bmatrix} x_1 \\ x_2 \end{bmatrix} = \begin{bmatrix} 1 & \dfrac{1-c}{s} \\ 0 & 1 \end{bmatrix} \begin{bmatrix} 1 & 0 \\ -s & 1 \end{bmatrix} \begin{bmatrix} 1 & \dfrac{1-c}{s} \\ 0 & 1 \end{bmatrix} \begin{bmatrix} x_1 \\ x_2 \end{bmatrix}. \tag{2.1}$$

Coefficient $(1-c)/s = \tan(\phi/2)$ can be calculated in advance and is not considered as an additional multiplication. The scheme of this three step lifting multiplication Wx is given in Figure 2.1. The lifting scheme for the

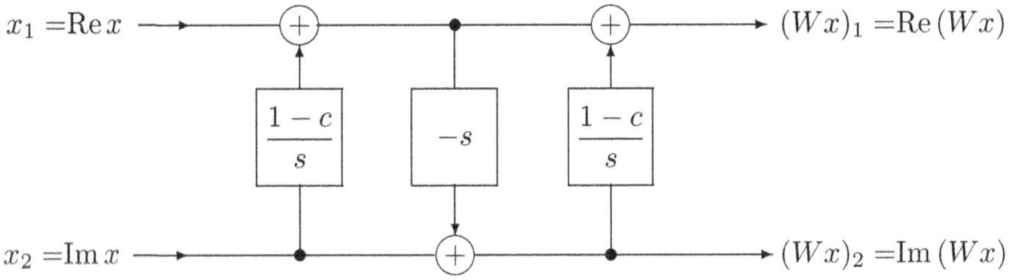

FIGURE 2.1
Three step lifting scheme for calculating Wx.

inverse transform $(Wx)_1, (Wx)_2 \rightarrow (x_1, x_2)$ is the same, except the sign of the coefficient $s = \sin(\phi)$ should be changed in the decomposition (2.1).

1 Step:

$$\begin{bmatrix} y_1 \\ y_2 \end{bmatrix} = \begin{bmatrix} 1 & \dfrac{1-c}{s} \\ 0 & 1 \end{bmatrix} \begin{bmatrix} x_1 \\ x_2 \end{bmatrix} = \begin{bmatrix} x_1 + \dfrac{1-c}{s}x_2 \\ x_2 \end{bmatrix}.$$

2 Step:

$$\begin{bmatrix} z_1 \\ z_2 \end{bmatrix} = \begin{bmatrix} 1 & 0 \\ -s & 1 \end{bmatrix} \begin{bmatrix} y_1 \\ y_2 \end{bmatrix} = \begin{bmatrix} y_1 \\ -sy_1 + y_2 \end{bmatrix}.$$

3 Step:

$$\begin{bmatrix} (Wx)_1 \\ (Wx)_2 \end{bmatrix} = \begin{bmatrix} 1 & \dfrac{1-c}{s} \\ 0 & 1 \end{bmatrix} \begin{bmatrix} z_1 \\ z_2 \end{bmatrix} = \begin{bmatrix} z_1 + \dfrac{1-c}{s}z_2 \\ z_2 \end{bmatrix}.$$

For integer approximation of the operation of multiplication Wx, a non-linear operation of quantization Q is used after each multiplication in the lifting scheme. For instance, we can consider the quantization to be rounding operation, $Q(a) = [a]$. This operation can also be flooring or ceiling, but the flooring and rounding operations are used frequently. The results of the three step multiplication are calculated as follows.

1 Step: (x_1 and x_2 are integers)

$$\begin{bmatrix} y_1 \\ y_2 \end{bmatrix} = \begin{bmatrix} 1 & \dfrac{1-c}{s} \\ 0 & 1 \end{bmatrix} \diamond \begin{bmatrix} x_1 \\ x_2 \end{bmatrix} = \begin{bmatrix} x_1 + Q\left(\dfrac{1-c}{s}x_2\right) \\ x_2 \end{bmatrix}.$$

2 Step: (y_1 and y_2 are integers)

$$\begin{bmatrix} z_1 \\ z_2 \end{bmatrix} = \begin{bmatrix} 1 & 0 \\ -s & 1 \end{bmatrix} \diamond \begin{bmatrix} y_1 \\ y_2 \end{bmatrix} = \begin{bmatrix} y_1 \\ -Q(sy_1) + y_2 \end{bmatrix}.$$

3 Step: (z_1 and z_2 are integers)

$$\begin{bmatrix} [Wx]_1 \\ [Wx]_2 \end{bmatrix} = \begin{bmatrix} 1 & \dfrac{1-c}{s} \\ 0 & 1 \end{bmatrix} \diamond \begin{bmatrix} z_1 \\ z_2 \end{bmatrix} = \begin{bmatrix} z_1 + Q\left(\dfrac{1-c}{s}z_2\right) \\ z_2 \end{bmatrix}.$$

$[Wx]_1$ and $[Wx]_2$ denote respectively the results of integer approximations of $(Wx)_1$ and $(Wx)_2$, by this lifting scheme. The symbol $\diamond$ is used for the multiplication with quantization, $a \diamond b = Q(ab)$.

In formulas and signal-flow graphs we also will use the following symbols for the lifting scheme and its integer implementation, respectively:

$$\begin{array}{|c|c|c|} \hline \dfrac{1-c}{s} & -s & \dfrac{1-c}{s} \\ \hline \end{array} \quad \text{and} \quad \begin{array}{|c|c|c|} \hline Q & -s & Q \\ \hline \frac{1-c}{s} & Q & \frac{1-c}{s} \\ \hline \end{array}.$$

For example, when $(x_1, x_2) = (2, 3)$ and $\phi = \pi/4$, the coefficients $c = s = 0.7071$ and

$$\begin{array}{|c|c|c|} \hline \dfrac{1-c}{s} & -s & \dfrac{1-c}{s} \\ \hline \end{array} \circ \begin{bmatrix} 2 \\ 3 \end{bmatrix} = \begin{array}{|c|c|c|} \hline \sqrt{2}-1 & -0.7071 & \sqrt{2}-1 \\ \hline \end{array} \circ \begin{bmatrix} 2 \\ 3 \end{bmatrix} = \begin{bmatrix} 3.5355 \\ 0.7071 \end{bmatrix},$$

$$\begin{array}{|c|c|c|} \hline Q & -0.7071 & Q \\ \hline \sqrt{2}-1 & Q & \sqrt{2}-1 \\ \hline \end{array} \circ \begin{bmatrix} 2 \\ 3 \end{bmatrix} = \begin{bmatrix} 3 \\ 1 \end{bmatrix}.$$

The method of lifting scheme is invertible. Each lifting step changes only one value of the input and it can be constructed by subtracting (or adding) the same quantized value $Q(\cdot)$ that has been added (or subtracted) to it. The diagram of the three step lifting multiplication with three quantizers for calculation of the integer approximation of the multiplication Wx is given in Figure 2.2.

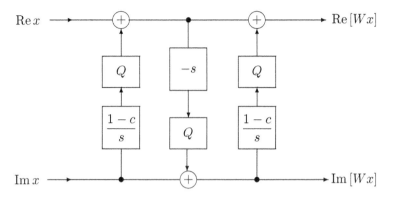

FIGURE 2.2
Three step lifting scheme with quantizers Q for approximation of Wx.

We also consider the lifting scheme for the rotation by angle ϕ,

$$\begin{bmatrix} c & -s \\ s & c \end{bmatrix}\begin{bmatrix} x_1 \\ x_2 \end{bmatrix} = \begin{bmatrix} 1 & \dfrac{c-1}{s} \\ 0 & 1 \end{bmatrix}\begin{bmatrix} 1 & 0 \\ s & 1 \end{bmatrix}\begin{bmatrix} 1 & \dfrac{c-1}{s} \\ 0 & 1 \end{bmatrix}\begin{bmatrix} x_1 \\ x_2 \end{bmatrix}$$

with the similar integer approximation by a quantizer Q of each stage of this scheme. This lifting scheme is inverse to the scheme given in (2.1). The following symbols are used for these three-step lifting schemes of the rotation:

$$\boxed{\begin{array}{c|c|c} \dfrac{c-1}{s} & s & \dfrac{c-1}{s} \end{array}} \quad \text{and} \quad \boxed{\begin{array}{c|c|c} Q & s & Q \\ \hline \frac{c-1}{s} & Q & \frac{c-1}{s} \end{array}}$$

For example, when $(x_1, x_2) = (2, 3)$ and the angle of rotation is $\phi = \pi/4$, the coefficients $c = s = 0.7071$ and

$$\boxed{\begin{array}{c|c|c} \frac{c-1}{s} & s & \frac{c-1}{s} \end{array}} \circ \begin{bmatrix} 2 \\ 3 \end{bmatrix} = \boxed{\begin{array}{c|c|c} 1-\sqrt{2} & 0.7071 & 1-\sqrt{2} \end{array}} \circ \begin{bmatrix} 2 \\ 3 \end{bmatrix} = \begin{bmatrix} -0.7071 \\ 3.5355 \end{bmatrix},$$

$$\boxed{\begin{array}{c|c|c} Q & 0.7071 & Q \\ \hline 1-\sqrt{2} & Q & 1-\sqrt{2} \end{array}} \circ \begin{bmatrix} 2 \\ 3 \end{bmatrix} = \begin{bmatrix} -1 \\ 4 \end{bmatrix}.$$

For the rotation by the angle and $\phi = \pi/8$, we have the following calculations:

$$\boxed{\begin{array}{c|c|c} \frac{c-1}{s} & s & \frac{c-1}{s} \end{array}} \circ \begin{bmatrix} 2 \\ 3 \end{bmatrix} = \boxed{\begin{array}{c|c|c} -0.1989 & 0.3827 & -0.1989 \end{array}} \circ \begin{bmatrix} 2 \\ 3 \end{bmatrix} = \begin{bmatrix} 0.6997 \\ 3.5370 \end{bmatrix},$$

$$\boxed{\begin{array}{c|c|c} Q & 0.3827 & Q \\ \hline -0.1989 & Q & -0.1989 \end{array}} \circ \begin{bmatrix} 2 \\ 3 \end{bmatrix} = \begin{bmatrix} 0 \\ 3 \end{bmatrix}.$$

2.2 Lifting schemes for DFT

In this section, we consider the implementation of the lifting schemes for calculating the discrete Fourier transform. In one of the known methods of the integer-to-integer DFT [29], it is proposed to use the lifting scheme for each nontrivial complex multiplication by coefficient $W^t = e^{-j\varphi_t}, \varphi_t = (2\pi/N)t$, $t = 1 : (N/2 - 1)$. Such multiplications are substituted by the three step lifting multiplications with quantizers Q. As an example, Figure 2.3 shows the signal-flow graph of the 8-point integer DFT when two lifting schemes with quantizers are used. Split-radix structure of the fast algorithm is chosen for the $N = 8$ case. Three real multiplications are used for each lifting scheme, or six multiplications for the integer approximation of the 8-point DFT.

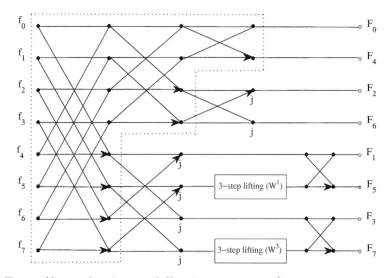

FIGURE 2.3 (See color insert following page 242.)
Signal-flow graph of calculation of the 8-point DFT by two lifting schemes.

Example 2.1
For the $N = 8$ case, we take the signal $\mathbf{x} = (1, 2, 3, 4, 5, 6, 7, 8)'$. Two multi-

plications by the coefficients $W = \exp(-j\pi/4)$ and $W^3 = \exp(-j3\pi/4)$ are required in the 8-point DFT after the 2nd stage of the algorithm.

Figure 2.4 shows the signal-flow graph of the 8-point DFT with two lifting schemes with quantizers, that approximate these multiplications. The numerical data of the signal and transform, as well as all intermediate calculations, are shown at corresponding knots of the signal-flow graph. The multiplication

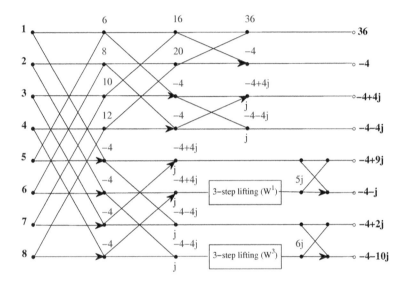

FIGURE 2.4

Signal-flow graph of calculation of the 8-point DFT by lifting schemes with quantizers.

of the complex number $-4 + 4j$ by W through the lifting scheme equals $5j$, and the multiplication of $-4 - 4j$ by W^3 through the lifting scheme equals 6. Indeed, the following holds:

$$5.6569j = (-4 + 4j) \cdot W \rightarrow \begin{array}{|c|c|c|} \hline \mathcal{Q} & -1/\sqrt{2} & \mathcal{Q} \\ \hline \sqrt{2}-1 & \mathcal{Q} & \sqrt{2}-1 \\ \hline \end{array} \circ \begin{bmatrix} -4 \\ 4 \end{bmatrix} = \begin{bmatrix} 0 \\ 5 \end{bmatrix},$$

$$5.6569j = (-4 - 4j) \cdot W^3 \rightarrow \begin{array}{|c|c|c|} \hline \mathcal{Q} & -1/\sqrt{2} & \mathcal{Q} \\ \hline 1+\sqrt{2} & \mathcal{Q} & 1+\sqrt{2} \\ \hline \end{array} \circ \begin{bmatrix} -4 \\ -4 \end{bmatrix} = \begin{bmatrix} 0 \\ 6 \end{bmatrix}.$$

One can notice that the same complex number, $5.6569j$, is approximated differently by these two lifting schemes. The values of this integer transform together with the DFT are given in the following table:

TABLE 2.1
Integer approximation of the 8-point DFT

p	F_p	$F_{p,3LS}$	error
0	36	36	
1	$-4 + 9.6569j$	$-4 + 9j$	$0.6569j$
2	$-4 + 4j$	$-4 + 4j$	
3	$-4 + 1.6569j$	$-4 + 2j$	$-0.3431j$
4	-4	-4	
5	$-4 - 1.6569j$	$-4 - j$	$-0.6569j$
6	$-4 - 4j$	$-4 - 4j$	
7	$-4 - 9.6569j$	$-4 - 10j$	$0.3431j$

The property of complex conjugate of the Fourier transform components F_p and F_{8-p}, $p = 1, 2, 3$, does not hold for this integer approximation of the DFT. Indeed, $F_{7,3LS} = -4 - 10j \neq \bar{F}_{1,3LS} = -4 + 9j$, and $F_{5,3LS} = -4 - j \neq \bar{F}_{3,3LS} = -4 + 2j$. The root-mean-square error of approximation equals

$$\varepsilon_{3ls} = \sqrt{\frac{1}{8} \sum_{p=0}^{7} |F_{p;3LS} - F_p|^2} = 0.3705.$$

⬜

We now compare the use of the lifting scheme in the above algorithm with the paired algorithm. The outputs of the paired transform are real when the input is real, and there is no need to use the three-step lifting scheme in the paired FFT for the $N = 8$ case. Indeed, the lifting scheme for the real number $x = x_1$ can be rewritten as follows.

1 Step: $(Wx)_2 = z_2 = -sx$.

2 Step:

$$(Wx)_1 = \left[1 \quad \frac{1-c}{s} \right] \left[\begin{matrix} x \\ z_2 \end{matrix} \right] = x + \frac{1-c}{s} z_2.$$

The integer approximation by this scheme is also performed in two steps as:

1 Step: $(Wx)_2 = z_2 = -Q(sx)$.

2 Step:

$$[(Wx)_1] = \left[1 \quad \frac{1-c}{s} \right] \diamond \left[\begin{matrix} x \\ z_2 \end{matrix} \right] = x + Q\left(\frac{1-c}{s} z_2 \right).$$

Thus the multiplication is approximated as

$$Wx \to \left(x + Q\left(\frac{c-1}{s} Q(sx) \right), -Q(sx) \right),$$

and two multiplications are used instead of three in the 3-step lifting. In addition, two operations of rounding are used.

We use the following symbols for the two-step lifting scheme and its integer implementation, respectively:

$$\boxed{\begin{array}{c|c} & \frac{1-c}{s} \\ \hline -s & \end{array}} \quad \text{and} \quad \boxed{\begin{array}{c|c} -s & Q \\ \hline Q & \frac{1-c}{s} \end{array}}.$$

For the inverse lifting operations, we use the symbols

$$\boxed{\begin{array}{c|c} c-1 & \\ \hline \frac{}{s} & s \end{array}} \quad \text{and} \quad \boxed{\begin{array}{c|c} Q & s \\ \hline \frac{c-1}{s} & Q \end{array}}.$$

We can use these two-step lifting schemes and modify the paired eight-point DFT by approximating the multiplications by factors $W^1 = 0.7071 - 0.7071j$ and $W^3 = -0.7071 - 0.7071j$. As an example, Figure 2.5 shows the signal-flow graph of the integer eight-point DFT of the signal $f = \{1, 2, 3, 4, 5, 6, 7, 8\}$. Two blocks of the two-step lifting schemes are used for computing the integer approximations of multiplications of -4 by factors W^1 and W^3, respectively.

The multiplications of -4 by W^1 and W^3 are calculated as follows:

$$-4 \cdot W \to \boxed{\begin{array}{c|c} -0.7071 & Q \\ \hline Q & 0.4142 \end{array}} \circ \begin{bmatrix} -4 \\ 0 \end{bmatrix} = \begin{bmatrix} -3 \\ 3 \end{bmatrix},$$

$$-4 \cdot W^3 \to \boxed{\begin{array}{c|c} -0.7071 & Q \\ \hline Q & 2.4142 \end{array}} \circ \begin{bmatrix} -4 \\ 0 \end{bmatrix} = \begin{bmatrix} 3 \\ 3 \end{bmatrix}.$$

It follows from Figure 2.5 that the implementation of the two-step lifting schemes in the paired algorithm results in the following approximation of the 8-point DFT of the discrete signal f :

$$\begin{array}{|ll|}
\hline
F_0' = 36 & \\
F_1' = -4 + 10j & F_7' = -4 - 10j \\
F_2' = -4 + 4j & F_6' = -4 - 4j \\
F_3' = -4 + 2j & F_5' = -4 - 2j \\
F_4' = -4 & \\
\hline
\end{array} \qquad (2.2)$$

where we denote by F_p', $p = 0 : 7$, the components of the integer paired DFT. The property of complex conjugate holds for this transform. Four multiplications (not six) are required and the root-mean-square error of approximation equals

$$\varepsilon' = \sqrt{\frac{1}{8} \sum_{p=0}^{7} |F_p' - F_p|^2} = 0.2426$$

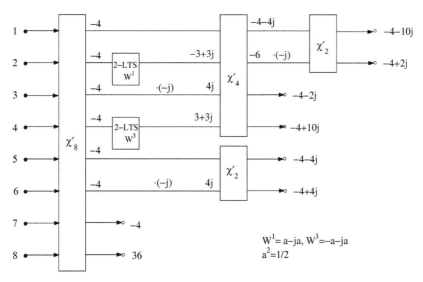

FIGURE 2.5
Signal-flow graph of the lifting scheme implementation for calculating the integer 8-point DFT of the signal $f(n) = \{1, 2, 3, 4, 5, 6, 7, 8\}$. Two-step lifting schemes are used to multiply two outputs -4 by twiddle factors W^1 and W^3.

which is less than the error ε_{3ls}. The pointwise errors, $F_p - F'_p$, of the integer transform are given in the following table:

p	0	1	2	3	4	5	6	7
error	0	$-0.3431j$	0	$-0.3431j$	0	$0.3431j$	0	$0.3431j$

Example 2.2
Consider the signal $\mathbf{x} = (1, 2, 4, 4, 3, 7, 5, 8)'$. The complete signal-flow graph of the integer eight-point DFT of this signal is given in Figure 2.6. Two two-step lifting schemes are used for computing the integer approximations of multiplications of -5 and -4 by the factors W^1 and W^3, respectively. The multiplication of the number -5 by W through the lifting scheme equals $-3 + 4j$,

$$-5 \cdot W \rightarrow \boxed{\begin{array}{c|c} -0.7071 & \mathcal{Q} \\ \hline \mathcal{Q} & 0.4142 \end{array}} \circ \begin{bmatrix} -5 \\ 0 \end{bmatrix} = \begin{bmatrix} -3 \\ 4 \end{bmatrix},$$

and the multiplication of -4 by W^3 through the lifting scheme equals $3 + 3j$.

Figure 2.7 shows for comparison the signal-flow graph with two three-step lifting schemes for integer approximation of the DFT of the signal $\mathbf{f}$ by the

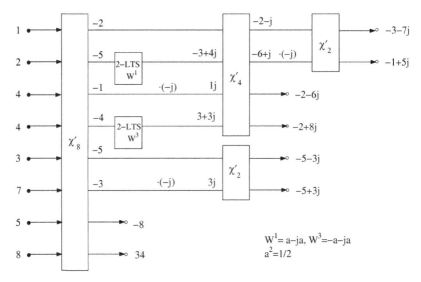

FIGURE 2.6

Signal-flow graph with two lifting scheme implementations for calculating the integer 8-point DFT of the signal $f(n) = \{1, 2, 4, 4, 3, 7, 5, 8\}$.

split-radix algorithm. The integer approximations of multiplications of complex numbers $-5 + 4j$ and $-5 - 4j$ by the factors W^1 and W^3 are calculated respectively as

$$-0.7071 + 6.3640j = (-5+4j) \cdot W \rightarrow \begin{array}{|c|c|c|} \hline Q & -0.7071 & Q \\ \hline 0.4142 & Q & 0.4142 \\ \hline \end{array} \circ \begin{bmatrix} -5 \\ 4 \end{bmatrix} = \begin{bmatrix} -1 \\ 6 \end{bmatrix}$$

$$0.7071 + 6.3640j = (-5-4j) \cdot W^3 \rightarrow \begin{array}{|c|c|c|} \hline Q & -0.7071 & Q \\ \hline 2.4142 & Q & 2.4142 \\ \hline \end{array} \circ \begin{bmatrix} -5 \\ -4 \end{bmatrix} = \begin{bmatrix} 2 \\ 7 \end{bmatrix}.$$

The values of these two integer transforms together with the DFT are given in Table 2.2, and the pointwise errors in Table 2.3.

We obtain

$$\sum_{p=0}^{7} e_{2LS}(p) = 0, \qquad \sum_{p=0}^{7} e_{3LS}(p) = 0.$$

and for the root-mean-square errors, $\varepsilon_{p;2LS} = 0.5298 < \varepsilon_{p;3LS} = 0.7574$.

The property of complex conjugate of the Fourier transform components F_p and F_{8-p}, $p = 1, 2, 3$, does not hold for these two integer approximations of the DFT. Indeed, for the algorithm with the 3-step lifting scheme, we have $F_{7,3LS} = -4 - 10j \neq \bar{F}_{1,3LS} = -4 + 9j$, and $F_{5,3LS} = -4 - j \neq \bar{F}_{3,3LS} = -4 + 2j$. This property $F_p = \bar{F}_{8-p}$ does not hold also for the paired algorithm, although two multiplications less are used in this algorithm and the error of

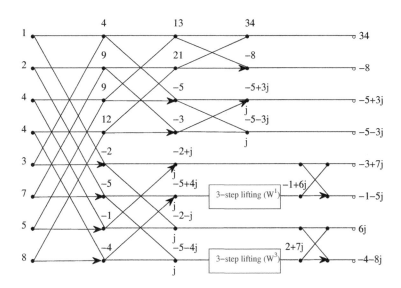

FIGURE 2.7

Signal-flow graph with two lifting scheme implementations for calculating the integer 8-point DFT of the signal $f(n) = \{1, 2, 4, 4, 3, 7, 5, 8\}$.

TABLE 2.2

Discrete Fourier transforms of $f(n)$

p	F_p	$F_{p,2LS}$	$F_{p,3LS}$	
0	34	34	34	
1	$-2.7071 + 7.3640j$	$-2 + 8j$	$-3 + 7j$	$F_1 \neq \bar{F}_7$
2	$-5 + 3j$	$-5 + 3j$	$-5 + 3j$	$F_2 = \bar{F}_6$
3	$-1.2929 + 5.3640j$	$-1 + 5j$	$6j$	$F_3 \neq \bar{F}_5$
4	-8	-8	-8	
5	$-1.2929 - 5.3640j$	$-2 - 6j$	$-1 - 5j$	
6	$-5 - 3j$	$-5 - 3j$	$-5 - 3j$	
7	$-2.7071 - 7.3640j$	$-3 - 7j$	$-4 - 8j$	

TABLE 2.3

Errors of approximation of DFT

p	1	3	5	7
e_{2LS}	$-0.7071 - 0.6360j$	$-0.2929 + 0.3640j$	$0.7071 + 0.6360j$	$0.2929 - 0.3640j$
e_{3LS}	$0.2929 + 0.3640j$	$-1.2929 - 0.6360j$	$-0.2929 - 0.3640j$	$1.2929 + 0.6360j$

approximation is smaller than when compared with the split-radix algorithm.
[]

The above examples show that the lifting schemes in both paired and split-radix algorithms do not provide good results. We next consider the concept of the reversible integer Fourier transform with control bits, which has been implemented for calculation of the DFT by the paired algorithm.

2.3   One-point integer transform

In this section, we describe one-point integer transforms which will be applied for calculating a new reversible integer-to-integer discrete transform which is an approximation of the Fourier transform. Our goal is to define reversible integer transforms for the multiplications of integer numbers by the twiddle factors which are used in the discrete Fourier transform.

We describe a one-point integer transformation which we call a *transformation with one additional bit (TOAB)*. Given a real factor $a \in (0.5, 1)$, the following transformation of integers $x \neq 0$ is considered

$$\mathcal{A} = \mathcal{A}_a : x \rightarrow \begin{cases} \vartheta_0 = [ax], \\ \vartheta_1 = \dfrac{1 + \text{sign}(ax - \vartheta_0)}{2}, \end{cases} \tag{2.3}$$

where $[.]$ denotes the round function and $\text{sign}(t)$ is the sign function which equals 1 when $t > 0$, and -1 when $t < 0$. It is assumed that $\text{sign}(0) = -1$. The transformation $\mathcal{A}$ is a one-to-one transformation and has the following property. For any integer ϑ_0, there are at most two different integer inputs x and $x + 1$ which may have the same image component ϑ_0, i.e., $\vartheta_0(x) = \vartheta_0(x + 1)$. In such a case, the second components of the transforms $\mathcal{A}[x]$ and $\mathcal{A}[x + 1]$ are different, i.e., $\vartheta_1(x) \neq \vartheta_1(x + 1)$. It should also be noted that in the general case, $\mathcal{A}(-x) \neq -\mathcal{A}(x)$. However, we can define the transform $\mathcal{A}(x)$ over negative integer numbers x as $\mathcal{A}(x) = -\mathcal{A}(-x)$. The product ax can be written as $[ax] + b$, where $b \in [-0.5, 0.5)$, and the transformation $\mathcal{A}$ as

$$\mathcal{A} : x \rightarrow \begin{cases} \vartheta_0 = [ax] \\ \begin{cases} 0, & \text{if } b \leq 0 \\ 1, & \text{if } b > 0. \end{cases} \end{cases} \tag{2.4}$$

Example 2.3
Let $a = \sqrt{2}/2 = 0.7071$. For integers equal $x = 1, 5$, and 6, we obtain the following TOAB:

$$\mathcal{A} : 1 \rightarrow \begin{cases} \vartheta_0 = [0.7071] = 1 \\ \vartheta_1 = 0 \end{cases}$$

$$\mathcal{A}: 5 \rightarrow \begin{cases} \vartheta_0 = [0.7071 \cdot 5] = [3.5355] = 4 \\ \vartheta_1 = 0 \end{cases}$$

$$\mathcal{A}: 6 \rightarrow \begin{cases} \vartheta_0 = [0.7071 \cdot 6] = [4.2426] = 4 \\ \vartheta_1 = 1. \end{cases}$$

⬜

It can be seen from these examples that when rounding the multiplication by factor a, we also save information about a way this rounding is performing, by the floor or ceiling function. The binary parameter ϑ_1 refers to as a *control parameter* or *control bit* that allows for performing the inverse integer transformation $\mathcal{A}^{-1}$. The inverse transformation is defined by

$$\mathcal{A}^{-1} = \mathcal{A}_a^{-1} : \vartheta_0 \rightarrow x = \begin{cases} \left\lfloor \left| \dfrac{\vartheta_0}{a} \right| \right\rfloor, & \text{if } \vartheta_1 = 0 \\ \left\lceil \left| \dfrac{\vartheta_0}{a} \right| \right\rceil, & \text{if } \vartheta_1 = 1 \end{cases} \tag{2.5}$$

where $\lfloor . \rfloor$ denotes the flooring function and $\lceil . \rceil$ denotes the ceiling function. When the control bit $\vartheta_1 = 1$, the rounding in $\mathcal{A}$ transform has been performed by the floor function, then $\mathcal{A}^{-1}$ inverse transform uses the ceiling function, and vice versa when $\vartheta_1 = 0$.

Example 2.4
Let $a = \sqrt{2}/2 = 0.7071$. For $x = 1$, we have the following pair of integer transforms:

$$\mathcal{A}: x = 1 \rightarrow \begin{cases} \vartheta_0 = 1 \\ \vartheta_1 = 0 \end{cases}$$

$$\mathcal{A}^{-1} : \vartheta_0 = 1 \rightarrow \left\lfloor \frac{1}{0.7071} \right\rfloor = \lfloor 1.4142 \rfloor = 1, \text{ since } \vartheta_1 = 0.$$

We now consider values of transforms for the cases when $x = 5$ and 6 :

$$\mathcal{A}: 5 \rightarrow \begin{cases} \vartheta_0 = 4 \\ \vartheta_1 = 0 \end{cases}, \qquad \mathcal{A}: 6 \rightarrow \begin{cases} \vartheta_0 = 4 \\ \vartheta_1 = 1 \end{cases}.$$

In both cases, the result of multiplication by factor a is approximated by $\vartheta_0 = 4$. Since $\vartheta_0/0.7071 = 5.6569$, the use of control bits 0 and 1 results in the following inverse transforms:

$$\mathcal{A}^{-1} : 4 \rightarrow \begin{cases} \lfloor 5.6569 \rfloor = 5, & \text{for } x = 5, \text{ since } \vartheta_1 = 0 \\ \lceil 5.6569 \rceil = 6, & \text{for } x = 6, \text{ since } \vartheta_1 = 1. \end{cases}$$

⬜

The $a = 0.5$ case can be considered separately. The transformation $\mathcal{A}_{0.5}$ can be defined as in (2.3), but under the condition that $[0.5] = 0$ and $[-0.5] = -1$. The inverse transformation $\mathcal{A}_{0.5}^{-1}$ can be defined as $\mathcal{A}_{0.5}^{-1}\{\vartheta_0, \vartheta_1\} = 2\vartheta_0 + \vartheta_1$. For instance, if $x = 3$, then

$$\mathcal{A}_{0.5} : 3 \to \begin{cases} \vartheta_0 = [1.5] = 1 \\ \vartheta_1 = 1 \end{cases} \quad \mathcal{A}_{0.5}^{-1} : \begin{bmatrix} 1 \\ 1 \end{bmatrix} \to 2\vartheta_0 + \vartheta_1 = 3.$$

The implementation of integer multiplication ax by a factor $a \in (0, 0.5)$ is not reversible. For instance, when $a = 0.3827$ and $x = 4$, we have $\mathcal{A}_a : 4 \to \{\vartheta_0 = [1.5308] = 2, \vartheta_1 = 0\}$, and $\mathcal{A}_a^{-1} : \{2, 0\} \to \lfloor 5.2260 \rfloor = 5 \neq 4$. However, the following should be noted. In the general $N > 2$ case, the absolute value of at least one of the components of twiddle factors $W_N^k = \exp(-j2\pi k/N) = w_1 - jw_2$, $k = 1 : (N-1)$ lies in the interval $[0.5, 1]$. This property allows us to use only one control bit for multiplication of the integer x by complex twiddle factors W_N^k.

As an example, we consider two twiddle factors $W_{16}^1 = 0.9239 - 0.3827j$ and $W_{16}^3 = 0.3827 - 0.9239j$. In the first case when $w_1 = 0.9239$ and $w_2 = 0.3827$, to approximate the pair of products $(w_1 x, w_2 x)$, we define the following one-to-three reversible transformation:

$$\mathcal{B} : x \to \begin{cases} \vartheta_0 = \lceil w_1 x \rceil \\ \vartheta_1 = \dfrac{1 + \text{sgn}(w_1 x - \vartheta_0)}{2} \\ \vartheta_2 = \lceil w_2 x \rceil \end{cases} = \begin{cases} \mathcal{A}_{w_1}(x) \\ \vartheta_2 = \lceil w_2 x \rceil \end{cases} \tag{2.6}$$

This transform uses two multiplications, one 'if' operation, and two rounding.

Example 2.5
For integer $x = 4$, the transform $\mathcal{B}(4) = [4, \underline{0}, 2]$, where the control bit is underlined. Indeed

$$\mathcal{B} : 4 \to \begin{cases} \vartheta_0 = \lceil 0.9239 \cdot 4 \rceil = \lceil 3.6956 \rceil = 4 \\ \vartheta_1 = \underline{0} \\ \vartheta_2 = \lceil 0.3827 \cdot 4 \rceil = \lceil 1.5308 \rceil = 2. \end{cases} \tag{2.7}$$

Thus, the integer approximation of multiplication

$$4W_{16}^1 = 4[0.9239 - 0.3827j] = 3.6955 - 1.5307j$$

by $\mathcal{B}$ transform equals $4 - 2j$. To perform the inverse transform, it is enough to use only the first two values of the output, i.e., $[4, \underline{0}]$. There is no need to process the output 2. The calculations can be fulfilled as follows:

$$\vartheta_0 \to \frac{\vartheta_0}{w_1} = \frac{4}{0.9239} = 4.3295 \to \lfloor 4.3295 \rfloor = 4 \ \ (\text{since } \vartheta_1 = 0).$$

The control bit shows that the first rounding in $\mathcal{B}$ transform has been performed by the ceiling function and for the inverse transform the floor function is to be used. Thus, the inverse transform is defined as $\mathcal{B}^{-1}\{\vartheta_0, \vartheta_1, \vartheta_2\} = \mathcal{A}_{w_1}^{-1}\{\vartheta_0, \vartheta_1\}$.   □

In a similar way, the case when $w_1 = 0.3827$ and $w_2 = 0.9239$, i.e., $w_1 < w_2$, is considered. The multiplication of an integer x by complex factor $a = w_1 - jw_2$ is approximated as the number $(\vartheta_0 - j\vartheta_1)$, which is calculated by

$$\mathcal{B}: x \rightarrow \begin{cases} \vartheta_0 = [w_1 x] \\ \vartheta_2 = [w_2 x] \\ \vartheta_1 = \dfrac{1 + \text{sgn}(w_2 x - \vartheta_0)}{2} \end{cases} = \begin{cases} \vartheta_0 = [w_1 x] \\ \mathcal{A}_{w_2}(x). \end{cases} \tag{2.8}$$

The inverse transform is defined as $\mathcal{B}^{-1}\{\vartheta_0, \vartheta_2, \vartheta_1\} = \mathcal{A}_{w_2}^{-1}\{\vartheta_2, \vartheta_1\}$. The control bit is used thus for saving information about the rounding of the multiplication of the integer x by the biggest real w_1 or imaginary w_2 part of the twiddle factor a. If $w_1, w_2 > 0.5$, then the control bit can be defined by either part.

Example 2.6

Let $x = 7$, $N = 15$, and $W_{15}^1 = 0.9135 - 0.4067j$. In this case, $w_1 = 0.9135$, $w_2 = 0.4067$, and the multiplication $7W_{15}^1 = 6.3948 - 2.8472j$ is approximated by $6 - 3j$ as follows:

$$\mathcal{B}: 7 \rightarrow \begin{cases} \vartheta_0 = 6 \\ \vartheta_1 = 1 \\ \vartheta_2 = 3 \end{cases} \text{ and } \mathcal{A}_{w_1}^{-1}: \begin{bmatrix} 6 \\ 1 \end{bmatrix} = \left\lfloor \frac{6}{w_1} = 6.5678 \right\rfloor = 7 \text{ (since } \vartheta_1 = 1).$$

We now take the factor $W_{15}^2 = 0.6691 - 0.7431j$. Then $w_1 = 0.6691$, $w_2 = 0.7431$, and the multiplication $7W_{15}^2 = 4.6839 - 5.2020j$ is approximated by $5 - 5j$ as follows:

$$\mathcal{B}: 7 \rightarrow \begin{cases} \vartheta_0 = 5 \\ \vartheta_1 = 0 \\ \vartheta_2 = 5 \end{cases} \text{ and } \mathcal{A}_{w_1}^{-1}: \begin{bmatrix} 5 \\ 0 \end{bmatrix} = \left\lfloor \frac{5}{w_1} = 7.4724 \right\rfloor = 7 \text{ (since } \vartheta_1 = 0).$$

If we define the control bit from the multiplication by $w_2 = 0.7431$, then the multiplication $7W_{15}^2$ is approximated by $5 - 5j$ as follows:

$$\mathcal{B}: 7 \rightarrow \begin{cases} \vartheta_0 = 5 \\ \vartheta_1 = 1 \\ \vartheta_2 = 5 \end{cases} \text{ and } \mathcal{A}_{w_2}^{-1}: \begin{bmatrix} 5 \\ 1 \end{bmatrix} = \left\lfloor \frac{5}{w_2} = 6.7282 \right\rfloor = 7 \text{ (since } \vartheta_1 = 1).$$

□

2.3.1   The eight-point integer Fourier transform

The N-point paired transform is an integer-to-integer transform that does not require multiplications, but $2N - 2$ additions and subtractions. As shown in

Figure 1.9, the calculation of the eight-point DFT by the paired transformations χ'_8, χ'_4, and χ'_2 requires only two noninteger operations of multiplication

$$(1 - j)ax = ax - jax, \quad \text{and} \quad (-1 - j)ax = -ax - jax, \quad (2.9)$$

where $a = \sqrt{2}/2$. The multiplications by factor a, which are required on the
first stage of the calculation, result in noninteger values of components F_1,
F_3, F_5, and F_7. Namely, such multiplications are performed with the second
and fourth outputs of the eight-point paired transform $\chi'_8 \circ f$.

We can modify the signal-flow graph of the eight-point DFT by approxima-
ting the multiplications by factor a in (2.9) with integer transforms $\mathcal{A} = \mathcal{A}_a$
defined in (2.3). By doing that, we obtain integer outputs composing together
the N-point discrete transform, which we call *the integer DFT*. The signal-
flow graph of the integer eight-point DFT is given in Figure 2.8. Two blocks
with the one-point integer transform $\mathcal{A}$ have been added to the signal-flow
graph of Figure 1.9. The multiplications of the values $f'_{1,1}$ and $f'_{1,3}$ of the
eight-point paired transform by factors of $(1 - j)a$ and $(-1 - j)a$ are replaced
respectively by approximations $(1 - j) \cdot \vartheta_0(f'_{1,1})$ and $(-1 - j) \cdot \vartheta_0(f'_{1,3})$. The
second binary outputs $\vartheta_1(f'_{1,1})$ and $\vartheta_1(f'_{1,3})$ of the applied transform $\mathcal{A}$ are
considered as additional outputs (two control bits) of the Fourier transform,
and we denote them by α_1 and α_2.

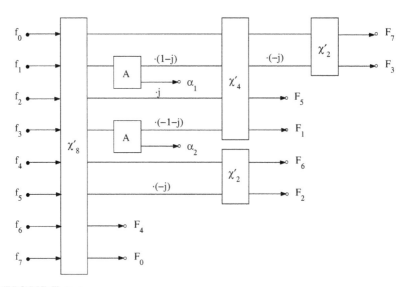

FIGURE 2.8
Signal-flow graph of calculation of the integer eight-point DFT with two control bits.

The transformation $\{f_0, f_1, ..., f_7\} \rightarrow \{F_0, F_1, ..., F_7, \alpha_1, \alpha_2\}$ is called the eight-point *integer discrete Fourier transform with two control bits* (2cb-IDFT).

Example 2.7

Let f be the sequence $\{1, 2, 4, 4, 3, 7, 5, 8\}$. The eight-point DFT of f consists of the following data:

$$
\begin{array}{|ll|}
\hline
F_0 = 34 & \\
F_1 = -2.7071 + 7.3640j & F_7 = -2.7071 - 7.3640j \\
F_2 = -5 + 3j & F_6 = -5 - 3j \\
F_3 = -1.2929 + 5.3640j & F_5 = -1.2929 - 5.3640j \\
F_4 = -8 & \\
\hline
\end{array}
\qquad (2.10)
$$

where we write the complex conjugate values of the DFT by pairs, because of the real input f. To calculate the eight-point integer DFT with two control bits, we first perform the paired transform of the sequence, which results in the output (see Figure 2.9)

$$
\chi'_8 \circ f = \Big\{ \underbrace{-2, -5, -1, -4}_{\text{signal } f'_1}, \underbrace{-5, -3}_{\text{signal } f'_2}, -8, 34 \Big\}.
$$

On this stage of calculation, we obtain $F_0 = 34$ and $F_4 = -8$. Then, the splitting-signal $f'_2 = \{-5, -3\}$ is multiplied by the weighted coefficients $\{1, -j\}$ and the two-point paired transform is calculated over the new signal $\{-5, 3j\}$, which results in the outputs $F_6 = -5 - 3j$ and $F_2 = -5 + 3j$.

On the next stage, the first splitting-signal $f'_1 = \{-2, -5, -1, -4\}$ should be multiplied by the weighted coefficients $\{1, (1 - j)a, -j, (-1 - j)a\}$ and processed then by the four-point paired transform. The multiplication of -5 by factor a and then by $(1 - j)$ is approximated by the TOAB as

$$
-5 \rightarrow \begin{cases} \vartheta_0 = -[0.7071 \cdot 5] = -4 \\ \vartheta_1 = 1 \end{cases} \rightarrow \begin{cases} (1 - j)\vartheta_0 = -4 + 4j \\ \alpha_1 = 1. \end{cases}
$$

In a similar way, the multiplication of -4 by factor a and then by $(-1 - j)$ is approximated by the TOAB as

$$
-4 \rightarrow \begin{cases} \vartheta_0 = -[0.7071 \cdot 4] = -3 \\ \vartheta_1 = 1 \end{cases} \rightarrow \begin{cases} (-1 - j)\vartheta_0 = 3 + 3j \\ \alpha_2 = 1. \end{cases}
$$

The new weighted splitting-signal becomes $f'_{1;new} = \{-2, -4 + 4j, j, 3 + 3j\}$ and the four-point paired transform of this signal results in the following data:

$$
\chi'_8 \circ f'_{1;new} = \{-2 - j, -7 + j, -1 - 6j, -3 + 8j\}.
$$

On this stage of calculation, we obtain $F_5 = -1 - 6j$ and $F_1 = -3 + 8j$. On the last stage, the first two components of the data, $\{-2 - j, -7 + j\}$, multiplied

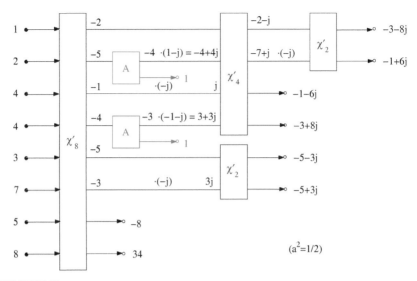

FIGURE 2.9

Signal-flow graph of calculation of the integer eight-point DFT of the signal $\{1, 2, 4, 4, 3, 7, 5, 8\}$.

by weighted coefficients $\{1, -j\}$ yield the components $F_7 = -3 - 8j$ and $F_3 = -1 + 6j$, after calculating the two-point paired transform.

The final result of the eight-point integer DFT with two control bits $\{\{F_p; p = 0 : 7\}, \alpha_1, \alpha_2\}$ equals

$$
\begin{vmatrix}
F_0' = 34 & \\
F_1' = -3 + 8j & F_7' = -3 - 8j \\
F_2' = -5 + 3j & F_6' = -5 - 3j \\
F_3' = -1 + 6j & F_5' = -1 - 6j \\
F_4' = -4 &
\end{vmatrix}
\tag{2.11}
$$

One can see that the property of the Fourier transform for a real input, which is expressed by $F_{N-p} = \bar{F}_p$, $p = 1 : 3$, holds for the considered integer discrete Fourier transform, too. Here $\bar{F}_p$ denotes the complex conjugate of F_p.

The pointwise errors of the integer transform are given in the following table together with the errors of the paired algorithm of the DFT with two-step lifting schemes:

p	1	3	5	7
e_{2cb}	$0.2929 - 0.6360j$	$-0.2929 - 0.6360j$	$-0.2929 + 0.6360j$	$0.2929 + 0.6360j$
e_{2LS}	$-0.7071 - 0.6360j$	$-0.2929 + 0.3640j$	$0.7071 + 0.6360j$	$0.2929 - 0.3640j$

The root-mean-square error equals $\varepsilon_{p;2cb} = 0.4951 < \varepsilon_{p;2LS} = 0.5298 < \varepsilon_{p;3LS} = 0.7574$. ☐

2.3.2 Eight-point inverse integer DFT

In this section, we describe a way to perform a transform that is inverse to the proposed integer DFT with two control bits. The inverse is not just the conjugate DFT with control bits.

Figure 2.10 shows the signal-flow graph of the inverse eight-point integer DFT with the inverse four-point paired transform χ_4', whose second and first outputs are processed by the inverse transform $\mathcal{A}^{-1}$. Two control bits α_1 and α_2 of the eight-point integer DFT are used for these transforms.

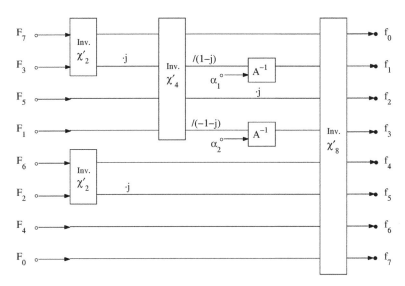

FIGURE 2.10
Signal-flow graph of calculation of the inverse integer eight-point DFT.

Example 2.8
We consider the eight-point integer DFT with two control bits that has been described in Example 2.7, when the input-to-output transformation is defined as (see 2.11)

$$\{1, 2, 4, 4, 3, 7, 5, 8\} \rightarrow \{-3 - 8j, -1 + 6j, -1 - 6j, \\ -3 + 8j, -5 - 3j, -5 + 3j, -8, 34, \underline{1}, \underline{1}\}.$$

The output of the transform is ordered in accordance with the paired transform.

The inverse eight-point integer DFT with two control bits of this output is calculated, as shown in the signal-flow graph of Figure 2.11, with two two-point inverse paired transforms on the first stage, and the four- and eight-point inverse paired transforms on the second and third stages of calculations, respectively. All values of inputs and outputs for these transforms are illustrated in the figure. Two inverse two-to-one integer transforms $\mathcal{A}^{-1}$ (each of which uses control bit 1) are calculated for dividing the second and fourth outputs $-4 + 4j$ and $3 + 3j$ of the inverse four-point paired transform by factor a.

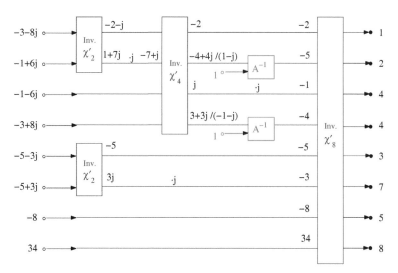

FIGURE 2.11
Signal-flow graph of calculation of the inverse integer eight-point DFT $\{1, 2, 4, 4, 3, 7, 5, 8\}$.

The matrices of the inverse 8-, 4-, and 2-point paired transformations are the following:

$$[(\chi_8')^{-1}] = \frac{1}{8}\begin{bmatrix} 4 & 0 & 0 & 0 & 2 & 0 & 1 & 1 \\ 0 & 4 & 0 & 0 & 0 & 2 & -1 & 1 \\ 0 & 0 & 4 & 0 & -2 & 0 & 1 & 1 \\ 0 & 0 & 0 & 4 & 0 & -2 & -1 & 1 \\ -4 & 0 & 0 & 0 & 2 & 0 & 1 & 1 \\ 0 & -4 & 0 & 0 & 0 & 2 & -1 & 1 \\ 0 & 0 & -4 & 0 & -2 & 0 & 1 & 1 \\ 0 & 0 & 0 & -4 & 0 & -2 & -1 & 1 \end{bmatrix}$$

$$[(\chi_4')^{-1}] = \frac{1}{4}\begin{bmatrix} 2 & 0 & 1 & 1 \\ 0 & 2 & -1 & 1 \\ -2 & 0 & 1 & 1 \\ 0 & -2 & -1 & 1 \end{bmatrix}, \quad [(\chi_2')^{-1}] = \frac{1}{2}\begin{bmatrix} 1 & 1 \\ -1 & 1 \end{bmatrix}.$$

▯

We now compare the result of calculating the 8-point IDFT with two control bits with the methods of integer entries (IE) of the transform matrix and the paired transform-based method of two-step lifting schemes (LS). The discrete signal $f = \{1, 2, 4, 4, 3, 7, 5, 8\}$ is considered. Results of these transforms are shown in Table 2.4.

TABLE 2.4
The 8-point DFT and integer DFTs of the signal $\{1, 2, 4, 4, 3, 7, 5, 8\}$

signal		DFT	IE-ITFT	2LS-IFFT	2cb-IDFT
1	F_0	34	34	34	34
2	F_1	$-2.7071 + 7.3640j$	$-5 + 11j$	$-2 + 8j$	$-3 + 8j$
4	F_2	$-5 + 3j$	$-5 + 3j$	$-5 + 3j$	$-5 + 3j$
4	F_3	$-1.2929 + 5.3640j$	$-3 + 7j$	$6j$	$-1 + 6j$
3	F_4	-8	-8	-8	-8
7	F_5	$-1.2929 - 5.3640j$	$-5 - 7j$	$-1 - 5j$	$-1 - 6j$
5	F_6	$-5 - 3j$	$-5 - 3j$	$-5 - 3j$	$-5 - 3j$
8	F_7	$-2.7071 - 7.3640j$	$-5 - 11j$	$-4 - 8j$	$-3 - 8j$
	a_1				1
	a_2				1
	ϵ		2.4530	0.5298	0.4951

The prototype matrix (8×8) of the eight-point integer Fourier transform (ITFT) with integer entries has been taken from [28], when the smallest integer solution for the unknowns is considered, i.e., $a_2 = c_2 = 1$ and $a_1 = a_3 = a_4 = c_1 = c_3 = c_4 = 2$. The matrix of this transform equals

$$
\begin{bmatrix}
1 & 1 & 1 & 1 & 1 & 1 & 1 & 1 \\
a_1 & a_2(1-j) & -ja_1 & a_2(-1-j) & -a_1 & a_2(-1+j) & ja_1 & a_2(1+j) \\
1 & -j & -1 & j & 1 & -j & -1 & j \\
c_1 & c_2(-1-j) & jc_1 & c_2(1-j) & -c_1 & c_2(1+j) & -jc_1 & c_2(-1+j) \\
1 & -1 & 1 & -1 & 1 & -1 & 1 & -1 \\
c_1 & c_2(-1+j) & -jc_1 & c_2(1+j) & -c_1 & c_2(1-j) & jc_1 & c_2(-1-j) \\
1 & j & -1 & -j & 1 & j & -1 & -j \\
a_1 & a_2(1+j) & ja_1 & a_2(-1+j) & -a_1 & a_2(-1-j) & -ja_1 & a_2(1-j)
\end{bmatrix}.
$$

Values of the ITFT at frequency-points $p = 1, 3$, and 5 differ much from the original values. For this example, the root-mean-square error of the approximation equals $\varepsilon_{IE} = 2.4530$ and exceeds five times the error provided by the 2cb-IDFT, $\varepsilon_{2cb} = 0.4951$. The 8-point ITFT uses 12 butterflies with 24 operations of addition. For the chosen set of unknowns, the ITFT uses 8 trivial multiplications. For other prototype matrices, the transform uses 8 multiplications. For instance, we can take the following set of unknowns: $a_1 = 7, a_2 = 5, a_3 = 18, a_4 = 13$, and $c_1 = 13, c_2 = 9, c_3 = 10, c_4 = 7$. The error of approximation by the ITFT in this case exceeds 2.4530. For real inputs, the 8-point ITFT requires 21 additions and at most four multiplications (plus a few factors for normalization).

The data of the 5th column in the table corresponds to the approximation of the 8-point split-radix DFT by the paired transform method of two-step lifting schemes. The error of approximation of the 8-point DFT of the given signal by this method equals $\varepsilon_{LS} = 0.5298$ and is greater than the 2cb-IDFT. The three-step lifting scheme in the split-radix algorithm, which is described in Example 2.2, uses 12 butterflies and two three-step lifting schemes for multiplications by factors W_8^1 and W_8^3. When the input is real, and when assuming that the complex conjugate property is valid for the lifting scheme-based integer approximation of the DFT, ten and a half butterflies and one lifting scheme can be used. The lifting scheme is performed by 3 multiplications and 3 additions. Therefore, the total number of operations could be reduced to $(10 \times 2 + 1) + 3 = 24$ additions and 3 multiplications. The use of the lifting scheme increases the number of bits for the output, and for $N = 8$ the least upper bound of this number is 3 bits [29]. The proposed IDFT with two control bits requires 2 multiplications, and 20 additions as the fast paired DFT, plus two 'if' operations and two additional bits.

2.3.3   General method of control bits

The method of control bits can also be applied for computing the DFT of higher orders $N \geq 8$, when different twiddle factors are used. To demonstrate this, we consider the paired algorithm of the 16-point DFT, which requires 12 multiplications by factors from the set $\{W_{16}^t; t = 1 : 7\}$. Values of these factors are the following (with accuracy of four digits after the point):

$$
\begin{array}{ll}
W_{16}^1 = 0.9239 - 0.3827j, & W_{16}^2 = 0.7071 - 0.7071j \\
W_{16}^3 = 0.3827 - 0.9239j, & W_{16}^5 = -0.3827 - 0.9239j \\
W_{16}^6 = -0.7071 - 0.7071j, & W_{16}^7 = -0.9239 - 0.3827j
\end{array}
$$

and $W_{16}^4 = -j$ is a "trivial" factor. The integer approximation of multiplications by factors $W_{16}^2 = W_8^1$ and $W_{16}^6 = W_8^3$ have been described above for the 8-point DFT.

We consider thus the remaining four factors, which can be written in the form $w = \pm w_1 - jw_2$, where numbers $w_2 \neq w_1 = 0.9239$ or 0.3827. The multiplication of an integer x by each of such a complex factor w and its inverse operation can be implemented by the integer transforms $\mathcal{B}$ and $\mathcal{B}^{-1}$ with one additional bit, which have been defined in (2.6)-(2.8).

2.3.4   16-point IDFT with 8 and 12 control bits

We now consider the implementation of the one-point integer transformation $\mathcal{B}$ for computing the 16-point integer DFT by the paired transforms. The approximation of each multiplication of an integer by the twiddle factor of the set $\{W_{16}^t; t = 1, 3, 5, 7\}$ requires two multiplications and one control bit. The integer approximation of four multiplications by two twiddle factors $W_{16}^t = (\pm 1 - j)a$, for $t = 2, 6$, requires one multiplication and one control bit each. Therefore, the 16-point DFT by the paired transforms can be approximated by the integer transform requiring 8 control bits. The number of multiplications equals $4(2) + 4 = 12$. Figure 2.12 shows the signal-flow graph of the 16-point integer DFT. Since the input is considered to be real, the following components are complex conjugates of each other: $F_{15} = \bar{F}_1$, $F_7 = \bar{F}_9$, $F_{11} = \bar{F}_5$, $F_3 = \bar{F}_{13}$, $F_{14} = \bar{F}_2$, $F_6 = \bar{F}_{10}$, and $F_4 = \bar{F}_{12}$. Three incomplete paired transforms χ_8', χ_4', and χ_2' are used in this algorithm. Together with the 16-paired transform χ_{16}' and other χ_2', these transforms require $30 + [2(9) + 2(3) + 0] + 2(2) = 58$ additions. The implementation of eight $\mathcal{B}$ and $\mathcal{A}$ operations uses 12 multiplications, 8 'if' operations and 8 control bits. Thus, the 16-point DFT when using this graph requires 12 multiplications, 58 additions, and 8 control bits.

If we use the method of signal-flow graph simplification for the 16-point DFT (which is given in Figure 1.14) with 12 real multiplications by real and imaginary parts of twiddle factors, then the integer approximations of these multiplications will lead to the 16-point fast IDFT with 12 multiplications, 58 additions, and 12 control bits (plus 12 'if' operations).

2.3.5 Inverse 16-point integer DFT

The following should be noted for the simplified block-diagram of Figure 2.12. This diagram can be used only for calculation of the forward 16-point integer DFT, when assuming that the property of the complex conjugate ($F_{16-p} = \bar{F}_p$) holds for the integer approximation of the Fourier transform by the proposed paired transform method. Thus we force this property to be true, regardless of the results of the omitted part in the block-scheme of the paired algorithm. To perform the inverse transform of the obtained integer complex data F_p, $p = 0 : 15$, we need to use the property of the complex conjugate and reconstruct a few steps in the calculation, which were omitted during the simplification of the graph.

Indeed, let us consider the inverse procedure, by examining the block diagram from the bottom. From the value of F_{12}, the inputs of the two-point paired transform can be reconstructed as shown in Figure 2.13. There is no need to consider the second output as $F_4 = \bar{F}_{12}$, since the 13th and 14th outputs of the 16-point paired transform are real.

Now we need to compute the first two outputs of the 4-point paired transform by knowing F_2 and F_{10}. Since we are assuming the property of the complex conjugate for the integer Fourier transform, the calculation of these two outputs can be done as shown in Figure 2.14. The outputs are A and jB.

Finally, we will define the first four outputs of the 8-point paired transform by knowing F_1, F_9, F_5, and F_{13}. For that, we first look again on the full block-diagram of the 16-point DFT with control bits, which is given in Figure 2.15. In this diagram, eight control bits are used on the first stage and two integer

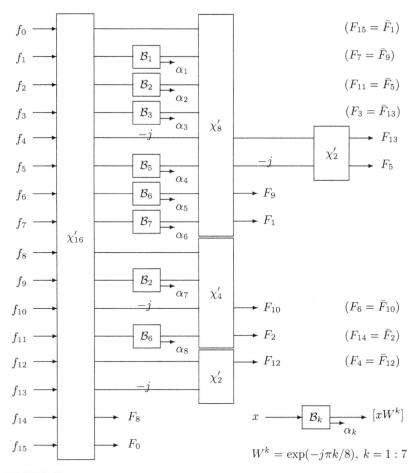

FIGURE 2.12
Block-diagram of the 16-point IDFT with 8 control bits, when the input is real.

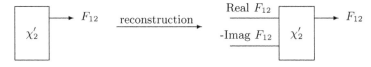

FIGURE 2.13
Reconstruction of the inputs of the 2-point paired transform.

rotations on the second stage. We need four additional control bits (α_9, α_{10}) and $(\alpha_{11}, \alpha_{12})$, which are missing from the multiplication of the 2nd and 4th outputs of the eight-point paired transform by factors W^2 and W^6, respectively. These two outputs are complex, and therefore four control bits are

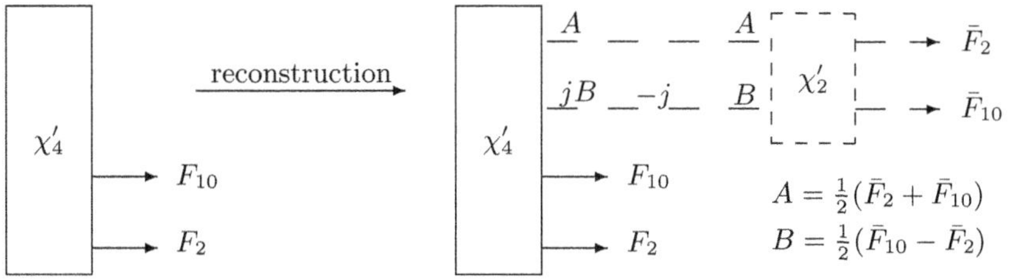

FIGURE 2.14
Reconstruction of the outputs of the 4-point paired transform.

required. These complex multiplications are operations of rotation, which we denote by C_2 and C_6, respectively. Four additional control bits are not shown in the figure.

The outputs of the 8-point paired transform can be calculated as shown in Figure 2.16. All outputs of the 4-point paired transform are calculated as for the previous stage,

$$\chi_4'[\text{input}] = \left\{ C = \frac{1}{2}(\bar{F}_1 + \bar{F}_9), Dj = \frac{j}{2}(\bar{F}_9 - \bar{F}_1), \bar{F}_5, \bar{F}_{13} \right\}.$$

We denote this input as a vector $\mathbf{z} = (z_0, z_1, z_2, z_3)'$. Then, by using the inverse paired transform, we can calculate these inputs as

$$\begin{bmatrix} z_0 \\ z_1 \\ z_2 \\ z_3 \end{bmatrix} = \frac{1}{4} \begin{bmatrix} 2 & 0 & 1 & 1 \\ 0 & 2 & -1 & 1 \\ -2 & 0 & 1 & 1 \\ 0 & -2 & -1 & 1 \end{bmatrix} \begin{bmatrix} \frac{1}{2}(\bar{F}_1 + \bar{F}_9) \\ \frac{j}{2}(\bar{F}_9 - \bar{F}_1) \\ \bar{F}_5 \\ \bar{F}_{13} \end{bmatrix} = \frac{1}{4} \begin{bmatrix} 1 & 0 & 1 & 1 \\ 0 & j & -1 & 1 \\ -1 & 0 & 1 & 1 \\ 0 & -j & -1 & 1 \end{bmatrix} \begin{bmatrix} \bar{F}_1 + \bar{F}_9 \\ \bar{F}_9 - \bar{F}_1 \\ \bar{F}_5 \\ \bar{F}_{13} \end{bmatrix}.$$

At this stage, we need four control bits, $\alpha_9, \alpha_{10}, \alpha_{11}$, and α_{12}, to reconstruct the 2nd and 4th values of the output of the 8-point paired transform, which we denote by χ_1 and χ_3. For that, we use the two-to-one inverse integer transformation $\mathcal{A}_a^{-1}$, where $a = 1/\sqrt{2} = 0.7071$, as follows:

$$z_1 \rightarrow \frac{z_1}{1 - j} = x_1 + jy_1 = \begin{bmatrix} x_1 \\ y_1 \end{bmatrix} \rightarrow \begin{bmatrix} \mathcal{A}_a^{-1}(x_1, \alpha_9) \\ \mathcal{A}_a^{-1}(y_1, \alpha_{10}) \end{bmatrix} \rightarrow \chi_1$$

where $\chi_1 = \mathcal{A}_a^{-1}(x_1, \alpha_9) + j\mathcal{A}_a^{-1}(y_1, \alpha_{10})$, and

$$z_3 \rightarrow -\frac{z_3}{1 + j} = x_3 + jy_3 = \begin{bmatrix} x_3 \\ y_3 \end{bmatrix} \rightarrow \begin{bmatrix} \mathcal{A}_a^{-1}(x_3, \alpha_{11}) \\ \mathcal{A}_a^{-1}(y_3, \alpha_{12}) \end{bmatrix} \rightarrow \chi_3$$

where $\chi_3 = \mathcal{A}_a^{-1}(x_3, \alpha_{11}) + j\mathcal{A}_a^{-1}(y_3, \alpha_{12})$. The integer transformations $\mathcal{A}_a^{-1}$ are used, since the transformations $\mathcal{B}_2$ and $\mathcal{B}_6$ are substituted by the transformations $\mathcal{A}_a$ followed by the multiplications by $(1-j)$ and $(-1-j)$, respectively.

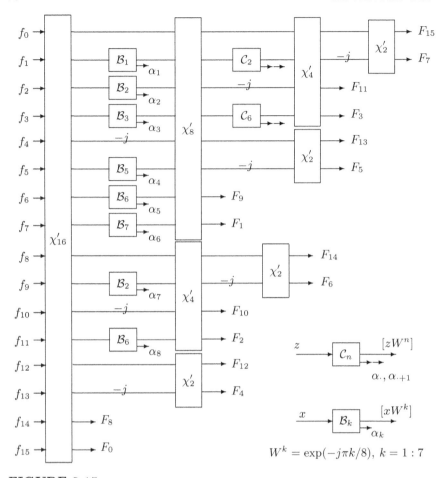

FIGURE 2.15

Block-scheme of the 16-point integer DFT with 8 control bits and two integer rotations.

Note also that two other outputs of the 8-point paired transform are equal to $\chi_0 = z_0$ and $\chi_2 = jz_2$. The full diagram of reconstruction of the outputs of the 8-point paired transform is given in Figure 2.16. Thus, with four additional control bits, the missing outputs of the 8-point paired transform can be defined, and therefore the inverse integer 16-point DFT can be calculated.

Summarizing the above reasoning, we can propose the block-diagram for calculating the reversible 16-point DFT with 12 control bits, which is shown in Figure 2.17. This realization requires four additional real multiplications to calculate the last four control bits. This realization assumes that the property of the complex conjugate of the proposed integer DFT does hold. In this case, when the property of the complex conjugate of the proposed integer

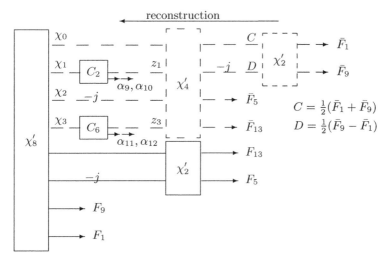

FIGURE 2.16
Block-scheme of the reconstruction of outputs of the 8-point discrete paired transform in the 16-point integer DFT.

DFT does not take place, the above procedure of calculating the values of the four inputs $z_0, z_1, z_2,$ and z_3 could not be very precise. That would lead to errors of calculation of the inverse 8-paired transform, and then the original signal f_n. An additional correction of the last four control bits or values of z_k could thus be required. Therefore, in the general case, for the invertible integer Fourier transform, the calculation by the complete block-diagram of the paired algorithm is needed. To analyze the problem of calculation of the reversible 16-point integer DFT, we describe in detail an example. At the same time, we check the property of the complex conjugate for the proposed 16-point integer DFT.

Example 2.9
For the signal $\mathbf{x} = (1, 2, 4, 4, 3, 7, 5, 8, 8, 5, 7, 3, 4, 4, 2, 1)'$, we first consider the integer Fourier transform by 8 control bits, when using the simplified block-diagram. The calculations of the transform can be divided into the following stages.

Stage 1: The paired transform of the signal equals

$$\chi'_{16}[\mathbf{x}] = (-7, -3, -3, 1, -1, 3, 3, 7, 2, -4, 4, -2, -2, 2, \underline{0}, \underline{68})'$$

and therefore, $F_8 = 0$ and $F_0 = 68$.

Stage 2: The first eight components of the paired transform, which represent the first integer splitting-signal, $f_{T'_1} = \{-7, -3, -3, 1, -1, 3, 3, 7\}$, are modified by integer approximations of multiplications by twiddles coefficients $\{1, W,$

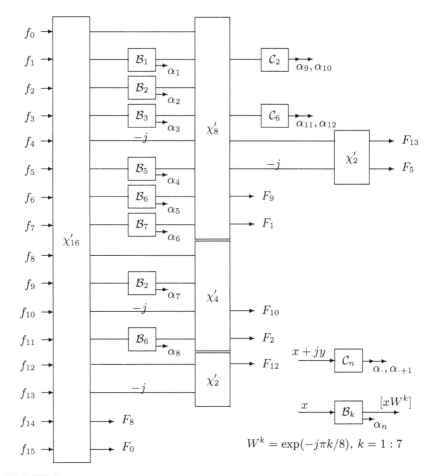

FIGURE 2.17
Block-scheme of the 16-point integer DFT with 12 control bits.

$W^2, W^3, -j, W^5, W^6, W^7\}$, where $W = \exp(-j\pi/8)$, as follows:

$$-7 \to -7$$

$$\mathcal{B}_1 : 3 \to \begin{cases} \vartheta_0 = -3 \\ \alpha_1 = \underline{0} \\ \vartheta_2 = 1 \end{cases} \to \begin{cases} -3 + j \\ \alpha_1 = 0 \end{cases}$$

$$\mathcal{A}_{0.7071} : -3 \to \begin{cases} \vartheta_0 = -2 \\ \alpha_2 = 1 \end{cases} \to \cdot(1-j) \to \begin{cases} -2 + 2j \\ \alpha_2 = 1 \end{cases}$$

$$\mathcal{B}_3 : 1 \to \begin{cases} \vartheta_0 = 0 \\ \alpha_3 = \underline{0} \\ \vartheta_2 = -1 \end{cases} \to \begin{cases} -j \\ \alpha_3 = 0 \end{cases}$$

$$-1 \rightarrow \cdot(-j) \rightarrow j$$

$$\mathcal{B}_5 : 3 \rightarrow \begin{cases} \vartheta_0 = -1 \\ \alpha_4 = \underline{0} \\ \vartheta_2 = -3 \end{cases} \rightarrow \begin{cases} -1 - 3j \\ \alpha_4 = 0 \end{cases}$$

$$\mathcal{A}_{0.7071} : 3 \rightarrow \begin{cases} \vartheta_0 = 2 \\ \alpha_5 = 1 \end{cases} \xrightarrow{\cdot(-1-j)} \begin{cases} -2 - 2j \\ \alpha_5 = 1 \end{cases}$$

$$\mathcal{B}_7 : 7 \rightarrow \begin{cases} \vartheta_0 = -6 \\ \alpha_6 = \underline{1} \\ \vartheta_2 = -3 \end{cases} \rightarrow \begin{cases} -6 - 3j \\ \alpha_6 = 1 \end{cases}$$

As a result, we obtain the following modified complex-integer splitting-signal:

$$g_{T_1'} = \{-7, -3 + j, -2 + 2j, -j, j, -1 - 3j, -2 - 2j, -6 - 3j\}$$

and the six control bits $\{\alpha_1, \alpha_2, \ldots, \alpha_6\} = \{0, 1, 0, 0, 1, 1\}$.

The next four components of the 16-point paired transform are modified as

$$2 \rightarrow 2$$

$$\mathcal{A}_{0.7071} : -4 \rightarrow \begin{cases} \vartheta_0 = -3 \\ \alpha_7 = 0 \end{cases} \xrightarrow{\cdot(1-j)} \begin{cases} -3 + 3j \\ \alpha_7 = 1 \end{cases}$$

$$4 \rightarrow \cdot(-j) \rightarrow -4j$$

$$\mathcal{A}_{0.7071} : -2 \rightarrow \begin{cases} \vartheta_0 = -1 \\ \alpha_8 = 1 \end{cases} \xrightarrow{\cdot(-1-j)} \begin{cases} 1 + j \\ \alpha_8 = 1 \end{cases}$$

The four-point modified complex-integer splitting-signal is thus equal to $g_{T_2'} = \{2, -3 + 3j, -4j, 1 + j\}$ and the two control bits $\{\alpha_7, \alpha_8\} = \{0, 1\}$.

The last two components of the 16-point paired transform are modified as $-2 \rightarrow -2$, $2 \rightarrow \cdot(-j) \rightarrow -2j$. From these two values, the value of the Fourier component F_{12} is defined, $F_{12} = -2 - (-2j) = -2 + 2j$.

Stage 3: The 8-point paired transform of the modified splitting-signal $g_{T_1'}$ equals

$$\chi_8'[g_{T_1'}] = \{-7 - j, -2 + 4j, 4j, 6 + 2j, -3 + j, 2 + 2j, \underline{-1 + 7j}, \underline{-21 - 5j}\}$$

and therefore, $F_9 = -1 + 7j$ and $F_1 = -21 - 5j$. The incomplete transform equals $\chi_{8;in}'[g_{T_1'}] = \{-3 + j, 2 + 2j, \underline{-1 + 7j}, \underline{-21 - 5j}\}$. The first two components of this transform are modified and then transformed by χ_2' as follows:

$$\begin{bmatrix} -3 + j \\ 2 + 2j \end{bmatrix} \rightarrow \begin{bmatrix} -3 + j \\ -j(2 + 2j) \end{bmatrix} \rightarrow \begin{bmatrix} 1 & -1 \\ 1 & 1 \end{bmatrix} \begin{bmatrix} -3 + j \\ 2 - 2j \end{bmatrix} = \begin{bmatrix} -5 + 3j \\ -1 - j \end{bmatrix}.$$

Therefore, $F_{13} = -5 + 3j$ and $F_5 = -1 - j$.

The incomplete 4-point paired transform equals $\chi_{4;in}'[g_{T_2'}] = \{4 - 8j, 0\}$ and therefore, $F_{10} = 4 - 8j$ and $F_2 = 0$. The full transform of the modified splitting-signal $g_{T_2'}$ equals $\chi_4'[g_{T_2'}] = \{2 + 4j, -4 + 2j, 4 - 8j, 0\}$.

And finally, the incomplete two-point paired transform of the signal $(-2, -2j)$ equals $-2 + 2j$, which means that $F_{12} = -2 + 2j$. Figure 2.18 shows the

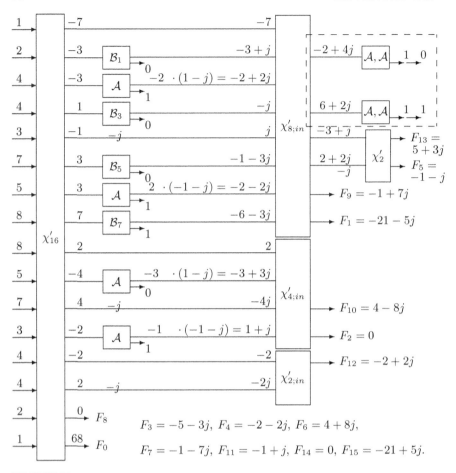

FIGURE 2.18
Block-scheme of the 16-point integer DFT with 8 (and/or 12) control bits.

block diagram of calculation of the direct integer 16-point DFT with eight control bits. The obtained nine values of the integer transform, namely, $F_0, F_1, F_2, F_5, F_8, F_9, F_{10}, F_{12}$, and F_{13}, together with the remaining seven values defined by using the complex conjugate property, are given in Table 2.5.

The values of the 16-point DFT of the signal are also given in the table, as well as the root-mean-square error of integer approximation

$$\varepsilon_{8cb} = \sqrt{\frac{1}{16} \sum_{p=0}^{15} |F_p - F_{p;8cb}|^2} = 0.5353.$$

If the inverse integer DFT is needed, the additional control bits $(\alpha_9, \alpha_{10}, \alpha_{11}, \alpha_{12}) = (1, 0, 1, 1)$ are calculated when multiplying the integers $2, 4, 6, 2$ by the factor $a = 0.7071$, as shown in the dashed box in the block-scheme of Figure 2.18.

TABLE 2.5
The 16-point DFT and integer DFTs of the signal by using the simplified block-scheme with 8 and 12 control bits, as well as two lifting schemes

f_n	F_p	FFT	8cb-FFT	12cb-FFT	8cb+2ls FFT
1	F_0	68	68	68	$\cdot$
2	F_1	$-21.2468 - 4.2263j$	$-21 - 5j$	$-21 - 5j$	$\cdot$
4	F_2	$0.5858 + 0.2426j$	0	0	$\cdot$
4	F_3	$-4.7364 - 3.1648j$	$-5 - 3j$	$-4 - 2j$	$-4 - 3j$
3	F_4	$-2 - 2j$	$-2 - 2j$	$-2 - 2j$	$\cdot$
7	F_5	$-0.7783 - 1.1648j$	$-1 - j$	$-1 - j$	$\cdot$
5	F_6	$3.4142 + 8.2426j$	$4 + 8j$	$4 + 8j$	$\cdot$
8	F_7	$-1.2385 - 6.2263j$	$-1 - 7j$	$-2 - 6j$	$-1 - 6j$
8	F_8	0	0	0	$\cdot$
5	F_9	$-1.2385 + 6.2263j$	$-1 + 7j$	$-1 + 7j$	$\cdot$
7	F_{10}	$3.4142 - 8.2426j$	$4 - 8j$	$4 - 8j$	$\cdot$
3	F_{11}	$-0.7783 + 1.1648j$	$-1 + j$	-2	$-2 + j$
4	F_{12}	$-2 + 2j$	$-2 + 2j$	$-2 + 2j$	$\cdot$
4	F_{13}	$-4.7364 + 3.1648j$	$-5 + 3j$	$-5 + 3j$	$\cdot$
2	F_{14}	$0.5858 - 0.2426j$	0	0	$\cdot$
1	F_{15}	$-21.2468 + 4.2263j$	$-21 + 5j$	$-20 + 4j$	$-21 + 4j$
	cont.bits		01001101	010011011011	01001101
	RMSE ε		0.5353	0.7938	0.5812
	#mult.	12	12	$12 + 2(2)$	$12 + 2(3)$

Stage 4 (Full transform): When performing all calculations in the considered paired algorithm, the first four components of the 8-point paired transform are modified as follows:

$$-7 - j \rightarrow -7 - j$$

$$\mathcal{A}_{0.7071} : -2 + 4j \rightarrow \begin{cases} \vartheta_0 = -1 \\ \alpha_8 = 1 \end{cases} +j \cdot \begin{cases} \vartheta_0 = 3 \\ \alpha_9 = 0 \end{cases} \rightarrow \cdot(1 - j) \rightarrow \begin{cases} 2 + 4j \\ \alpha_8, \alpha_9 = 1, 0 \end{cases}$$

$$4j \rightarrow \cdot(-j) \rightarrow 4$$

$$\mathcal{A}_{0.7071} : 6 + 2j \rightarrow \begin{cases} \vartheta_0 = 4 \\ \alpha_{10} = 1 \end{cases} +j \cdot \begin{cases} \vartheta_0 = 1 \\ \alpha_{11} = 1 \end{cases} \rightarrow \cdot(-1 - j) \rightarrow \begin{cases} -3 - 5j \\ \alpha_{10}, \alpha_{11} = 1, 1 \end{cases}$$

The four-point modified complex-integer splitting-signal on this stage thus equals $h_{T'_2} = \{-7 - j, 2 + 4j, 4, -3 - 5j\}$ and the four control bits $\{\alpha_9, \alpha_{10}, \alpha_{11}, \alpha_{12}\} = \{1, 0, 1, 1\}$. The four-point paired transform of this signal equals $\chi'_4[h_{T'_1}] = \{-11 - j, 5 + 9j, \underline{-2}, \underline{-4 - 2j}\}$ and therefore, $F_{11} = -2$ and $F_3 = -4 - 2j$. The next calculation of the two-point paired transform over the obtained data $\{-11 - j, -j(5 + 9j)\}$ results in the components $F_{15} = -20 + 4j$ and $F_7 = -2 - 6j$.

To complete the calculations, we can perform another two-point paired transform over the modified outputs of the four-point paired transform on the first stage, $\{2+4j, -j(-4+2j)\}$. That will result in the components $F_{14} = 0$ and $F_6 = 4 + 8j$. The property of the complex conjugate always holds for these components in the algorithm, i.e., $F_{14} = \bar{F}_2$ and $F_6 = \bar{F}_{10}$. Therefore, these calculations can be omitted. Values of the obtained integer discrete Fourier transform are given in the 5th column of Table 2.5. Figure 2.19 shows all data including the intermediate ones, when performing those calculations on Stage 3. The property of the complex conjugate does not hold for this

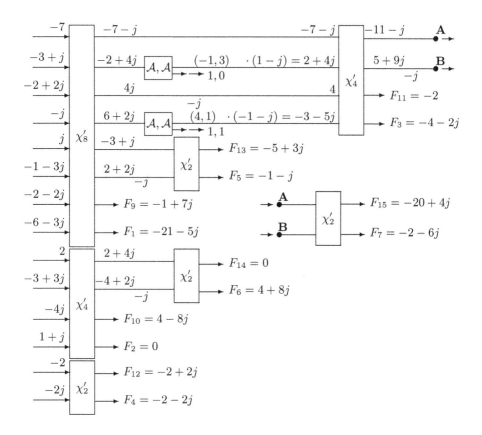

FIGURE 2.19
Block-scheme of the part of the 16-point integer DFT after Stage 2.

transform, since four vales of the transform have been changed slightly at the frequency-points $p = 3, 7, 11$, and 15. The real and imaginary parts of these transforms differ by ± 1. The root-mean-square error of integer approximation

$$\varepsilon_{12cb} = \sqrt{\frac{1}{16} \sum_{p=0}^{15} |F_p - F_{p;12cb}|^2} = 0.7938 > \varepsilon_{8cb} = 0.5353.$$

The pointwise difference between the components of the integer DFTs with 8 and 12 control bits at the frequency-points $3, 7, 11$, and 15 is given in the following table:

p	3	7	11	15
$F_{p;8cb} - F_{p;12cb}$	$-1 - j$	$1 - j$	$1 + j$	$-1 + j$

Stage 4 (Lifting scheme): We now consider the application of the three-step lifting schemes for integer approximation of two rotations $\mathcal{C}_2$ and $\mathcal{C}_6$, instead of the multiplications with control bits. Each integer lifting scheme requires three multiplications instead of two multiplications when using two integer transforms $\mathcal{A}$.

The integer approximations of multiplication of complex numbers $-2 + 4j$ and $6 + 2j$ by the factors W^2 and W^6 are calculated respectively as

$$1.4142 + 4.2426j = (-2+4j) \cdot W^2 \rightarrow \begin{array}{|c|c|c|} \hline \mathcal{Q} & -0.7071 & \mathcal{Q} \\ \hline 0.4142 & \mathcal{Q} & 0.4142 \\ \hline \end{array} \circ \begin{bmatrix} -2 \\ 4 \end{bmatrix} = \begin{bmatrix} 2 \\ 4 \end{bmatrix}$$

and

$$-2.8284 - 5.6569j = (6+2j) \cdot W^6 \rightarrow \begin{array}{|c|c|c|} \hline \mathcal{Q} & -0.7071 & \mathcal{Q} \\ \hline 2.4142 & \mathcal{Q} & 2.4142 \\ \hline \end{array} \circ \begin{bmatrix} 6 \\ 2 \end{bmatrix} = \begin{bmatrix} -3 \\ -6 \end{bmatrix}.$$

Here, the quantizing operation is the rounding, i.e., $\mathcal{Q}(x) = [x]$.

The part of the block-scheme of the 16-point DFT where two integer lifting schemes are implemented is shown in Figure 2.20. The new value $-3 - 6j$ of

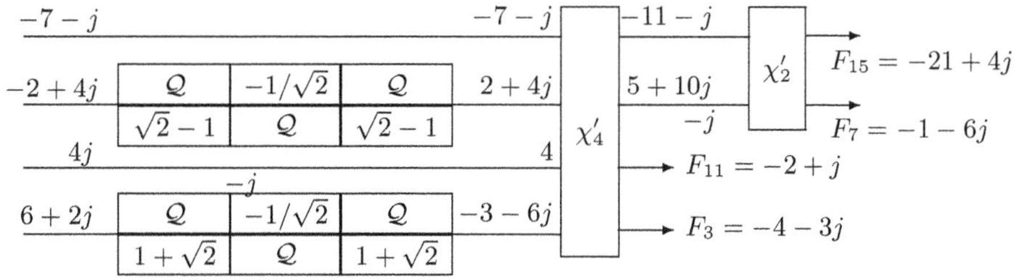

FIGURE 2.20

The implementation of two integer lifting schemes on Stage 3 of calculation of the 16-point DFT with eight control bits.

the second lifting scheme, instead of $-3 - 5j$ in the transforms $(\mathcal{A}_a, \mathcal{A}_a)$ with two control bits $(0, 1)$, changes the three outputs of the following four-point paired transform, as well as the next two-point transform. As a result, the difference of the 16-point integer DFT occurs at frequency-points $3, 7, 11$, and 15. The pointwise difference between the components of the integer DFTs at these points, when using two integer lifting schemes and the above integer transforms with 8 and 12 control bits are given in the following table:

p	3	7	11	15
$F_{p;8cb+2ls} - F_{p;8cb}$	1	j	-1	$-j$
$F_{p;8cb+2ls} - F_{p;12cb}$	$-j$	1	j	-1

The only four values of the 16-point integer DFT with two lifting schemes which differ from the transforms with only control bits are shown in the last column of Table 2.5. The property of complex conjugate of the Fourier transform components F_p and F_{8-p}, $p = 3, 7$ does not hold for the integer approximations of the 16-point DFT by the lifting schemes. The root-mean-square error of integer approximation by lifting schemes equals

$$\varepsilon_{2ls} = \sqrt{\frac{1}{16} \sum_{p=0}^{15} |F_p - F_{p;8cb+2ls}|^2} = 0.5812 > \varepsilon_{8cb} = 0.5353.$$

⬜

2.3.6   Codes for the forward 16-point integer FFT

Below are the MATLAB-based codes for computing the integer discrete Fourier transform by the paired transform with eight control bits. The test signal in the main program is the signal **x** from Example 2.9. Code "intfft16_cb8f.m" is for calculation of the forward integer DFT; however, the calculation of an additional four control bits is also given on Stage 4A of the code. Stage 4B for calculation of two operations of multiplication, C_2 and C_6, by three-step lifting schemes is also added to the code. This code can be extended easily for the full calculation of the 16-point DFT, as shown in the comments in Stage III in the code. Therefore, the eight- and four-point paired transforms have not been changed by the incomplete transforms, which could save six operations of addition.

```
% -----------------------------------------------------------------
% demo_pfft16.m file of programs (library of codes of Grigoryans)
% List of codes for 16-point integer FFT with control bits:
% 16-point integer FFT with 8 control bits - "intfft16_cb8f.m"
% 1-D fast direct paired transform  - "fastpaired_1d.m"
% integer A operation - "it_1bit.m"
% integer B operation - "it_1bitB.m"
%
    x=[1  2  4  4  3  7  5  8  8  5  7  3  4  4  2  1];
    [F_int,F_cb]=intfft16_cb8f(x);
    % 68,-21-5i,0,-5-3i,-2-2i,-1-i,4+8i,-1-7i,0,-1+7i,4-8i,-1+i,
```

```
%   -2+2i,-5+3i,0,-21+5i,0,1,0,0,1,1,0,1
%   ----------------------------------------------------------------
    function [F_int,F_cb]=intfft16_cb8f(x)
    F_int=zeros(1,16);      % integer DFT
    F_cb=zeros(1,8);        % control bits
    p2=pi/8;                % case is N=16
%   Stage I:
    y_1=fastpaired_1df(x);
    F_int(9)=y_1(15);
    F_int(1)=y_1(16);
%   Stage II: Modification of the splitting-signals
    % 1. Trivial multiplications
    y_paired=y_1;
    f_points=[5,11,14];
    y_paired(f_points) =y_1(f_points)*(-j);
    % 2. Integer multiplications by W1,W3,W5, and W7
    f_points=[1,3,5,7];
    n_bits=  [1,3,4,6];
    for nn=1:4
        fp=f_points(nn);
        wp=exp(-j*p2*fp);
        y1=y_1(fp+1);       sn=sign(y1);
        x2=sn*y1;
        [r0,r2,r1]=it_1bitB(x2,wp);   % 2 multiplications
        y_paired(fp+1)=sn*(r0+j*r2);
        nb=n_bits(nn);
        F_cb(nb)=r1;
    end;
    % 3. Integer multiplications by W2 and W6
    %    for the 1st and 2nd splitting-signals
    f_points=[3,7,10,12];
    n_bits=  [2,5,7,8];
    cj=[1-j,-1-j,1-j,-1-j];
    for nn=1:4
        fp=f_points(nn);
        y1=y_1(fp);         sn=sign(y1);
        x3=sn*y1;
        [xi,r1]=it_1bit(x3);          % 1 multiplication
        y_paired(fp)=sn*cj(nn)*xi;
        nb=n_bits(nn);
        F_cb(nb)=r1;
    end;
%   Stage III:
    % 1. 8-point paired transform of the 1st modified signal
    x_2=y_paired(1:8);
    y_2=fastpaired_1d(x_2);
    F_int(10)=y_2(7);
    F_int(2) =y_2(8);
    % 1.1 The following 2-point paired transform
```

```
        x_5=[y_2(5) -j*y_2(6)];
        y_5=fastpaired_1d(x_5);
        F_int(14)=y_5(1);
        F_int(6) =y_5(2);
        % Stage IV (A or B) can be added here for the full DFT, with the
        % followed 4- and 2-point paired transforms over the modified
        % signal y_5 (calculated on Stage IV) as in the next steps 2 and 3.
        %
        % 2. 4-point paired transform of the 2nd modified signal
        x_3=y_paired(9:12);
        y_3=fastpaired_1d(x_3);
        F_int(11)=y_3(3);
        F_int(3) =y_3(4);
        % 3. 2-point incomplete paired transform of the 3rd modified signal
        F_int(13)=y_paired(13)-y_paired(14);
%       Making the property of the complex conjugate
        p=[4,6,14,3,7,11,15];       % use p=[4,6,14] for the full DFT
        p1=16-p;
        F_int(p+1)=conj(F_int(p1+1));
%       End of calculations for the forward 16-point integer DFT
% ----------------------------------------------------------------------
%       For the inverse 16-point DFT use one of these stages in Stage III:
%       Stage IVA (Additional four control bits):
        % When multiplying the complex inputs y_2(2) and y_2(4)
        % by the vectors W2 and W6. Integer A-operation is used.
          xia=zeros(1,4);
          x2=y_2(2); x4=y_2(4);
          xy=[real(x2),imag(x2),real(x4),imag(x4)];
          for k=1:4
              x1=xy(k);  if x1<0 x1=-x1; end;
              [xi,r1]=it_1bit(x1);        % 1 multiplication
              xia(k)=xi;
              F_cb(8+k)=r1;
          end;
          y_5=[y_2(1),xia(1)+j*xia(2),-j*y_2(3),xia(3)+j*xia(4)];
%       Stage IVB (Additional two lifting schemes):
        % Integer liftings for the inputs y_2(2) and y_2(4) when
        % multiplying them by W2 and W6 (3 multiplications each)
        y_lifts=zeros(1,2);
        c=1/sqrt(2); s=c;    % c=0.7071
        A2=[ 1 0
            -s 1];
        for k=1:2
            cs=(1-c)/s;
            A1=[1 cs
                0  1];
            x=y_2(2*k);
            y1=A1*[real(x) imag(x)]';   y1=round(y1);
            y2=A2*y1;                    y2=round(y2);
```

```
        y3=A1*y2;                           y3=round(y3);
        y_lifts(k)=y3(1)+y3(2)*j;
        c=-c;
    end;
    y_5=[y_2(1),y_lifts(1),-j*y_2(3),y_lifts(2)];
% ------------------------------------------------------------------
    function [r,r0]=it_1bit(x1)
        a=0.7071;  % a=sqrt(2)/2;
        xa=x1*a;
        r=round(xa);
        r0=0;
        if r0<xa  r0=1; end;
% ------------------------------------------------------------------
    function [r0,r2,r1]=it_1bitB(x1,w)
        x=sign(x1)*x1;
        wr=real(w);  wi=imag(w);
        w1=wr; w2=wi;
        k=0;                            % addition on Dec. 1, 2007
        if abs(wi)>abs(wr) w1=wi; w2=wr; k=1; end;
        xw1=x*w1;
        r0=round(xw1);
        r1=(1+sign(abs(xw1)-abs(r0)))/2; % the control bit
        xw2=x1*w2;
        r2=round(xw2);
        r22=r2;                         % addition on Dec. 1, 2007
        if k==1 r2=r0; r0=r22; end;
% ------------------------------------------------------------------
```

All data of calculations used in Example 2.9 have been obtained from this program.

2.3.6.1 General case $N = 2^r$, $r \geq 4$.

As for the $N = 16$ case discussed above in detail, the integer discrete Fourier transform of a high order $N = 2^r \geq 16$ can be analyzed, too. In the case when an integer and reversible approximation of the N-points DFT are needed in one universal program, we can use the recursive form of the paired algorithm, which is based on the following matrix formula:

$$[\mathcal{F}_N] = P\left(\bigoplus_{n=1}^{r}[\mathcal{F}_{N/2^n}] \oplus 1\right)[\bar{W}][\chi'_N],$$

where P is a permutation, and $[\bar{W}]$ is the diagonal matrix with twiddle coefficients given by

$$\bigoplus_{n=1}^{r} \text{diag}\left\{1, W_{N/2^{n-1}}^1, W_{N/2^{n-1}}^2, \ldots, W_{N/2^{n-1}}^{N/2^n-1}\right\}.$$

For instance, it is not difficult to modify for this purpose the recursive code "paired_1dfft.m" given in section 1.4. Indeed, for that in Stage 2 of that code, the command "z=z.*w" should be substituted by the following set of commands:

```
% -------------------------------------------------------------------
%    F_cb=[];
     M=length(z);
     for k=2:M-3
        wp=w(k);
        if wp==1 ;
        elseif wp==-j z(k)=-j*z(k);
        else
            x1=real(z(k)); sn1=sign(x1);
            x2=imag(z(k)); sn2=sign(x2);
            x1=sn1*x1;      x2=sn2*x2;
            [r0,r2,r1]=it_1bitB(x1,wp);        % 2 multiplications
            [s0,s2,s1]=it_1bitB(x2,wp);        % 2 multiplications
            z(k)=sn1*(r0+j*r2) + sn2*(s0*j-s2);
            F_cb=[F_cb r1 s1];                 % control bits
        end;
     end
% -------------------------------------------------------------------
```

The implementation of the method of control bits does not increase the number of operations of multiplication and addition, but bits and 'if' operations. For $N = 2^r$, $r > 4$, the calculation of the N-point DFT by the paired transform requires multiplications by $2^{r-1}(r-3) + 2$ twiddle factors [8, 13]. The total number of real multiplications in this implementation equals four times the number of operations of multiplication, $m(N)$, used in the paired fast algorithm, i.e., $\mu_T(N) = 4m(N) = 4 \cdot [N/2(\log_2 N - 3) + 2]$, since each $\mathcal{B}$ operation over the real and imaginary parts of the input requires two multiplications. This number can be reduced essentially and the following simplification could be done in this code.

On the first stage of the paired algorithm, the N-point paired transform is real, and the operations $\mathcal{B}$ should be used only on real data. On this stage, we thus can change the above part as follows:

```
% -------------------------------------------------------------------
%    M=N;
     for k=2:M-3
        wp=w(k);
        if wp==1 ;
        elseif wp==-j z(k)=-j*z(k);
        else
            x1=z(k);    sn1=sign(x1);
            x1=sn1*x1;
            [r0,r1]=it_1bitB(x1,wp);           % 2 multiplications
            z(k)=sn1*(r0+j*r2);
```

```
        F_cb=[F_cb r1];                    % control bits
      end;
   end
% -----------------------------------------------------------------
```

That will save

$$\mu_1(N) = 2[(N/2 - 2) + (N/4 - 2) + (N/8 - 2) + \cdots + (4 - 2)] = 2(N - 2r)$$

multiplications, or 16 multiplications for the $N = 16$ case.

Note also that in the recursive realization of the N-point integer DFT, only the integer $\mathcal{B}$ operations are used, which required two multiplications each. Therefore, when the twiddle coefficients "wp=w(k)" equal $\pm 0.7071 - 0.7071j$, the integer operations $\mathcal{B}$ can be substituted by $\mathcal{A}_{0.7071}$, which saves one multiplication for each substitution for real data and two multiplications for the complex data. The number of such twiddle coefficients in the paired algorithm equals $\mu_2(N) = 2[(r - 2) + (r - 3) + \cdots + 1)] = (r - 1)(r - 2)$. Therefore the total number of real multiplications can be reduced by

$$\mu(N) = \mu_T(N) - [\mu_1(N) + 2\mu_2(N)] = 2N(r - 4) - 2r^2 + 10r + 4,$$

and in particular, $\mu(16) = 12$ and $\mu(32) = 68$.

The implementation of the proposed method for approximation of the DFT requires one or two control bits for each such factor, depending on the inputs. In the first stage of the paired algorithm, when the 2^r-point DFT is decomposed by DFTs of orders $N/2, N/4, ..., 4, 2, 1, 1$, the multiplication of integer outputs of the paired transform by $N - 2r$ twiddle factors requires one control bit for each factor. For the next stages of the paired algorithm, each such factor requires at maximum two control bits since many inputs are complex. The total number of control bits is calculated by $c(N) = N(r - 4) + 2r + 4$. A few values of numbers of real multiplications and control bits are given in Table 2.6.

TABLE 2.6
The number of multiplications and control bits for computing the
N-point integer DFTs

N	8	16	32	64	128	256	512	1024	2048	4096
$C(N)$	2	12	46	144	402	1044	2582	6168	14362	32796
$\mu(N)$	2	12	68	248	744	2004	5052	12192	28544	65372
$\Delta(N)$			6	34	124	372	1002	2526	6096	14272
$\mu_{cb+ls}(N)$			62	214	620	1632	4050	9666	22448	51100
$\mu_{3ls}(N)$	6	30	102	294	774	1926	4614	10758	24582	55302

For comparison, the number $\Delta(N)$ of multiplications that can be saved in the integer approximation of the integer DFT by using the integer lifting schemes after the second stage is also given in the table. The three-step lifting scheme uses one less multiplication than four integer $\mathcal{B}$ (not $\mathcal{A}$) transforms of complex integers. This saving is calculated by

$$\Delta(N) = \left[\frac{N}{2}(r-3) + 2 \right] - \left[(N - 2r) + (r-2)(r-3) \right]$$
$$= \frac{N}{2}(r-5) - r^2 + 7r - 4.$$

Therefore, the number of multiplications in the integer DFT with control bits and lifting schemes equals

$$\mu_{cb+ls}(N) = \mu(N) - \Delta(N) = N/2(3r - 11) - r^2 + 3r + 8,$$

when $r > 4$. The number of multiplications for the integer DFT by the split-radix algorithm with three-step lifting schemes can be calculated by

$$\mu_{3ls}(N) = 3m(N) = 3 \left[N/2(r-3) + 2 \right], \quad r \geq 3.$$

These numbers for $r = 4:12$ are also given in the table, for comparison with the algorithms with control bits. One can notice that $\mu_{3ls}(N) \leq \mu(N)$ when $N = 2^r > 128$.

The paired algorithm of the fast Fourier transform exists for many other orders $N \neq 2^r$ of the transform as well. For each such order N, the concept of the integer transformations $\mathcal{A}$ and $\mathcal{B}$ with one control bit can thus be used for the integer approximation of the N-point DFT in a similar way. For instance, the 12-point DFT by the paired transform (see [9], p. 254) uses four twiddle factors (one control bit to be given to each) followed by three 3-point DFTs with two factors each. For real inputs, the integer 12-point DFT requires $4 + 3(4) = 16$ control bits.

Thus, with the aid of control bits, the problem of calculating the integer-to-integer DFT and its inverse transform has been solved. The integer DFT with control bits is fast and its computational complexity is comparative with the fast DFT. The implementation of the integer approximation of the discrete Fourier transform (DFT) runs into the problem of conservation of one of the most important properties of the transform, the property of the complex conjugate for real data. Therefore, two cases for integer implementation of the DFT have been considered, when inputs are real. In the first case, this property is assumed and that leads to the saving of many operations as in the case of the traditional DFT. In the general case, when this property does not hold, additional computations and control bits are required, to make the integer transformation reversible.

2.4   DFT in vector form

In Section 1.3, we have considered the method of graph simplification for calculating the fast Fourier transform by paired transforms. The calculations have been split in such a way that real and imaginary parts of the transform on each stage of the algorithm are calculated separately. This algorithm has been demonstrated in detail for the 8- and 16-point DFT. We now generalize this method and consider the N-point DFT in real space.

2.4.1 DFT in real space

The N-point discrete Fourier transform is defined as the decomposition of the signal by N the roots of the unit

$$W^k = e^{-\frac{2\pi j}{N}k} = c_k - js_k = \cos(\frac{2\pi}{N}k) - j\sin(\frac{2\pi}{N}k), \quad k = 0 : (N-1),$$

which are located on the unit circle. During the fast calculation of the transform, data on different stages of calculation are multiplied on such twiddle factors. These multiplications define the arithmetical complexity of the transform, and therefore, in the fast transform the number of these operations is reduced to a minimum.

We now consider in matrix form the multiplication of the complex number $\mathbf{x}$, which we consider as the vector $(x_1, x_2)'$, by twiddle coefficients W^k, where $k = 0 : (N-1)$,

$$\mathbf{x} = \begin{pmatrix} x_1 \\ x_2 \end{pmatrix} \rightarrow W^k \mathbf{x} = (c_k - js_k)(x_1 + jx_2) = c_k x_1 + s_k x_2 + j(c_k x_2 - s_k x_1)$$

In matrix form this multiplication can be written as

$$T^k \mathbf{x} = \begin{pmatrix} c_k & s_k \\ -s_k & c_k \end{pmatrix} \begin{pmatrix} x_1 \\ x_2 \end{pmatrix} = \begin{pmatrix} \cos\varphi_k & \sin\varphi_k \\ -\sin\varphi_k & \cos\varphi_k \end{pmatrix} \begin{pmatrix} x_1 \\ x_2 \end{pmatrix}$$

where $W^k = (\cos\varphi_k, -\sin\varphi_k) = \cos\varphi_k - j\sin\varphi_k$, and the angle $\varphi_k = 2\pi k/N$. The matrix of rotation by this angle is denoted by T^k.

Thus we transfer the complex space into the 2-D real space, $C \rightarrow R^2$, and consider each operation of multiplication by twiddle coefficient as the elementary rotation, or the Givens transformation,

$$W^k \rightarrow T^k, \quad k = 0 : (N-1).$$

The discrete Fourier transform of the vector $\mathbf{f} = (f_0, f_1, f_2, ..., f_{N-1})'$

$$F_p = \sum_{n=0}^{N-1} W^{np} f_n, \quad p = 0 : (N-1),$$

in the complex space C^N has the following matrix:

$$[\mathcal{F}_N] = \begin{bmatrix} 1 & 1 & 1 & 1 & 1 & 1 \\ 1 & W^1 & W^2 & W^3 & \cdots & W^{N-1} \\ 1 & W^2 & W^4 & W^6 & \cdots & W^{N-2} \\ 1 & \cdots & \cdots & \cdots & \cdots & \cdots \\ 1 & & & & & \\ 1 & W^{N-1} & W^{N-2} & W^{N-3} & & W^1 \end{bmatrix}.$$

Consider the transform of the complex plane C^N into the real plane R^{2N} by

$$\mathbf{f} = (f_0, f_1, f_2, ..., f_{N-1})' \rightarrow \bar{\mathbf{f}} = (r_0, i_0, r_1, i_1, r_2, i_2, ..., r_{N-1}, i_{N-1})',$$

where we denote $r_k = \text{Re} f_k$ and $i_k = \text{Im} f_k$, when $k = 0 : (N-1)$. The vector $\bar{\mathbf{f}}$ is composed from the original vector $\mathbf{f}$, and its vector-component is denoted by $\bar{\mathbf{f}}_n = (\bar{f}_{2n}, \bar{f}_{2n+1})' = (r_n, i_n)'$. The N-point DFT is thus represented in the real space R^{2N} as the $2N$-point transform

$$\bar{\mathbf{F}}_p = \begin{bmatrix} R_p \\ I_p \end{bmatrix} = \sum_{n=0}^{N-1} T^{np} \bar{\mathbf{f}}_n = \sum_{n=0}^{N-1} T^{np} \begin{bmatrix} r_n \\ i_n \end{bmatrix}, \quad p = 0 : (N-1). \qquad (2.12)$$

In matrix form, the Fourier transform in the space R^{2N} is described by the following matrix $(2N \times 2N)$:

$$[\mathcal{F}_{N\text{-}b}] = X = \begin{bmatrix} I & I & I & I & I & I \\ I & T^1 & T^2 & T^3 & \cdots & T^{N-1} \\ I & T^2 & T^4 & T^6 & \cdots & T^{N-2} \\ I & \cdots & \cdots & \cdots & \cdots & \cdots \\ I & T^{N-1} & T^{N-2} & T^{N-3} & & T^1 \end{bmatrix}. \qquad (2.13)$$

The (n, p)-th blocks 2×2 of this matrix are defined as $X_{(n,p)} = T^{np}$, where $n, p = 0 : (N-1)$. Note that according to the definition, the rotation matrices T^k compose the one-parametric group with period N. Indeed,

$$T^{k_1 + k_2} = T^{k_1} T^{k_2}, \quad (T^0 = T^N = I),$$

for any $k_1, k_2 = 0 : (N-1)$.

Example 2.10
The six-point DFT in the real space R^{12} has the following matrix:

$$[\mathcal{F}_{6\text{-}b}] = X = \begin{bmatrix} 1 & 0 & 1 & 0 & 1 & 0 & 1 & 0 & 1 & 0 & 1 & 0 \\ 0 & 1 & 0 & 1 & 0 & 1 & 0 & 1 & 0 & 1 & 0 & 1 \\ 1 & 0 & c_1 & s_1 & c_2 & s_2 & c_3 & s_3 & c_4 & s_4 & c_5 & s_5 \\ 0 & 1 & -s_1 & c_1 & -s_2 & c_2 & -s_3 & c_3 & -s_4 & c_4 & -s_5 & c_5 \\ 1 & 0 & c_2 & s_2 & c_4 & s_4 & 1 & 0 & c_2 & s_2 & c_4 & s_4 \\ 0 & 1 & -s_2 & c_2 & -s_4 & c_4 & 0 & 1 & -s_2 & c_2 & -s_4 & c_4 \\ 1 & 0 & c_3 & s_3 & 1 & 0 & c_3 & s_3 & 1 & 0 & c_3 & s_3 \\ 0 & 1 & -s_3 & c_3 & 0 & 1 & -s_3 & c_3 & 0 & 1 & -s_3 & c_3 \\ 1 & 0 & c_4 & s_4 & c_2 & s_2 & 1 & 0 & c_4 & s_4 & c_2 & s_2 \\ 0 & 1 & -s_4 & c_4 & -s_2 & c_2 & 0 & 1 & -s_4 & c_4 & -s_2 & c_2 \\ 1 & 0 & c_5 & s_5 & c_4 & s_4 & c_3 & s_3 & c_2 & s_2 & c_1 & s_1 \\ 0 & 1 & -s_5 & c_5 & -s_4 & c_4 & -s_3 & c_3 & -s_2 & c_2 & -s_1 & c_1 \end{bmatrix}$$

and the determinant $\det(X) = 6^6$. For the cosine and sine coefficients of this matrix, we have $c_{6-k} = c_k$ and $s_{6-k} = -s_k$, when $k = 1, 2$, and $c_3 = -1$ and

$s_3 = 0$. By using only the coefficients $0, \pm 1, c_1, c_2, s_1$, and s_2, this matrix can be written as

$$
X =
\begin{bmatrix}
1 & 0 & 1 & 0 & 1 & 0 & 1 & 0 & 1 & 0 & 1 & 0 \\
0 & 1 & 0 & 1 & 0 & 1 & 0 & 1 & 0 & 1 & 0 & 1 \\
1 & 0 & c_1 & s_1 & c_2 & s_2 & -1 & 0 & c_2 & -s_2 & c_1 & -s_1 \\
0 & 1 & -s_1 & c_1 & -s_2 & c_2 & 0 & -1 & s_2 & c_2 & s_1 & c_1 \\
1 & 0 & c_2 & s_2 & c_2 & -s_2 & 1 & 0 & c_2 & s_2 & c_2 & -s_2 \\
0 & 1 & -s_2 & c_2 & s_2 & c_2 & 0 & 1 & -s_2 & c_2 & s_2 & c_2 \\
1 & 0 & -1 & 0 & 1 & 0 & -1 & 0 & 1 & 0 & -1 & 0 \\
0 & 1 & 0 & -1 & 0 & 1 & 0 & -1 & 0 & 1 & 0 & -1 \\
1 & 0 & c_2 & -s_2 & c_2 & s_2 & 1 & 0 & c_2 & -s_2 & c_2 & s_2 \\
0 & 1 & s_2 & c_2 & -s_2 & c_2 & 0 & 1 & s_2 & c_2 & -s_2 & c_2 \\
1 & 0 & c_1 & -s_1 & c_2 & -s_2 & -1 & 0 & c_2 & s_2 & c_1 & s_1 \\
0 & 1 & s_1 & c_1 & s_2 & c_2 & 0 & -1 & -s_2 & c_2 & -s_1 & c_1
\end{bmatrix}.
$$

For the real vector $\mathbf{f}$, we consider $\bar{\mathbf{f}} = \mathbf{f}$, and the DFT in the real space R^{12} can be written as

$$
X\mathbf{f} =
\begin{bmatrix}
1 & 1 & 1 & 1 & 1 & 1 \\
0 & 0 & 0 & 0 & 0 & 0 \\
1 & c_1 & c_2 & -1 & c_2 & c_1 \\
0 & -s_1 & -s_2 & 0 & s_2 & s_1 \\
1 & c_2 & c_2 & 1 & c_2 & c_2 \\
0 & -s_2 & s_2 & 0 & -s_2 & s_2 \\
1 & -1 & 1 & -1 & 1 & -1 \\
0 & 0 & 0 & 0 & 0 & 0 \\
1 & c_2 & c_2 & 1 & c_2 & c_2 \\
0 & s_2 & -s_2 & 0 & s_2 & -s_2 \\
1 & c_1 & c_2 & -1 & c_2 & c_1 \\
0 & s_1 & s_2 & 0 & -s_2 & -s_1
\end{bmatrix}
\begin{bmatrix}
f_0 \\ f_1 \\ f_2 \\ f_3 \\ f_4 \\ f_5
\end{bmatrix}
=
\begin{bmatrix}
R_0 \\ I_0 \\ R_1 \\ I_1 \\ R_2 \\ I_2 \\ R_3 \\ I_3 \\ R_4 \\ I_4 \\ R_5 \\ I_5
\end{bmatrix}.
$$

Since the property of the complex conjugate holds for the DFT, i.e.,

$$
R_4 = R_2, \ I_4 = -I_2, \ \text{and} \ R_5 = R_1, \ I_5 = -I_1,
$$

we can remove the last four rows from the matrix and write the transform as

$$
\hat{X}\mathbf{f} =
\begin{bmatrix}
1 & 1 & 1 & 1 & 1 & 1 \\
0 & 0 & 0 & 0 & 0 & 0 \\
1 & c_1 & c_2 & -1 & c_2 & c_1 \\
0 & -s_1 & -s_2 & 0 & s_2 & s_1 \\
1 & c_2 & c_2 & 1 & c_2 & c_2 \\
0 & -s_2 & s_2 & 0 & -s_2 & s_2 \\
1 & -1 & 1 & -1 & 1 & -1 \\
0 & 0 & 0 & 0 & 0 & 0
\end{bmatrix}
\begin{bmatrix}
f_0 \\ f_1 \\ f_2 \\ f_3 \\ f_4 \\ f_5
\end{bmatrix}
=
\begin{bmatrix}
R_0 \\ I_0 \\ R_1 \\ I_1 \\ R_2 \\ I_2 \\ R_3 \\ I_3
\end{bmatrix}.
$$

or, after removing two zero imaginary components I_0 and I_3, in the following form:

$$
\hat{X}\mathbf{f} = \begin{bmatrix}
1 & 1 & 1 & 1 & 1 & 1 \\
1 & c_1 & -c_1 & -1 & -c_1 & c_1 \\
0 & -s_1 & -s_1 & 0 & s_1 & s_1 \\
1 & -c_1 & -c_1 & 1 & -c_1 & -c_1 \\
0 & -s_1 & s_1 & 0 & -s_1 & s_1 \\
1 & -1 & 1 & -1 & 1 & -1
\end{bmatrix}
\begin{bmatrix} f_0 \\ f_1 \\ f_2 \\ f_3 \\ f_4 \\ f_5 \end{bmatrix}
= \begin{bmatrix} R_0 \\ R_1 \\ I_1 \\ R_2 \\ I_2 \\ R_3 \end{bmatrix},
$$

since $c_2 = -c_1 = -0.5$ and $s_2 = s_1 = 0.8660$. This matrix can be decomposed as $\hat{X} = A_1 + c_1 A_{c_1} + s_1 A_{s_1}$, where

$$
A_1 = \begin{bmatrix}
1 & 1 & 1 & 1 & 1 & 1 \\
1 & 0 & 0 & -1 & 0 & 0 \\
0 & 0 & 0 & 0 & 0 & 0 \\
1 & 0 & 0 & 1 & 0 & 0 \\
0 & 0 & 0 & 0 & 0 & 0 \\
1 & -1 & 1 & -1 & 1 & -1
\end{bmatrix}
= \begin{bmatrix}
1 & 0 & 0 & 1 & 0 & 0 \\
1 & 0 & 0 & -1 & 0 & 0 \\
0 & 0 & 0 & 0 & 0 & 0 \\
1 & 0 & 0 & 1 & 0 & 0 \\
0 & 0 & 0 & 0 & 0 & 0 \\
1 & 0 & 0 & -1 & 0 & 0
\end{bmatrix}
+ \begin{bmatrix}
0 & 1 & 1 & 0 & 1 & 1 \\
0 & 0 & 0 & 0 & 0 & 0 \\
0 & 0 & 0 & 0 & 0 & 0 \\
0 & 0 & 0 & 0 & 0 & 0 \\
0 & 0 & 0 & 0 & 0 & 0 \\
0 & -1 & 1 & 0 & 1 & -1
\end{bmatrix}
$$

and

$$
A_{c_1} = \begin{bmatrix}
0 & 0 & 0 & 0 & 0 & 0 \\
0 & 1 & -1 & 0 & -1 & 1 \\
0 & 0 & 0 & 0 & 0 & 0 \\
0 & -1 & -1 & 0 & -1 & -1 \\
0 & 0 & 0 & 0 & 0 & 0 \\
0 & 0 & 0 & 0 & 0 & 0
\end{bmatrix},
\qquad
A_{s_1} = \begin{bmatrix}
0 & 0 & 0 & 0 & 0 & 0 \\
0 & 0 & 0 & 0 & 0 & 0 \\
0 & -1 & -1 & 0 & 1 & 1 \\
0 & 0 & 0 & 0 & 0 & 0 \\
0 & -1 & 1 & 0 & -1 & 1 \\
0 & 0 & 0 & 0 & 0 & 0
\end{bmatrix}.
$$

We now calculate the number of operations when calculating the transform

$$
\hat{X}\mathbf{f} = A_1\mathbf{f} + c_1(A_{c_1}\mathbf{f}) + s_1(A_{s_1}\mathbf{f}).
$$

Eight additions/subtractions are required when using the matrix A_1, four such operations for calculating $A_{s_1}\mathbf{f}$. The calculation of $A_{c_1}\mathbf{f}$ does not require any additions; the 2nd and 4th components of this product can be defined during the calculation of $A_1\mathbf{f}$ in its second addend. The sum of these three products $A_1\mathbf{f}$, $c_1(A_{c_1}\mathbf{f})$, and $s_1(A_{s_1}\mathbf{f})$ requires two additions (the summation is only along the 2nd and 4th rows). Thus the number of additions for the 6-point DFT equals $a(6) = (8+0+4)+2 = 14$. The number of multiplications (only by s_1) equals 2, i.e., $m(6) = 2$. Two multiplications by $c_1 = -0.5$ are considered as the shift of one bit each. It is interesting to note for comparison that the paired algorithm of the six-point DFT for real data uses six multiplications [9].

The determinant of the matrix $\hat{X}$ equals -54 and the inverse matrix is

$$
\hat{X}^{-1} = \frac{1}{6}\begin{bmatrix}
1 & 1 & 0 & 1 & 0 & 1 \\
1 & 1 & -2s_1 & -1 & -2s_1 & -1 \\
1 & -1 & -2s_1 & -1 & 2s_1 & 1 \\
1 & -1 & 0 & 1 & 0 & -1 \\
1 & -1 & 2s_1 & -1 & -2s_1 & 1 \\
1 & 1 & 2s_1 & -1 & 2s_1 & -1
\end{bmatrix},
\qquad (s_1 = 0.8660).
$$

The multiplication by this matrix also requires two multiplications. □

The DFT in the real space R^{2N} is linear as in the complex plane C^N. Multiplication of the DFTs of two vectors $\mathbf{f}$ and $\mathbf{g} \in R^N$ corresponds to the aperiodic convolution of these vectors, $\mathbf{f} * \mathbf{g}$. In vector representation, the multiplication of the components F_p and G_p of these vectors can be written as

$$\mathbf{F}_p \bullet \mathbf{G}_p = \begin{bmatrix} R_{p;g} & -I_{p;g} \\ I_{p;g} & R_{p;g} \end{bmatrix} \begin{bmatrix} R_{p;f} \\ I_{p;f} \end{bmatrix} = |G_p| \begin{bmatrix} c_{p;g} & -s_{p;g} \\ s_{p;g} & c_{p;g} \end{bmatrix} \begin{bmatrix} R_{p;f} \\ I_{p;f} \end{bmatrix}$$

where $c_{p;g}$ and $s_{p;g}$ are the cosine and sine components of the complex number G_p, respectively.

$$c_{p;g} = \cos(\phi_p), \quad s_{p;g} = \sin(\phi_p), \quad \phi_p = Arg(G_p) = \tan^{-1}(R_{p;g}/I_{p;g}).$$

Thus the block-wise rotation of the DFT of the signal $\mathbf{f}$ in the space R^{2N} by angles defined from the polar form of the complex numbers G_p, $p = 0 : (N-1)$ results in the aperiodic convolution of vectors $\mathbf{f}$ and $\mathbf{g}$. Such block-wise rotation can be performed over the DFT of the signal $\mathbf{g}$, when using angles from the polar forms of F_p, $p = 0 : (N-1)$.

For example, when the vectors $\mathbf{f}$ and $\mathbf{g}$ are real and six-dimensional, $N = 6$, we can calculate the values of the six-point DFT of the convolution $\mathbf{f} * \mathbf{g}$ by

$$\hat{X}(\mathbf{f} * \mathbf{g}) = \begin{bmatrix} |G_0| \begin{bmatrix} 1 & 0 \\ 0 & 1 \end{bmatrix} \\ & |G_1| \begin{bmatrix} c_{1;g} & -s_{1;g} \\ s_{1;g} & c_{1;g} \end{bmatrix} \\ & & \ddots \\ & & & |G_5| \begin{bmatrix} c_{5;g} & -s_{5;g} \\ s_{5;g} & c_{5;g} \end{bmatrix} \end{bmatrix} \hat{X}(\mathbf{f}).$$

Since the signals $\mathbf{f}$ and $\mathbf{g}$ are real, this six-block matrix can be reduced to a four-block matrix with rotations only by the first four angles ϕ_p, $p = 0 : 3$.

The inner product is preserved for the extended DFT, i.e.,

$$(\bar{\mathbf{f}}, \bar{\mathbf{g}}) = \sum_{n=0}^{2N-1} \bar{f}_n \bar{g}_n = (X\bar{\mathbf{f}}, X\bar{\mathbf{g}}) = \frac{1}{N} \sum_{k=0}^{2N-1} [F]_k [G]_k,$$

where we denote by $[F]_k$ and $[G]_k$ the components of the extended DFTs of the vectors $\bar{\mathbf{f}}$ and $\bar{\mathbf{g}}$, respectively. For example, when signals are real and $\mathbf{f} = (1, 2, 4, 7, 5, 6)'$ and $\mathbf{g} = (3, 1, 2, 1, 4, 7)'$, the 12-point images of the extended 6-point DFTs of these vectors equal, respectively,

$$X\bar{\mathbf{f}} = (25, 0, -6.5, 4.3301, -0.5, 2.5981, -5, 0, -0.5, -2.5981, -6.5, -4.3301)'$$
$$X\bar{\mathbf{g}} = (18, 0, 3, 6.9282, -3, 3.4641, 0, 0, -3, -3.4641, 3, -6.9282)'$$

and $(X\bar{\mathbf{f}}, X\bar{\mathbf{g}}) = (\bar{\mathbf{f}}, \bar{\mathbf{g}}) = 82$. If we consider the complex case, for example, two complex vectors $\mathbf{f} = (1 + 2j, 4 + 7j, 5 + 6j, 3 + 2j, 1 + 4j, 2 + 5j)'$ and $\mathbf{g} = (3 + j, 2 + j, 4 + 7j, 4 + 5j, 2 + j, 2 + j)'$, then we obtain

$$(X\bar{\mathbf{f}}, X\bar{\mathbf{g}}) = \frac{1}{6} \sum_{k=0}^{11} [F]_k [G]_k = (\bar{\mathbf{f}}, \bar{\mathbf{g}}) = \sum_{n=0}^{11} \bar{f}_n \bar{g}_n = 119$$

$$= \operatorname{Re}\left((\mathbf{f}, \mathbf{g}) = \sum_{n=0}^{5} f_n g_n^* = 119 + 12j \right).$$

Here $\bar{\mathbf{f}} = (1, 2, 4, 7, 5, 6, 3, 2, 1, 4, 2, 5)'$ and $\bar{\mathbf{g}} = (3, 1, 2, 1, 4, 7, 4, 5, 2, 1, 2, 1)'$.

2.4.2 Integer representation of the DFT

The definition of the DFT in the real space R^{2N} by (2.12) and (2.13) can be generalized by using the two-point transforms different from the rotations. In the real space R^{2N}, we define the $2N$-point transform over the vector $\bar{\mathbf{f}} = (f_0, f_1, f_2, f_3, ..., f_{2N-2}, f_{2N-1})'$ by

$$\bar{\mathbf{F}}_p = \begin{bmatrix} F_{2p} \\ F_{2p+1} \end{bmatrix} = \sum_{n=0}^{N-1} T^{np} \bar{\mathbf{f}}_n = \sum_{n=0}^{N-1} T^{np} \begin{bmatrix} f_{2n} \\ f_{2n+1} \end{bmatrix}, \quad p = 0 : (N-1), \quad (2.14)$$

where T is a matrix 2×2 with determinant one, $\det T = 1$. It is assumed also that such a matrix defines the one-parametric group with period N. We call the transformation $X : \mathbf{f} \to \mathbf{F}$ *the T-generated N-block discrete transform,* or *N-block T-GDT*. In the case when each pair (f_{2n}, f_{2n+1}) represents the complex component of the vector $\mathbf{f}$ and the matrix T is the matrix of the Givens rotation by angle $2\pi/N$, this definition leads to the N-point DFT extended into the real space R^{2N}. We will call this Fourier transform *the N-block discrete Fourier transform,* or *the N-block W-GFT*.

2.4.2.1 The $N = 6$ case

In the real space R^{12}, we define the integer transformation X, which is based on the integer two-point transformation, with the following matrix:

$$T = \begin{bmatrix} 1 & -1 \\ 1 & 0 \end{bmatrix}, \quad (\det(T) = 1),$$

which composes a one-parametric group with period $N = 6$, i.e., $T^6 = I$.

The six-block T-GDT is defined by the matrix whose (n, p)-th blocks 2×2 are calculated by $X_{(n,p)} = T^{np}$, where $n, p = 0 : 5$. The transform has the following binary matrix (12×12) (or block-matrix (6×6)):

$$X = X(T) = \begin{bmatrix} I & I & I & I & I & I \\ I & T^1 & T^2 & T^3 & T^4 & T^5 \\ I & T^2 & T^4 & I & T^2 & T^4 \\ I & T^3 & I & T^3 & I & T^3 \\ I & T^4 & T^2 & I & T^4 & T^2 \\ I & T^5 & T^4 & T^3 & T^2 & T^1 \end{bmatrix}.$$

The determinant of this matrix $\det(X) = 6^6$, as for the Fourier transformation. The block matrices are

$$T^2 = \begin{bmatrix} 0 & -1 \\ 1 & -1 \end{bmatrix}, T^3 = \begin{bmatrix} -1 & 0 \\ 0 & -1 \end{bmatrix} = -I, T^4 = \begin{bmatrix} -1 & 1 \\ -1 & 0 \end{bmatrix} = -T, T^5 = \begin{bmatrix} 0 & 1 \\ -1 & 1 \end{bmatrix} = -T^2.$$

The inverse matrix is defining by inverting all blocks 2×2,

$$X^{-1} = \frac{1}{6} \begin{bmatrix} I & I & I & I & I & I \\ I & T^{-1} & T^{-2} & T^{-3} & T^{-4} & T^{-5} \\ I & T^{-2} & T^{-4} & I & T^{-2} & T^{-4} \\ I & T^{-3} & I & T^{-3} & I & T^{-3} \\ I & T^{-4} & T^{-2} & I & T^{-4} & T^{-2} \\ I & T^{-5} & T^{-4} & T^{-3} & T^{-2} & T^{-1} \end{bmatrix} = \frac{1}{6} \begin{bmatrix} I & I & I & I & I & I \\ I & T^5 & T^4 & T^3 & T^2 & T^1 \\ I & T^4 & T^2 & I & T^4 & T^2 \\ I & T^3 & I & T^3 & I & T^3 \\ I & T^2 & T^4 & I & T^2 & T^4 \\ I & T^1 & T^2 & T^3 & T^4 & T^5 \end{bmatrix},$$

or $X^{-1}(T) = X(T^{-1})/6 = X(T^5)/6$. Thus,

$$X^{-1}(n,p) = (X(n,p))^{-1}/6 = T^{-np}/6 = T^{6-np}/6, \quad n, p = 0 : 5.$$

The following also holds for this binary transform (as for the DFT): $X \cdot X(T^{-1}) = 6I$, $X^4 = X(T^{-1})^4 = 6^2 I$. The normalized matrices $X/\sqrt{6}$ and $X(T^{-1})/\sqrt{6}$ are the fourth roots of the identity matrix.

We consider one of the main properties of the DFT, the cyclic shift. Let for example, the vector $\bar{\mathbf{f}} = (f_0, f_1, f_2, f_3, f_4, f_5)'$ be shifted to the right cyclicly by one position, $\bar{\mathbf{f}} \to \overrightarrow{\mathbf{f}} = (f_5, f_0, f_1, f_2, f_3, f_4)'$. Then we obtain the following for the transform ($p = 0 : 5$):

$$\left(X \overrightarrow{\mathbf{f}} \right)_p = \sum_{n=0}^{5} T^{np} \mathbf{f}_{n-1} = \sum_{n=0}^{5} T^{(n+1)p} \mathbf{f}_n = T^p \sum_{n=0}^{5} T^{np} \mathbf{f}_n = T^p \left(X \bar{\mathbf{f}} \right)_p.$$

The property of shifting holds for the six-block T-GDT in the general case of shifting, too. The inner product is not preserved for the N-block T-GDT, i.e.,

$$(\bar{\mathbf{f}}, \bar{\mathbf{g}}) = \sum_{n=0}^{11} \bar{f}_n \bar{g}_n \neq (X\bar{\mathbf{f}}, X\bar{\mathbf{g}}) = \frac{1}{6} \sum_{k=0}^{11} [F]_k [G]_k,$$

where $[F]_k$ and $[G]_k$ denote the component of the six-block T-GDT of the vectors $\bar{\mathbf{f}}$ and $\bar{\mathbf{g}}$, respectively. For example, if we consider the vectors $\bar{\mathbf{f}} =$

$(1, 2, 4, 7, 5, 6, 3, 2, 1,\ 4, 2, 5)'$ and $\bar{\mathbf{g}} = (3, 1, 2, 1, 4, 7, 4, 5, 2, 1, 2, 1)'$, then we obtain

$$(X\bar{\mathbf{f}}, X\bar{\mathbf{g}}) = \frac{1}{6}\sum_{k=0}^{11}[F]_k[G]_k = 107 \neq (\bar{\mathbf{f}}, \bar{\mathbf{g}}) = 119,$$

as well as $|X\bar{\mathbf{f}}|^2 = (X\bar{\mathbf{f}}, X\bar{\mathbf{f}}) = 128.33 \neq |\bar{\mathbf{f}}|^2 = (\bar{\mathbf{f}}, \bar{\mathbf{f}}) = 131$.

That was expected, because of the matrix T. Therefore, we will try to change the inner product of the vectors and define it as $(\bar{\mathbf{f}}, \bar{\mathbf{g}})_A = \bar{\mathbf{f}}'A\bar{\mathbf{g}}$, $\bar{\mathbf{f}}, \bar{\mathbf{g}} \in R^{12}$, where A is a non-identity matrix (12×12) to be found. For that, we first define the matrix (or matrices) R, such that $T'RT = R$, and then define the matrix A as the block-diagonal matrix with the matrix R in the blocks, i.e., $A = I_6 \otimes R$. It is not difficult to see that the equation

$$\begin{bmatrix} 1 & 1 \\ -1 & 0 \end{bmatrix}\begin{bmatrix} a & b \\ c & d \end{bmatrix}\begin{bmatrix} 1 & -1 \\ 1 & 0 \end{bmatrix} = \begin{bmatrix} a & b \\ c & d \end{bmatrix}$$

leads to the following system of equations with unknown coefficients a, b, c, and $d: c = -b - a$, $d = a$. Thus, we obtain the family of solutions which are parameterized by a and b,

$$R = R(a, b) = \begin{bmatrix} a & b \\ -a - b & a \end{bmatrix}. \tag{2.15}$$

Let us consider, for instance, the case when $a = 0$ and $b = 1$, i.e.,

$$R = R(0, 1) = \begin{bmatrix} 0 & 1 \\ -1 & 0 \end{bmatrix}. \tag{2.16}$$

The "inner" product of two vectors $\bar{\mathbf{f}}$ and $\bar{\mathbf{g}}$ is calculated by

$$(\bar{\mathbf{f}}, \bar{\mathbf{g}})_A = \sum_{n=0}^{5}\bar{\mathbf{f}}'_n R\bar{\mathbf{g}}_n = \sum_{n=0}^{5}\begin{bmatrix} f_{2n} & f_{2n+1} \end{bmatrix}\begin{bmatrix} 0 & 1 \\ -1 & 0 \end{bmatrix}\begin{bmatrix} g_{2n} \\ g_{2n+1} \end{bmatrix}$$

$$= \sum_{n=0}^{5}\begin{bmatrix} f_{2n} & f_{2n+1} \end{bmatrix}\begin{bmatrix} g_{2n+1} \\ -g_{2n} \end{bmatrix} = \sum_{n=0}^{5}[f_{2n}g_{2n+1} - f_{2n+1}g_{2n}].$$

In the transform space, we define this product as

$$(X\bar{\mathbf{f}}, X\bar{\mathbf{g}})_A = \frac{1}{6}\sum_{p=0}^{5}(X\bar{\mathbf{f}})'_p R(X\bar{\mathbf{g}})_p = \frac{1}{6}\sum_{p=0}^{5}\begin{bmatrix} F_{2p} & F_{2p+1} \end{bmatrix}\begin{bmatrix} 0 & 1 \\ -1 & 0 \end{bmatrix}\begin{bmatrix} G_{2p} \\ G_{2p+1} \end{bmatrix}.$$

Note that with respect to the introduced "inner" product $(\bar{\mathbf{f}}, \bar{\mathbf{g}})_A$, the rows (and columns) of the matrices X are orthogonal. We leave the details of this proof to the reader.

Example 2.11

For the considered vectors $\bar{\mathbf{f}} = (1, 2, 4, 7, 5, 6, 3, 2, 1, 4, 2, 5)'$ and $\bar{\mathbf{g}} = (3, 1, 2, 1, 4, 7, 4, 5, 2, 1, 2, 1)'$, we obtain the following: $(\bar{\mathbf{f}}, \bar{\mathbf{g}})_A = (X\bar{\mathbf{f}}, X\bar{\mathbf{g}})_A = -12$. ⬚

It is not difficult to see that this "inner" product makes the norms of all vectors equal zero. Indeed, according to the definition, we have $(\bar{\mathbf{f}}, \bar{\mathbf{g}})_A = -(\bar{\mathbf{g}}, \bar{\mathbf{f}})_A$, and as a result, $(\bar{\mathbf{f}}, \bar{\mathbf{f}})_A = (\bar{\mathbf{g}}, \bar{\mathbf{g}})_A = (X\bar{\mathbf{f}}, X\bar{\mathbf{f}})_A = (X\bar{\mathbf{g}}, X\bar{\mathbf{g}})_A = 0$. Thus in the vector space with this product the last axiom of the inner product does not hold, namely, "4°. $(\bar{\mathbf{f}}, \bar{\mathbf{f}})_A = 0$ then $\bar{\mathbf{f}} = 0$." Therefore, we consider other matrices for R.

We describe the case when $a = 1$ and $b = 0$, i.e.,

$$R = R(0, 1) = \begin{bmatrix} 1 & 0 \\ -1 & 1 \end{bmatrix}.$$

With respect to the product defined by this matrix, the norm of a vector $\mathbf{x} = (x_0, x_1)'$ is calculated as

$$|\mathbf{x}|^2 = (\mathbf{x}, \mathbf{x})_R = \begin{bmatrix} x_0 & x_1 \end{bmatrix} \begin{bmatrix} 1 & 0 \\ -1 & 1 \end{bmatrix} \begin{bmatrix} x_0 \\ x_1 \end{bmatrix} = x_0^2 + x_1^2 - x_0 x_1.$$

This norm can be used for the metric $(d(\mathbf{x}, \mathbf{y}) = |\mathbf{x} - \mathbf{y}|^2)$, since it is always nonnegative,

$$|\mathbf{x}|^2 = \frac{1}{2}(x_0^2 + x_1^2) + \frac{1}{2}(x_0 - x_1)^2 \geq 0.$$

The product of two vectors $\mathbf{x} = (x_0, x_1)'$ and $\mathbf{y} = (y_0, y_1)'$ is calculated as

$$(\mathbf{x}, \mathbf{y})_R = \begin{bmatrix} x_0 & x_1 \end{bmatrix} \begin{bmatrix} 1 & 0 \\ -1 & 1 \end{bmatrix} \begin{bmatrix} y_0 \\ y_1 \end{bmatrix} = x_0 y_0 + x_1 y_1 - x_1 y_0,$$

which again is not equal to $(\mathbf{y}, \mathbf{x})_R$, when $\mathbf{y} \neq \mathbf{x}$. The "inner" product of two vectors $\bar{\mathbf{f}}$ and $\bar{\mathbf{g}}$ is calculated by

$$(\bar{\mathbf{f}}, \bar{\mathbf{g}})_A = \sum_{n=0}^{5} \begin{bmatrix} f_{2n} & f_{2n+1} \end{bmatrix} \begin{bmatrix} 1 & 0 \\ -1 & 1 \end{bmatrix} \begin{bmatrix} g_{2n} \\ g_{2n+1} \end{bmatrix} = \sum_{n=0}^{5} [f_{2n}g_{2n} + f_{2n+1}g_{2n+1} - f_{2n+1}g_{2n}],$$

and the norm of the vectors is defined as

$$|\bar{\mathbf{f}}|^2 = (\bar{\mathbf{f}}, \bar{\mathbf{f}})_A = \sum_{n=0}^{5} [f_{2n}^2 + f_{2n+1}^2 - f_{2n}f_{2n+1}] = \sum_{n=0}^{11} f_n^2 - \sum_{n=0}^{5} f_{2n}f_{2n+1}.$$

This norm can be used for the metric in the vector space R^{12}

$$d(\mathbf{f}, \mathbf{g}) = |\mathbf{f} - \mathbf{g}|^2 = (\bar{\mathbf{f}} - \mathbf{g}, \bar{\mathbf{f}} - \mathbf{g})_A \geq 0, \quad \mathbf{f}, \mathbf{g} \in R^{12}.$$

Consider first the basic functions of the six-point DFT,

$$f_n = f(n; p) = \cos(\frac{2\pi}{6}np), \quad g_n = g(n; p) = \sin(\frac{2\pi}{6}np), \quad n = 0 : 5,$$

for $p = 1$ and $p = 2$. Figure 2.21 shows these two discrete signals in part a, along with the "real" and "imaginary" parts of the T-GDT in b and c, respectively. One can see from (c) that the transform distinguishes the main frequency $p = 1$, as well as frequency 5, since $f(n; 5) = f(n; 1)$ and $g(n; 5) = -g(n; 1)$. In the "real" part of the transform, the frequency 5 is observed not 1. The similar results are shown in Figure 2.22, when the cosine and sine waves are calculated for the frequency $p = 2$. The transform distinguishes the main frequencies $p = 2$ and 4, since $g(n; 2) = -g(n; 4)$.

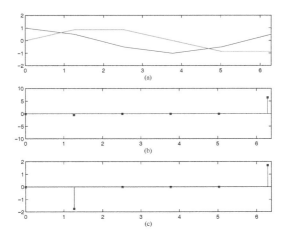

FIGURE 2.21
Cosine and sine waves of length 6 with the six-point T-GDTs for $p = 1$.

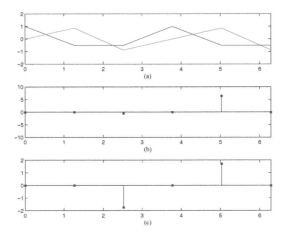

FIGURE 2.22
Cosine and sine waves of length 6 with the six-point T-GDTs for $p = 2$.

Note that we use the name "inner" product in quotes, since in both above cases the condition $(\bar{\mathbf{f}}, \bar{\mathbf{g}})_A = (\bar{\mathbf{g}}, \bar{\mathbf{f}})_A$ does not hold when $\bar{\mathbf{f}} \neq \bar{\mathbf{g}}$. In order to obtain this property, the coefficients of the matrix $R(a, b)$ in (2.15) should satisfy the following condition: $b = -a/2$. Thus we consider the following matrix:

$$R = R(a, -a/2) = \begin{bmatrix} a & -a/2 \\ -a/2 & a \end{bmatrix} = a \begin{bmatrix} 1 & -1/2 \\ -1/2 & 1 \end{bmatrix}.$$

We denote this matrix by $R(a)$ and consider the integer normalized matrix

$$R = \frac{1}{\sqrt{3}} R(2) = \frac{1}{\sqrt{3}} \begin{bmatrix} 2 & -1 \\ -1 & 2 \end{bmatrix} = \begin{bmatrix} 1.1547 & -0.5774 \\ -0.5774 & 1.1547 \end{bmatrix}, \quad \det R = 1. \quad (2.17)$$

The inner product of vectors $\mathbf{x} = (x_0, x_1)'$ and $\mathbf{y} = (y_0, y_1)'$ is calculated by this matrix as

$$(\mathbf{x}, \mathbf{y})_R = \frac{1}{\sqrt{3}} [x_0 \; x_1] \begin{bmatrix} 2 & -1 \\ -1 & 2 \end{bmatrix} \begin{bmatrix} y_0 \\ y_1 \end{bmatrix} = \frac{1}{\sqrt{3}} [2(x_0 y_0 + x_1 y_1) - (x_0 y_1 + y_0 x_1)].$$

The norm of a vector $\mathbf{x} = (x_0, x_1)'$ equals

$$|\mathbf{x}|^2 = \frac{2}{\sqrt{3}} [2(x_0^2 + x_1^2) - x_0 x_1] \geq 0,$$

and in the space R^{12}, the norm is calculated by

$$|\bar{\mathbf{f}}|^2 = (\bar{\mathbf{f}}, \bar{\mathbf{f}})_A = \frac{2}{\sqrt{3}} \sum_{n=0}^{5} [2(f_{2n}^2 + f_{2n+1}^2) - f_{2n} f_{2n+1}]$$

$$= \frac{4}{\sqrt{3}} \sum_{n=0}^{11} f_n^2 - \frac{2}{\sqrt{3}} \sum_{n=0}^{5} f_{2n} f_{2n+1}.$$

The inner product of two vectors $\bar{\mathbf{f}}$ and $\bar{\mathbf{g}} \in R^{12}$ is calculated by

$$(\bar{\mathbf{f}}, \bar{\mathbf{g}})_A = \frac{1}{\sqrt{3}} \sum_{n=0}^{5} [f_{2n} \; f_{2n+1}] \begin{bmatrix} 2 & -1 \\ -1 & 2 \end{bmatrix} \begin{bmatrix} g_{2n} \\ g_{2n+1} \end{bmatrix},$$

and in the transform space, the inner product is

$$(X\bar{\mathbf{f}}, X\bar{\mathbf{g}})_A = \frac{1}{6\sqrt{3}} \sum_{p=0}^{5} [F_{2p} \; F_{2p+1}] \begin{bmatrix} 2 & -1 \\ -1 & 2 \end{bmatrix} \begin{bmatrix} G_{2p} \\ G_{2p+1} \end{bmatrix}.$$

Example 2.12

For the $N = 6$ case, and vectors $\mathbf{f} = (1, 2, 4, 9, 5, 1, 3, 2, 5, 4, 6, 8)'$ and $\mathbf{g} = (2, 1, 8, 3, 4, 5, 4, 3, 2, 6, 4, 2)'$, we obtain the following values of the inner products:

$$(\bar{\mathbf{f}}, \bar{\mathbf{g}})_A = (X\bar{\mathbf{f}}, X\bar{\mathbf{g}})_A = \frac{1}{\sqrt{3}} \cdot 143 = 82.5611,$$

$$|\bar{\mathbf{f}}|_A^2 = (\bar{\mathbf{f}}, \bar{\mathbf{f}})_A = \frac{1}{\sqrt{3}} \cdot 330 = 190.5256,$$

$$|\bar{\mathbf{g}}|_A^2 = (\bar{\mathbf{g}}, \bar{\mathbf{g}})_A = \frac{1}{\sqrt{3}} \cdot 252 = 145.4923,$$

$$|\bar{\mathbf{f}} - \bar{\mathbf{g}}|_A^2 = (\bar{\mathbf{f}} - \bar{\mathbf{g}}, \bar{\mathbf{f}} - \bar{\mathbf{g}})_A = \frac{1}{\sqrt{3}} \cdot 296 = 170.8957.$$

Note for comparison that for the traditional inner product and the Fourier transform, W-GFT, we have the following data: $(\bar{\mathbf{f}}, \bar{\mathbf{g}}) = (F\bar{\mathbf{f}}, F\bar{\mathbf{g}}) = 180$, $|\bar{\mathbf{f}}|^2 = (\bar{\mathbf{f}}, \bar{\mathbf{f}}) = 282$, $|\bar{\mathbf{g}}|^2 = (\bar{\mathbf{g}}, \bar{\mathbf{g}}) = 204$, $|\bar{\mathbf{f}} - \bar{\mathbf{g}}|^2 = (\bar{\mathbf{f}} - \bar{\mathbf{g}}, \bar{\mathbf{f}} - \bar{\mathbf{g}}) = 126$. $\Box$

By using different matrices R parameterized by two values of a and b, we can define different metrics in the space of vectors R^{12}. The selection of these parameters can be done among only the integer numbers, which leads to an integer-valued metric. Thus, the six-point DFT in the complex space or in the traditional form has been transferred to the integer transformation in the space R^{12}. The simple binary matrix T for this purpose has been selected. Other matrices T may also be used and corresponding integer representations of the six-point DFT obtained. This matrix composes a cyclic group with period six. By selecting the generator T^2, we can consider the subgroup in this group and describe the integer presentation of the DFT in the subgroup. Namely, we can define the integer representation of the three-point DFT with the matrix $X(T^2)$.

2.4.2.2   The $N = 3$ case

In the real space R^6, we define the integer transformation X which is based on the integer two-point transformation with the following matrix:

$$S = T^2 = \begin{bmatrix} 0 & -1 \\ 1 & -1 \end{bmatrix}, \qquad (\det(S) = 1).$$

This matrix composes a one-parametric group with period $N = 3$, i.e., $S^3 = I$. The three-block S-GDT is defined by the matrix whose (n, p)-th blocks 2×2 are calculated by $X_{(n,p)} = S^{np}$, where $n, p = 0, 1, 2$,

$$X = X(S) = \begin{bmatrix} I & I & I \\ I & S^1 & S^2 \\ I & S^2 & S^4 \end{bmatrix} = \begin{bmatrix} I & I & I \\ I & S^1 & S^2 \\ I & S^2 & S^1 \end{bmatrix} = \begin{bmatrix} 1 & 0 & 1 & 0 & 1 & 0 \\ 0 & 1 & 0 & 1 & 0 & 1 \\ 1 & 0 & 0 & -1 & -1 & 1 \\ 0 & 1 & 1 & -1 & -1 & 0 \\ 1 & 0 & -1 & 1 & 0 & -1 \\ 0 & 1 & -1 & 0 & 1 & -1 \end{bmatrix}.$$

The determinant of this matrix $\det(X) = 3^3$, as for the three-point Fourier transform.

The inverse matrix is defined by inverting all blocks 2×2,

$$X^{-1} = \frac{1}{3}X(S^{-1}) = \frac{1}{3}\begin{bmatrix} I & I & I \\ I & S^{-1} & S^{-2} \\ I & S^{-2} & S^{-1} \end{bmatrix} = \frac{1}{3}\begin{bmatrix} 1 & 0 & 1 & 0 & 1 & 0 \\ 0 & 1 & 0 & 1 & 0 & 1 \\ 1 & 0 & -1 & 1 & 0 & -1 \\ 0 & 1 & -1 & 0 & 1 & -1 \\ 1 & 0 & 0 & -1 & -1 & 1 \\ 0 & 1 & 1 & -1 & -1 & 0 \end{bmatrix}.$$

Thus, $X^{-1}(n,p) = (X(n,p))^{-1}/3 = S^{-np}/3 = S^{3-np}/3$ for $n, p = 0, 1, 2$. The following also holds for this binary transform (as for the DFT): $X^4 = X(S^{-1})^4 = 3^2 I$. The normalized matrices $X/\sqrt{3}$ and $X^{-1}/\sqrt{3}$ are the fourth roots of the identity matrix.

The inner product of two vectors $\bar{\mathbf{f}}$ and $\bar{\mathbf{g}} \in R^6$ is defined by the same matrix R in (2.17) and calculated by

$$(\bar{\mathbf{f}}, \bar{\mathbf{g}})_A = \frac{1}{\sqrt{3}} \sum_{n=0}^{2} [f_{2n} \ f_{2n+1}] \begin{bmatrix} 2 & -1 \\ -1 & 2 \end{bmatrix} \begin{bmatrix} g_{2n} \\ g_{2n+1} \end{bmatrix},$$

and in the transform space, the inner product is

$$(X\bar{\mathbf{f}}, X\bar{\mathbf{g}})_A = \frac{1}{6\sqrt{3}} \sum_{p=0}^{2} [F_{2p} \ F_{2p+1}] \begin{bmatrix} 2 & -1 \\ -1 & 2 \end{bmatrix} \begin{bmatrix} G_{2p} \\ G_{2p+1} \end{bmatrix}.$$

The inner product is preserved after transformation, i.e., $(X\bar{\mathbf{f}}, X\bar{\mathbf{g}})_A = (\bar{\mathbf{f}}, \bar{\mathbf{g}})_A$, and the metric in the space R^6 can be defined as

$$d(\bar{\mathbf{f}}, \bar{\mathbf{g}})_A = |\bar{\mathbf{f}} - \bar{\mathbf{g}}|^2 = (\bar{\mathbf{f}} - \bar{\mathbf{g}}, \bar{\mathbf{f}} - \bar{\mathbf{g}})_A = (X\bar{\mathbf{f}} - X\bar{\mathbf{g}}, X\bar{\mathbf{f}} - X\bar{\mathbf{g}})_A = |X\bar{\mathbf{f}} - X\bar{\mathbf{g}}|^2.$$

Thus, we have described the isometric integer representation of the three-point DFT in the extended space R^6. This representation is called the three-block S-GDT, or the three-block T^2-GDT.

Example 2.13

Consider vectors $\mathbf{f} = (1, 2, 4, 9, 5, 1)'$ and $\mathbf{g} = (2, 1, 8, 3, 4, 5)'$. The S-GDTs of the vectors equal $X(\mathbf{f}) = (10, 12, -12, -8, 5, 2)'$, $X(\mathbf{g}) = (14, 9, 0, 2, -8, -8)'$, and for the difference of these vectors we have the following transform: $\mathbf{f} - \mathbf{g} = (-1, 1, -4, 6, 1, -4)' \rightarrow X(\bar{\mathbf{f}} - \bar{\mathbf{g}}) = X(\bar{\mathbf{f}}) - X(\bar{\mathbf{g}}) = (-4, 3, -12, -10, 13, 10)'$. The values of the inner products and norms for these vectors equal

$$(\bar{\mathbf{f}}, \bar{\mathbf{g}})_A = (X\bar{\mathbf{f}}, X\bar{\mathbf{g}})_A = \tfrac{1}{\sqrt{3}} \cdot 58, \ |\bar{\mathbf{f}}|_A^2 = (\bar{\mathbf{f}}, \bar{\mathbf{f}})_A = \tfrac{1}{\sqrt{3}} \cdot 170,$$

$$|\bar{\mathbf{g}}|_A^2 = (\bar{\mathbf{g}}, \bar{\mathbf{g}})_A = \tfrac{1}{\sqrt{3}} \cdot 146, \quad |\bar{\mathbf{f}} - \bar{\mathbf{g}}|_A^2 = (\bar{\mathbf{f}} - \bar{\mathbf{g}}, \bar{\mathbf{f}} - \bar{\mathbf{g}})_A = \tfrac{1}{\sqrt{3}} \cdot 200.$$

The three-block discrete Fourier transform has the matrix

$$[\mathcal{F}_{3\text{-b}}] = \begin{bmatrix} 1 & 0 & 1 & 0 & 1 & 0 \\ 0 & 1 & 0 & 1 & 0 & 1 \\ 1 & 0 & -0.5 & 0.8660 & -0.5 & -0.8660 \\ 0 & 1 & -0.8660 & -0.5 & 0.8660 & -0.5 \\ 1 & 0 & -0.5 & -0.8660 & -0.5 & 0.8660 \\ 0 & 1 & 0.8660 & -0.5 & -0.8660 & -0.5 \end{bmatrix}.$$

The Fourier transforms of these vectors $\mathbf{f} = (1, 2, 4, 9, 5, 1)' = (1 + 2j, 4 + 9j, 5 + j)'$ and $\mathbf{g} = (2, 1, 8, 3, 4, 5)' = (2 + j, 8 + 3j, 4 + 5j)'$ equal

$$
\begin{aligned}
F(\mathbf{f}) &= (10, 12, 3.4282, -2.1340, -10.4282, -3.8660)' \\
&= (10 + 12j, 3.4282 - 2.1340j, -10.4282 - 3.8660j)', \\
F(\mathbf{g}) &= (14, 9, -5.7321, -6.4641, -2.2679, 0.4641)' \\
&= (14 + 9j, -5.7321 - 6.4641j, -2.2679 + 0.4641j)'.
\end{aligned}
$$

In the vector space R^6 with a traditional inner product, we obtain the following data: $(\bar{\mathbf{f}}, \bar{\mathbf{g}}) = (F\bar{\mathbf{f}}, F\bar{\mathbf{g}}) = 88$, $|\bar{\mathbf{f}}|^2 = (\bar{\mathbf{f}}, \bar{\mathbf{f}}) = 128$, $|\bar{\mathbf{g}}|^2 = (\bar{\mathbf{g}}, \bar{\mathbf{g}}) = 119$, $|\bar{\mathbf{f}} - \bar{\mathbf{g}}|^2 = (\bar{\mathbf{f}} - \bar{\mathbf{g}}, \bar{\mathbf{f}} - \bar{\mathbf{g}}) = 71$. Thus, we obtain the integer representation of the Fourier transform of each vector in the space R^6,

$$
\mathbf{f} = \begin{bmatrix} 1 + 2j \\ 4 + 9j \\ 5 + j \end{bmatrix} \rightarrow \bar{\mathbf{f}} = \begin{bmatrix} 1 \\ 2 \\ 4 \\ 9 \\ 5 \\ 1 \end{bmatrix} \rightarrow F(\bar{\mathbf{f}}) = \begin{bmatrix} 10 \\ 12 \\ 3.4282 \\ -2.1340 \\ -10.4282 \\ -3.8660 \end{bmatrix} \rightarrow X(\bar{\mathbf{f}}) = \begin{bmatrix} 10 \\ 12 \\ -12 \\ -8 \\ 5 \\ 2 \end{bmatrix}
$$

and

$$
\mathbf{g} = \begin{bmatrix} 2 + j \\ 8 + 3j \\ 4 + 5j \end{bmatrix} \rightarrow \bar{\mathbf{g}} = \begin{bmatrix} 2 \\ 1 \\ 8 \\ 3 \\ 4 \\ 5 \end{bmatrix} \rightarrow F(\bar{\mathbf{g}}) = \begin{bmatrix} 14 \\ 9 \\ -5.7321 \\ -6.4641 \\ -2.2679 \\ 0.4641 \end{bmatrix} \rightarrow X(\bar{\mathbf{g}}) = \begin{bmatrix} 14 \\ 9 \\ 0 \\ 2 \\ -8 \\ -8 \end{bmatrix}.
$$

The integer representation $F(\cdot) \rightarrow X(\cdot)$ is described by the following matrix:

$$
Z = X \cdot [\mathcal{F}_{3\text{-}b}]^{-1} = \begin{bmatrix} 1 & 0 & 0 & 0 & 0 & 0 \\ 0 & 1 & 0 & 0 & 0 & 0 \\ 0 & 0 & -0.0774 & -0.2887 & 1.0774 & 0.2887 \\ 0 & 0 & -0.2887 & -0.0774 & 0.2887 & 1.0774 \\ 0 & 0 & 1.0774 & 0.2887 & -0.0774 & -0.2887 \\ 0 & 0 & 0.2887 & 1.0774 & -0.2887 & -0.0774 \end{bmatrix} = \begin{bmatrix} 1 & 0 & 0 \\ 0 & 1 & 0 \\ 0 & 0 & Z_{4\times4} \end{bmatrix}.
$$

The block-matrix (4×4) in Z can be decomposed as follows:

$$
Z_{4\times4} = 0.0774 \begin{bmatrix} -1 & 0 & 1 & 0 \\ 0 & -1 & 0 & 1 \\ 1 & 0 & -1 & 0 \\ 0 & 1 & 0 & -1 \end{bmatrix} + 0.2887 \begin{bmatrix} 0 & -1 & 0 & 1 \\ -1 & 0 & 1 & 0 \\ 0 & 1 & 0 & -1 \\ 1 & 0 & -1 & 0 \end{bmatrix} + \begin{bmatrix} 0 & 0 & 1 & 0 \\ 0 & 0 & 0 & 1 \\ 1 & 0 & 0 & 0 \\ 0 & 1 & 0 & 0 \end{bmatrix}.
$$

The multiplication of a vector by this matrix requires 4 multiplications (two multiplications by the factors of 0.0774 and 0.2887 each) and 10 additions.
□

It is interesting to note that in the $N = 6$ case considered above, the integer representation $F(\cdot) \to X(\cdot)$ is described by the similar block-diagonal matrix

$$Z = X \cdot [\mathcal{F}_{6\text{-b}}]^{-1} = 1 \oplus 1 \oplus Z_{10 \times 10},$$

where the matrix $Z_{10 \times 10}$ equals

$$
\begin{bmatrix}
-0.0774 & -0.2887 & 0 & 0\;0\;0 & 0 & 0 & 1.0774 & 0.2887 \\
-0.2887 & -0.0774 & 0 & 0\;0\;0 & 0 & 0 & 0.2887 & 1.0774 \\
0 & 0 & -0.0774 & -0.2887\;0\;0 & 1.0774 & 0.2887 & 0 & 0 \\
0 & 0 & -0.2887 & -0.0774\;0\;0 & 0.2887 & 1.0774 & 0 & 0 \\
0 & 0 & 0 & 0\;1\;0 & 0 & 0 & 0 & 0 \\
0 & 0 & 0 & 0\;0\;1 & 0 & 0 & 0 & 0 \\
0 & 0 & 1.0774 & 0.2887\;0\;0 & -0.0774 & -0.2887 & 0 & 0 \\
0 & 0 & 0.2887 & 1.0774\;0\;0 & -0.2887 & -0.0774 & 0 & 0 \\
1.0774 & 0.2887 & 0 & 0\;0\;0 & 0 & 0 & -0.0774 & -0.2887 \\
0.2887 & 1.0774 & 0 & 0\;0\;0 & 0 & 0 & -0.2887 & -0.0774
\end{bmatrix}.
$$

This matrix is defined by two nontrivial coefficients $a_1 = 0.2887$, $a_2 = 0.0774$, and $1 + a_2 = 1.0774$. If we define the matrix

$$Z_2 = \begin{bmatrix} 0.0774 & 0.2887 \\ 0.2887 & 0.0774 \end{bmatrix}, \qquad \det(Z_2) = -0.0774,$$

then the matrix $Z_{10 \times 10}$ can be written as follows:

$$
Z_{10 \times 10} =
\begin{bmatrix}
-Z_2 & 0\;0 & 0 & I + Z_2 \\
0 & -Z_2\;0 & I + Z_2 & 0 \\
0 & 0\;I & 0 & 0 \\
0 & I + Z_2\;0 & -Z_2 & 0 \\
I + Z_2 & 0\;0 & 0 & -Z_2
\end{bmatrix}
= J +
\begin{bmatrix}
-1 & 0\;0 & 0 & 1 \\
0 & -1\;0 & 1 & 0 \\
0 & 0\;0 & 0 & 0 \\
0 & 1\;0 & -1 & 0 \\
1 & 0\;0 & 0 & -1
\end{bmatrix}
\otimes Z_2.
$$

The symbol $\otimes$ denotes the direct product of matrices, and J is the opposite identity matrix 10×10.

The same two coefficients a_1 and a_2 compose the matrix of the integer representation of the discrete Fourier transform for the $N = 3$ case. The block-matrix $Z_{4 \times 4}$ for the three-block GDT can be written as

$$Z_{4 \times 4} = \begin{bmatrix} 0 & 1 \\ 1 & 0 \end{bmatrix} \otimes I + \begin{bmatrix} -1 & 1 \\ 1 & -1 \end{bmatrix} \otimes Z_2.$$

One can also note that the matrix $Z_{4 \times 4}$ can be seen in the middle and in the corners of the matrix $Z_{10 \times 10}$.

2.4.2.3 The $N = 4$ case

The DFT in the $N = 4$ case represents itself the integer transform. The four-point DFT is defined by four trivial twiddle factors ± 1 and $\pm j$. The rotation matrix in this case is binary

$$T = W = \begin{bmatrix} 0 & -1 \\ 1 & 0 \end{bmatrix}, \qquad (T^4 = I).$$

Therefore the matrix of the 4-block WFT is the integer matrix

$$F = X(T) = \begin{bmatrix} I & I & I & I \\ I & T^1 & T^2 & T^3 \\ I & T^2 & I & T^2 \\ I & T^3 & T^2 & T^1 \end{bmatrix} = \begin{bmatrix} 1 & 0 & 1 & 0 & 1 & 0 & 1 & 0 \\ 0 & 1 & 0 & 1 & 0 & 1 & 0 & 1 \\ 1 & 0 & 0 & 1 & -1 & 0 & 0 & -1 \\ 0 & 1 & -1 & 0 & 0 & -1 & 1 & 0 \\ 1 & 0 & -1 & 0 & 1 & 0 & -1 & 0 \\ 0 & 1 & 0 & -1 & 0 & 1 & 0 & -1 \\ 1 & 0 & 0 & -1 & -1 & 0 & 0 & 1 \\ 0 & 1 & 1 & 0 & 0 & -1 & -1 & 0 \end{bmatrix}.$$

The determinant of this matrix $\det(F) = 4^4$ and the inverse matrix equals the matrix $X(T^{-1})$ up to the normalized coefficient 4,

$$X^{-1} = X' = \frac{1}{4}X(T^{-1}) = \frac{1}{4}X(T').$$

We can define another representation of the four-point DFT in the space R^8 by considering, for instance, the matrix

$$T = \begin{bmatrix} -1 & 2 \\ -1 & 1 \end{bmatrix}, \qquad (\det(T) = 1).$$

This matrix composes a one-parametric group with period 4, i.e., $T^4 = I$,

$$T^2 = \begin{bmatrix} -1 & 0 \\ 0 & -1 \end{bmatrix}, \quad T^3 = \begin{bmatrix} 1 & -2 \\ 1 & -1 \end{bmatrix}, \quad T^4 = \begin{bmatrix} 1 & 0 \\ 0 & 1 \end{bmatrix} = I.$$

The corresponding 4-block T-GDT is defined by the following matrix (8×8) (or block-matrix (4×4)):

$$X = X(T) = \begin{bmatrix} I & I & I & I \\ I & T^1 & -I & T^3 \\ I & -I & I & -I \\ I & T^3 & -I & T^1 \end{bmatrix} = \begin{bmatrix} 1 & 0 & 1 & 0 & 1 & 0 & 1 & 0 \\ 0 & 1 & 0 & 1 & 0 & 1 & 0 & 1 \\ 1 & 0 & -1 & 2 & -1 & 0 & 1 & -2 \\ 0 & 1 & -1 & 1 & 0 & -1 & 1 & -1 \\ 1 & 0 & -1 & 0 & 1 & 0 & -1 & 0 \\ 0 & 1 & 0 & -1 & 0 & 1 & 0 & -1 \\ 1 & 0 & 1 & -2 & -1 & 0 & -1 & 2 \\ 0 & 1 & 1 & -1 & 0 & -1 & -1 & 1 \end{bmatrix}.$$

The determinant of this matrix $\det(X) = 4^4$.

2.5 Roots of the unit

When considering the $N = 5$ case, we face the problem of finding the integer matrix (2×2) which is the fifth root of the identity matrix. Probably such a matrix does not exit; at least we do not know about that. The same problem stands for many other values of N as well. We propose considering the representation of the discrete Fourier transform in the real space R^{2N}. This representation is not integer, and is based on the matrix (2×2) which is not a rotation, but a root of the identity matrix. For that, we first describe the $N = 5$ case.

For the angle $\varphi = 2\pi/5$ with $c_1 = \cos(\varphi) = 0.3090$ we consider the following matrix:

$$C_1 = C_1(\varphi) = \begin{bmatrix} -c_1 & c_1 - 1 \\ c_1 + 1 & -c_1 \end{bmatrix} = \begin{bmatrix} -0.3090 & -0.6910 \\ 1.3090 & -0.3090 \end{bmatrix}, \qquad (2.18)$$

as well as the matrix

$$C_2 = C_1(2\varphi) = \begin{bmatrix} -c_2 & c_2 - 1 \\ c_2 + 1 & -c_2 \end{bmatrix} = \begin{bmatrix} 0.8090 & -1.8090 \\ 0.1910 & 0.8090 \end{bmatrix} \neq C_1^2(\varphi), \qquad (2.19)$$

where $c_2 = \cos(2\varphi) = -0.8090 = -(c_1 + 0.5)$.

These matrices have the unique form:

$$C = cU + V, \qquad (2.20)$$

where $c = c_1$ or c_2, and matrices U and V equal

$$U = \begin{bmatrix} -1 & 1 \\ 1 & -1 \end{bmatrix}, \qquad V = \begin{bmatrix} 0 & -1 \\ 1 & 0 \end{bmatrix}. \qquad (2.21)$$

The matrix C satisfies the following equations:

$$C^5 = -I, \quad C^{10} = I, \quad \det(C) = 1,$$

and the multiplication of a vector $\mathbf{x}$ by matrix C requires two multiplications by c. By using this matrix, we can construct the 5-block C-GDT with the matrix

$$X(C) = \begin{bmatrix} I & I & I & I & I \\ I & C^1 & C^2 & C^3 & C^4 \\ I & C^2 & C^4 & -C^1 & -C^3 \\ I & C^3 & -C^1 & -C^4 & C^2 \\ I & C^4 & -C^3 & C^2 & -C^1 \end{bmatrix}.$$

For instance, when $C = C_1$, the matrix $X(C)$ equals

$$
\begin{bmatrix}
1\ 0 & 1 & 0 & 1 & 0 & 1 & 0 & 1 & 0 \\
0\ 1 & 0 & 1 & 0 & 1 & 0 & 1 & 0 & 1 \\
1\ 0 & -0.3090 & -0.6910 & -0.8090 & 0.4271 & 0.8090 & 0.4271 & 0.3090 & -0.6910 \\
0\ 1 & 1.3090 & -0.3090 & -0.8090 & -0.8090 & -0.8090 & 0.8090 & 1.3090 & 0.3090 \\
1\ 0 & -0.8090 & 0.4271 & 0.3090 & -0.6910 & 0.3090 & 0.6910 & -0.8090 & -0.4271 \\
0\ 1 & -0.8090 & -0.8090 & 1.3090 & 0.3090 & -1.3090 & 0.3090 & 0.8090 & -0.8090 \\
1\ 0 & 0.8090 & 0.4271 & 0.3090 & 0.6910 & -0.3090 & 0.6910 & -0.8090 & 0.4271 \\
0\ 1 & -0.8090 & 0.8090 & -1.3090 & 0.3090 & -1.3090 & -0.3090 & -0.8090 & -0.8090 \\
1\ 0 & 0.3090 & -0.6910 & -0.8090 & -0.4271 & -0.8090 & 0.4271 & 0.3090 & 0.6910 \\
0\ 1 & 1.3090 & 0.3090 & 0.8090 & -0.8090 & -0.8090 & -0.8090 & -1.3090 & 0.3090
\end{bmatrix}.
$$

The determinant of this matrix $\det(X(C)) = 448.6068$, and the inverse matrix $X^{-1}(C) \neq X(C^{-1})$.

For comparison with the traditional five-point DFT, we consider the orbit of a point $\mathbf{x} = (x_1, x_2)'$ with respect to the group of motion $\mathbf{C} = \{C^k; k = 0 : 9\}$. In other words, we consider ten points $\mathbf{y}_k$ in the plane, which are defined by

$$\mathbf{y}_0 = \mathbf{x} \rightarrow \mathbf{y}_1 = C\mathbf{x} \rightarrow \mathbf{y}_2 = C\mathbf{y}_1 \rightarrow \mathbf{y}_3 = C\mathbf{y}_2 \rightarrow \cdots \rightarrow \mathbf{y}_9 = C\mathbf{y}_8 \rightarrow \mathbf{x}.$$

If the matrix C would be equal to the matrix of rotation W by the angle $\varphi_0 = 2\pi/5$, the point $\mathbf{x}$ would be moving around the unit circle. Let us first consider the case when the matrix $C = C_1$ and the point $\mathbf{x} = (1, 0)'$. Figure 2.23 in part a shows the locations of all ten points $\mathbf{y}_k$, which are identified as k, and the drawing of the full path of the point $\mathbf{x}$ during its movement in b. As we see, this drawing is not a circle. All the points lie on the ellipse

$$x^2 + \frac{y^2}{b^2} = 1, \qquad \left(b = \frac{1+c}{1-c} = 1.3764\right) \tag{2.22}$$

but they move along the perimeter of the ellipse with the angle-step which we consider conditionally equal to $\varphi_0 = 3\varphi/2 = 3\pi/5$.

We now consider the case when the matrix $C = C_2$. The matrix $X(C_2)$ of the corresponding 5-block C-GDT equals

$$
\begin{bmatrix}
1\ 0 & 1 & 0 & 1 & 0 & 1 & 0 & 1 & 0 \\
0\ 1 & 0 & 1 & 0 & 1 & 0 & 1 & 0 & 1 \\
1\ 0 & 0.8090 & -1.8090 & 0.3090 & -2.9271 & -0.3090 & -2.9271 & -0.8090 & -1.8090 \\
0\ 1 & 0.1910 & 0.8090 & 0.3090 & 0.3090 & 0.3090 & -0.3090 & 0.1910 & -0.8090 \\
1\ 0 & 0.3090 & -2.9271 & -0.8090 & -1.8090 & -0.8090 & 1.8090 & 0.3090 & 2.9271 \\
0\ 1 & 0.3090 & 0.3090 & 0.1910 & -0.8090 & -0.1910 & -0.8090 & -0.3090 & 0.3090 \\
1\ 0 & -0.3090 & -2.9271 & -0.8090 & 1.8090 & 0.8090 & 1.8090 & 0.3090 & -2.9271 \\
0\ 1 & 0.3090 & -0.3090 & -0.1910 & -0.8090 & -0.1910 & 0.8090 & 0.3090 & 0.3090 \\
1\ 0 & -0.8090 & -1.8090 & 0.3090 & 2.9271 & 0.3090 & -2.9271 & -0.8090 & 1.8090 \\
0\ 1 & 0.1910 & -0.8090 & -0.3090 & 0.3090 & 0.3090 & 0.3090 & -0.1910 & -0.8090
\end{bmatrix}.
$$

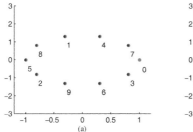
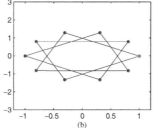

FIGURE 2.23

(a) Location of all points $\mathbf{y}_k$ and (b) scheme of the movement of the point $(1,0)$.

and its determinant $\det(X(C)) = 1.3932$. Figure 2.24 shows the locations of ten points $\mathbf{y}_k = C^k\mathbf{x}$, $k = 0:9$, in part a, along with the drawing of the full path of the point $\mathbf{x}$ during its movement in b. This drawing is an ellipse. All points are moving sequentially and anti-clock-wise along the ellipse with the step equal to $\varphi_0 = \pi/5$. The ellipse is described by the equation $x^2 + y^2/b^2 = 1$,

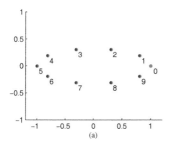
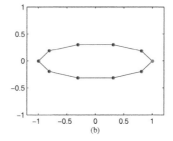

FIGURE 2.24 (See color insert following page 242.)

(a) Location of all points $\mathbf{y}_k$ and (b) the path of the point $(1,0)$.

$(b^2 = 0.3249^2)$.

To define the product with respect to which the transformation with the matrix $X(C)$ is isometric, we define the matrix A from the following equation: $C'AC = A$, or $C'A = AC^{-1}$. Since $U' = U$ and $V' = -V$, it is not difficult to obtain the matrix A,

$$A = \begin{bmatrix} 0 & 1 \\ -1 & 0 \end{bmatrix}.$$

The product of vectors $\mathbf{x} = (x_0, x_1)'$ and $\mathbf{y} = (y_0, y_1)'$ is calculated by this matrix as

$$(\mathbf{x}, \mathbf{y})_A = \mathbf{x}'A\mathbf{y} = \begin{bmatrix} x_0 & x_1 \end{bmatrix} \begin{bmatrix} 0 & 1 \\ -1 & 0 \end{bmatrix} \begin{bmatrix} y_0 \\ y_1 \end{bmatrix} = x_0 y_1 - x_1 y_0.$$

This product leads to the zero metric $|\mathbf{x}|^2 = (\mathbf{x}, \mathbf{x})_A = 0$.

We now prove that the locus of the points $\mathbf{y}_k$ is an ellipse, when considering the group of motion $\mathbf{C}$ of the N-block C-GDT, in the general $N > 1$ case. For that we find a function

$$F(\mathbf{x}) = \text{const}, \tag{2.23}$$

such that

$$F(C\mathbf{x}) = \text{const}. \tag{2.24}$$

Assuming that the ellipse is indeed the locus of the points $\mathbf{x} = (x, y)$ in (2.23), and taking this function in the canonical form $F(\mathbf{x}) = ax^2 + by^2 = 1$, we will try to solve equation (2.24),

$$F(C\mathbf{x}) = a[-cx + (c-1)y]^2 + b[(c+1)x - cy]^2 = 1.$$

When opening this equation, we obtain the following:

$$c^2(ax^2 + by^2) + a(c-1)^2 y^2 + b(c+1)^2 x^2 + xy[-2ac(c-1) - 2c(c+1)b] = 1. \tag{2.25}$$

By assuming that the factor of xy equals zero, $-2ac(c-1) - 2c(c+1)b = 0$, we obtain the following:

$$a = b\frac{1+c}{1-c} = b\frac{1 + \cos(\phi)}{1 - \cos(\phi)} = b \operatorname{ctg}^2 \frac{\phi}{2}.$$

Using this relation between a and b, it is not difficult to see that the first part of equation (2.25) can be written as

$$c^2(ax^2 + by^2) + a(c-1)^2 y^2 + b(c+1)^2 x^2 = ax^2 + by^2 = 1.$$

Thus the invariance of the function F has been proved. Note that, since $\phi = 2\pi/N$, the relation between a and b can be written as $a/b = \operatorname{ctg}^2 \pi/N$, or $\operatorname{ctg}\pi/N = \sqrt{a/b}$, and therefore the integer number N is expressed by

$$\frac{\pi}{N} = \operatorname{arcctg}\sqrt{\frac{a}{b}} + \pi k, \quad \text{or} \quad N = \frac{\pi}{\operatorname{arcctg}\sqrt{\frac{a}{b}} + \pi k},$$

where k is an integer.

2.5.1   Elliptic DFT

In this section we consider the Nth roots of the identity matrix (2×2), which are not defined by the Givens rotations, but transformations similar to the transformations $X(C)$ described above for the $N = 5$ case.

Given integer $N > 1$, we consider the angle $\varphi = \varphi_N = 2\pi/N$ and the matrix

$$C = C(\varphi) = \begin{bmatrix} -\cos\varphi & \cos\varphi - 1 \\ \cos\varphi + 1 & -\cos\varphi \end{bmatrix} = \cos\varphi \cdot U + V \qquad (2.26)$$

where the matrices U and V equal

$$U = \begin{bmatrix} -1 & 1 \\ 1 & -1 \end{bmatrix}, \qquad V = \begin{bmatrix} 0 & -1 \\ 1 & 0 \end{bmatrix}. \qquad (2.27)$$

The calculations show that the following holds for the matrix C :

$$C^N = \begin{cases} I, & \text{if } N = 2k; \\ -I, & \text{if } N = 2k + 1. \end{cases} \qquad (2.28)$$

In other words, the matrix C defines a one-parametric group with period N or $2N$, depending on the evenness of N. This group moves the point $\mathbf{x} = (1, 0)'$ along an ellipse as for the $N = 5$ case considered above. To illustrate this fact, we consider a few examples.

For the case $N = 7$, Figure 2.25 in part a shows the locations of 14 points $\mathbf{y}_k$, $k = 0 : 13$, and the drawing of the full path of the point $\mathbf{x}$ during its movement in b. The group of motion $\mathbf{C}$ has the period $2N = 14$, the angle-step of

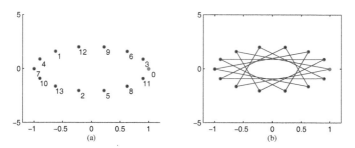

FIGURE 2.25 (See color insert following page 242.)
(a) Location of all points $\mathbf{y}_k$ and (b) scheme of the movement of the point $(1, 0)$, when $N = 7$.

rotation of the point along the perimeter of the ellipse $x^2 + y^2/2.0765^2 = 1$ equals $5/2\varphi_7 = 5/7\pi$.

When N is odd, we can define other roots of the identity matrix, by using the angle $\phi = \phi_N = 2\varphi_N$. The root-matrix equals

$$C = C(\phi) = \begin{bmatrix} -\cos\phi & \cos\phi - 1 \\ \cos\phi + 1 & -\cos\phi \end{bmatrix} = \cos\phi \cdot U + V, \qquad (2.29)$$

and for this matrix we have $C^N = -I$ and $C^{2N} = I$. The group of motion in this case leads also to the movement of the point $\mathbf{x}$ along the perimeter of an ellipse which differs from the ellipse corresponding to the case $C(\varphi)$. As an example, Figure 2.26 shows a similar picture of movement of the point $\mathbf{x}$, when considering the matrix $C = C(2\varphi_7)$. The period of the group of

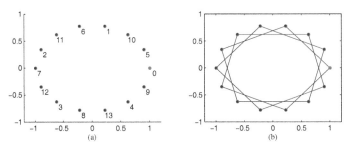

FIGURE 2.26 (See color insert following page 242.)
(a) Location of all points $\mathbf{y}_k$ and (b) scheme of the movement of the point $(1, 0)$, when $N = 7$.

motion $\mathbf{C}$ in this case also equals 14. The point $\mathbf{x}$ moves along the perimeter of the following ellipse $x^2 + y^2/0.7975^2 = 1$ with the angle-step equal to $3/2\varphi_7 = 3/7\pi$. Figure 2.27 shows the locations of eight points $\mathbf{y}_k$, $k = 0 : 7$, and the drawing of the full path of the point $\mathbf{x}$ during its movement in b. This is the $N = 8$ case.

For the same point $\mathbf{x}$, Figure 2.28 shows the diagram of location of 16 points $\mathbf{y}_k$, $k = 0 : 15$, and the full path of the point in part a for the case $N = 16$. In part b, a similar drawing is shown for the $N = 17$ case, when the matrix C is defined by the angle $\varphi_{17}/2 = \pi/17$.

It also should be mentioned that the inverse matrix $C(\varphi)^{-1}$ is calculated by

$$S(\varphi) = C(\varphi)^{-1} = \begin{bmatrix} -\cos\varphi & -\cos\varphi + 1 \\ -\cos\varphi - 1 & -\cos\varphi \end{bmatrix} = -\cos\varphi \cdot U_1 + V_1$$

where the matrices U_1 and V_1 equal

$$U_1 = \begin{bmatrix} 1 & 1 \\ 1 & 1 \end{bmatrix}, \qquad V_1 = V^{-1} = \begin{bmatrix} 0 & 1 \\ -1 & 0 \end{bmatrix}.$$

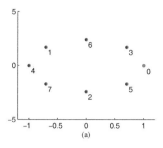

FIGURE 2.27 (See color insert following page 242.)

(a) Location of all points $\mathbf{y}_k$ and (b) scheme of the movement of the point $(1,0)$, when $N = 8$.

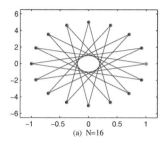 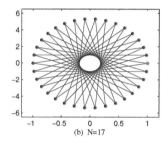

FIGURE 2.28 (See color insert following page 242.)

Schemes of movement of the point $(1,0)$ for the cases when (a) $N = 16$ and (b) $N = 17$.

The determinant $\det(S) = 1$ and $S^N = I$ if N is an even number, and $S^N = -I$ and $S^{2N} = I$ if N is an odd number. Thus, we have another class of roots $S(\varphi)$ of the identity matrix, which also describe the movement of the point $\mathbf{x}$ along the same ellipses which are defined by the direct matrices $C(\varphi)$. The diagrams of the full paths of $\mathbf{x}$ are also the same, but the point backs. The corresponding groups of motion $\mathbf{S}$ consist of the same matrices as $\mathbf{C}$ they are only ordered in the inverse way. As an example, for the $N = 8$ case, Figure 2.29 shows the locations of 8 points and the drawing of the full path of the point $\mathbf{x}$ in part a for the matrix $C(\varphi_7)$, and for $S(\varphi_7)$ in b.

By using the matrix $C = C(\varphi_N)$, we can construct the N-block C-GDT with the matrix

$$
X(C) = \begin{bmatrix}
I & I & I & I & \dots & I \\
I & C^1 & C^2 & C^3 & \dots & C^{N-1} \\
I & C^2 & C^4 & C^6 & \dots & C^{2N-2} \\
I & \dots & & & \dots & \dots \\
I & C^{N-1} & C^{2N-2} & C^{3N-3} & \dots & C^1
\end{bmatrix}, \tag{2.30}
$$

and consider the property $C^{kN+m} = \pm C^m$, when $k, m = 0, 1, ..., N - 1$. We

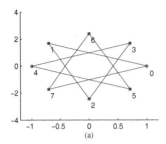 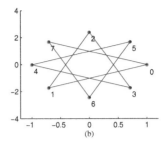

(a)   (b)

FIGURE 2.29
Schemes of the movement of the point $(1,0)$ by the groups of motion (a) **C**
and (b) **S**.

will call this transform *the N-block elliptic Fourier transformation*, or *the N-block elliptic FT (EFT)*. The matrix of the N-block EFT is parameterized by
the angle φ_N, as the N-point DFT with the matrix of rotation $T(\varphi_N)$. For
the case when N is odd, we also can define a similar concept of the N-block
EFT, by using the angle ϕ_N, i.e., the matrix $C = C(\phi_N)$ in composition of
$X(C)$. We here focus on the φ-based N-block EFT.

When N is even, the following property holds for the N-block EFT:

$$R(\mathbf{x})_{N-k} = R(\mathbf{x})_k, \quad I(\mathbf{x})_{N-k} = -I(\mathbf{x})_k, \quad k = 1 : (N/2 - 1),$$

where we denote by $R(\mathbf{x})$ and $I(\mathbf{x})$ the "real" and "imaginary" parts of the
N-block EFT of a real input $\mathbf{x}$, respectively.

Note that $C^{-1} = C^{N-1}$, when N is even, and $C^{-1} = -C^{N-1}$, when N is
odd. The N-block EFT based on the C^{-1} defines the inverse transform in the
$N = 2^r$ case, when $r > 1$. Indeed, $X(C) \cdot X(C^{-1}) = NI$. In the case when
$N \neq 2^r$ but even, the following holds:

$$X(C) \cdot X(C^{-1}) = N \begin{bmatrix} I_{N/2} & I_{N/2} \\ I_{N/2} & I_{N/2} \end{bmatrix}.$$

It is interesting to see if the elliptic Fourier transform can distinguish the
frequencies of the cosine or sine waves as the traditional transform does. The
EFT is given in the matrix form of (2.30) and in the real space R^{2N}. The
action of the matrix C on a vector $(x_1, x_2)'$ is not expressed as a complex
multiplication, as in the case of the matrix of rotation. It is clear that we
have two different concepts of the discrete Fourier transform. However, the
brief analysis of a few examples shows that these transforms have similar
properties, which we illustrate below.

Example 2.14
Let $N = 32$ and let the vector $\mathbf{x}_r$ be the following discrete-time cosine wave
in the time interval $[0, 2\pi] : x_r(t) = \cos(\omega_1 t)$, $\omega_1 = 2\pi/N$, where time-points

$t = t_n = n(2\pi/(N-1))$, $n = 0 : (N-1)$. The $2N$-dimensional vector is composed as $\bar{\mathbf{x}} = (x_r(0), 0, x_r(t_1), 0, x_r(t_2), 0, \ldots, 0, x_r(t_{N-1}), 0)'$. We denote by $\mathbf{y}_r$ and $\mathbf{y}_i$ the "real" and "imaginary" parts of the N-block EFT of $\mathbf{x}$, i.e.,

$$\mathbf{y}_r = (y(0), y(2), y(4), y(6), \ldots, y(2N-4), y(2N-2))',$$
$$\mathbf{y}_i = (y(1), y(3), y(5), y(7), \ldots, y(2N-3), y(2N-1))',$$

where $\mathbf{y} = X(C)\mathbf{x}$.

Figure 2.30 shows the discrete-time signal $\mathbf{x}_r$ in part a, along with the "real" part $\mathbf{y}_r$ of the N-block EFT in b. The transform recognizes the frequency ω_1, but in the frequency-points $p = 15$ and 17, instead of frequency-points $p = 1$ and 31 in the DFT case. There is a cyclic shift of the frequency-

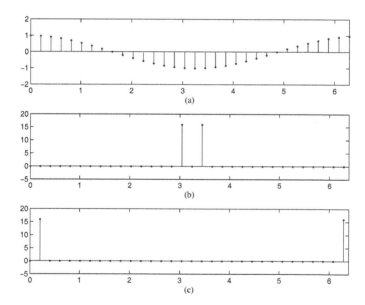

FIGURE 2.30
(a) The cosine signal with frequency ω_1, (b) the 32-block EFT, and (c) the permutation of the 32-block EFT.

points, namely, the cyclic permutation of the transform. Indeed, there are 32 points $\mathbf{y}_k$ on the ellipse, which divide the perimeter of the ellipse by 31 intervals, which we consider conditionally equal. If we consider how the group of motion $\mathbf{C}$ moves the point $\mathbf{x} = (1, 0)$ along the perimeter of the ellipse, we can see that this point runs all 32 points $\mathbf{y}_k$ on the ellipse with the speed of 15 intervals. Therefore, to make the movement sequential, we can change the order of the transform in accordance with the following permutation: $k \to P(k) = (15k) \bmod N$, $k = 0 : (N-1)$. The "real" part $\mathbf{y}_r$ of the N-block

EFT after this permutation is shown in c. This result is the exact result of the traditional DFT; the elliptic Fourier transform allows for distinguishing the main frequency of the cosine wave. It should be noted that the EFT distinguishes the main frequency of the signal not only in its real part, but in the imaginary part as well. The traditional DFT does not have this property. Figure 2.31 shows the "imaginary" part y_i of the N-block EFT in part a and b, respectively, before and after the permutation of the transform. The maximums of the imaginary part of the transform are at the same frequency-points. For comparison, the imaginary part of the DFT is given in c; it does not have maximums on the carrying frequency-points $p = 1$ and 31. In addition, the imaginary part of the DFT by amplitude is smaller than the EFT.

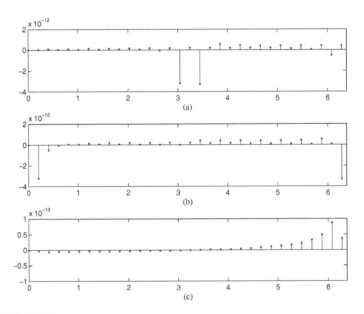

FIGURE 2.31

(a) The "imaginary" part of the 32-block EFT, (b) the "imaginary" part after the permutation, and (c) "imaginary" part of the 32-block DFT.

We now consider the cosine wave $x_r(t) = \cos(\omega_3 t)$ with the frequency equal $\omega_3 = 3\omega_1$. Figure 2.32 shows the discrete-time signal $\mathbf{x}_r$ in part a, along with the "real" part $\mathbf{y}_r$ of the N-block EFT before and after permutation of components in b and c, respectively. The transform recognizes the frequency ω_3 at the frequency-points $p = 13$ and 19, or at the frequency-points $3 = P(13) = 13 \cdot 15 \bmod 32$ and $29 = P(19) = 19 \cdot 15 \bmod 32$ after the permutation, as in the DFT case. We can see again that the EFT distinguishes the carrying

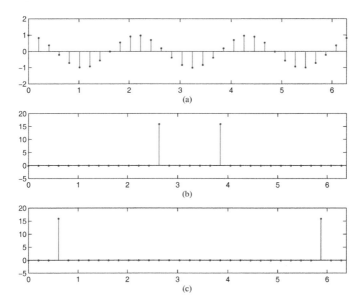

FIGURE 2.32
(a) The cosine signal with frequency ω_3, (b) the 32-block EFT, and (c) the permutation of the 32-block EFT.

frequency-points $p = 3$ and 29 in the imaginary part as well. Figure 2.33 shows the "imaginary" part $\mathbf{y}_i$ of the N-block EFT in part a and b, respectively, before and after the permutation of the transform. The imaginary part of the DFT is given in c, for comparison.

Figure 2.34 shows the discrete-time signal $x_r(t) = \cos(\omega_2 t)$ with the frequency equal to $\omega_2 = 2\omega_1$ in part a, and the "real" part $\mathbf{y}_r$ of the N-block EFT after the permutation of components in b. The discrete-time signal $x_r(t) = \cos(\omega_6 t)$ with the frequency equal $\omega_6 = 6\omega_1$ is shown in part c, along with the "real" part of the N-block EFT after the permutation of components in d. The transform recognizes the frequency ω_2 at the frequency-points $p = 2$ and 30, and the frequency ω_6 at the frequency-points $p = 6$ and 28. □

We now consider the example with a nonsinusoidal signal. Figure 2.35 shows the discrete-time signal of length $N = 128$ in the time interval $[0, 2\pi]$ in part a, along with the "real" part $\mathbf{y}_r$ of the 128-block EFT in b, and after the permutation of components in c. According to the group of motion in this case, the permutation of components is performed by $k \rightarrow P(k) = (63k) \bmod 128$, $k = 0 : 127$. The result in c is very close to the result of the 128-point DFT; the difference between these two transforms is shown in d. The error occurs in low frequencies and it is of order 10^{-11}.

Figure 2.36 shows the "imaginary" part $\mathbf{y}_i$ of the 128-block EFT in part a,

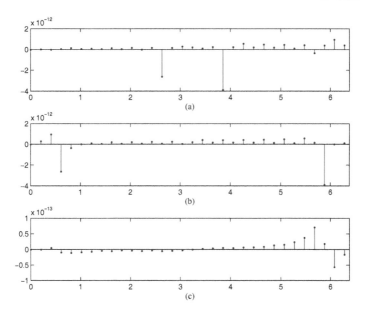

FIGURE 2.33
The "imaginary" part of the 32-block EFT (a) before and (b) after the permutation, and (c) "imaginary" part of the 32-block DFT.

as well as in b after the permutation of its components. For comparison, the imaginary part, $\mathbf{F}_i$, of the 128-block DFT is shown in c. The signals in b and c are similar; they approximately are equal up to the normalized coefficient 64, i.e., $\mathbf{y}_i \approx 64\mathbf{F}_i$.

Figure 2.37 shows the amplitude spectrum of the 128-point DFT in part a, along with the amplitude spectrum of the 128-block EFT after the permutation in b, which is calculated by

$$|\bar{\mathbf{y}}| = \sqrt{|\mathbf{y}_r|^2 + |\mathbf{y}_i|^2} = \sqrt{\sum_{k=0}^{127} \mathbf{y}_r^2(k) + \sum_{k=0}^{127} \mathbf{y}_i^2(k)}.$$

These transforms have been shifted to the center of the interval $[0, 2\pi]$. In part c, the amplitude spectrum of the 128-block EFT is shown after normalizing the imaginary part of the transform, $\mathbf{y}_i = \mathbf{y}_i/64$,

$$|\bar{\mathbf{y}}|_n = \sqrt{|\mathbf{y}_r|^2 + \left|\frac{\mathbf{y}_i}{64}\right|^2} = \sqrt{\sum_{k=0}^{127} \mathbf{y}_r^2(k) + \sum_{k=0}^{127} \left(\frac{\mathbf{y}_i(k)}{64}\right)^2}.$$

One can see that the elliptic and traditional Fourier transforms lead to similar results when the input signal is "real", i.e., each second component of the signal is zero.

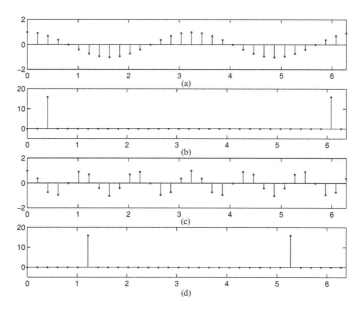

FIGURE 2.34

(a) The cosine signal with frequency ω_2 and (b) the 32-block EFT. (c) The cosine signal with frequency ω_6 and (d) the 32-block EFT. (The transforms are shown with the permutation.)

We now consider the example with a "complex" signal, when the second component of the signal, which represents the "imaginary" part of the signal is not zero. Figure 2.38 shows the discrete-time signal $\mathbf{z} = \{z_n; n = 0 : 255\}$ in part a. We denote by $\mathbf{z}_1$ and $\mathbf{z}_2$ the first and the second parts of this signal, i.e., $\mathbf{z}_1 = \{z_n; n = 0 : 127\}$ and $\mathbf{z}_2 = \{z_n; n = 128 : 255\}$. These parts are referred to as the "real" and "imaginary" parts of the complex signal $\bar{\mathbf{y}} = \{z_0, z_{128}, z_1, z_{129}, ..., z_{127}, z_{255}\} = \mathbf{z}_1 + j\mathbf{z}_2$. The real part of the 128-block EFT of $\bar{\mathbf{y}}$ is shown in b, and the imaginary part in c, after the permutation of their components. The transform is shifted to the center. The amplitude spectrum of the transform is shown in d. One can observe the symmetry of the amplitude spectrum with respect to the center, although the signal $\bar{\mathbf{y}}$ is not real. The property of symmetry does not hold for the traditional Fourier transform, when the signal is complex. In the considered case of $\bar{\mathbf{y}}$, this fact is illustrated in Figure 2.39, where the spectrum of the DFT is shown. The symmetry holds for the N-block EFT, not by DFT when the signal is complex. Another example is given in Figure 2.40, where the signal is shown in part a, along with the real part and imaginary part of the 128-block EFT in b and c, respectively. The amplitude spectrum of the transform is shown in d, and the spectrum of the DFT is shown in Figure 2.41. Summarizing the above reasoning and results we can state the following. In the theory

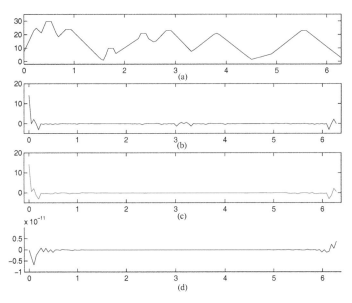

FIGURE 2.35
(a) Discrete-time signal of length 128, the real part of the 128-block EFT (b) and (c) DFT, and (d) the difference of the 128-block EFT and DFT.

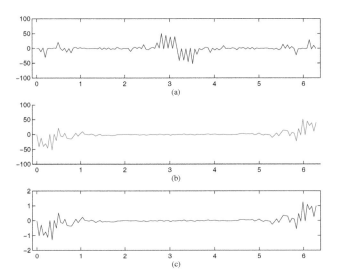

FIGURE 2.36
The imaginary part of the 128-block EFT before (a) and (b) after the permutation, and (c) the imaginary part of the 128-point DFT.

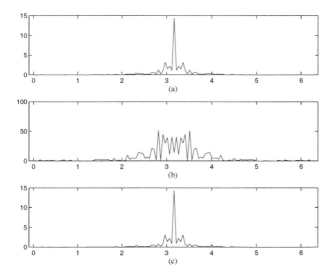

FIGURE 2.37

The amplitude spectrums of (a) the 128-block DFT and (b) the 128-block EFT after the permutation, and (c) after normalizing the imaginary part.

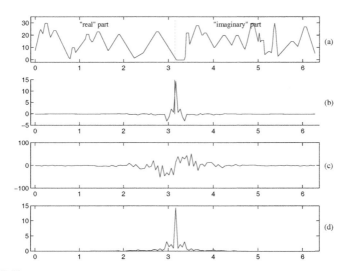

FIGURE 2.38

(a) Two parts of the complex signal of length 256, (b) the real and (c) imaginary parts the 128-block EFT after the permutation. (d) The amplitude spectrums of the 128-block EFT. (The transform has been shifted cyclicly to the center.)

of the discrete Fourier transform, the multiplication by the complex roots

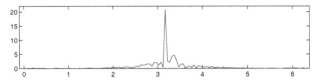

FIGURE 2.39
The amplitude spectrum of the 128-block DFT, after shifting cyclicly to the center.

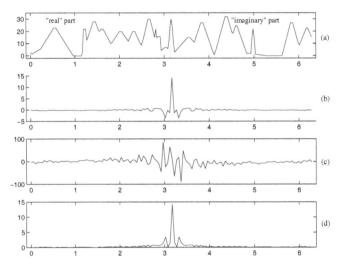

FIGURE 2.40
(a) Two parts of the complex signal of length 256, (b) the real and (c) imaginary parts the 128-block EFT after the permutation. (c) The amplitude spectrum of the 128-block EFT. (The transform has been shifted cyclicly to the center.)

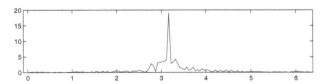

FIGURE 2.41
The amplitude spectrum of the 128-block DFT, after shifting cyclicly to the center.

$W = \cos(\varphi) - j\sin(\varphi)$ of the unit corresponds to the rotation $T(\varphi)$ in the 2-D space R^2. This transformation

$$W \rightarrow T(\varphi) = \begin{pmatrix} \cos(\varphi) & \sin(\varphi) \\ -\sin(\varphi) & \cos(\varphi) \end{pmatrix}$$

has the group property $W^k \rightarrow T(\varphi^k) = T^k(\varphi)$, for any integer k. This iso-morphism of the complex space C on the real space R^2 is reduced to the multiplication

$$T(\varphi)\begin{pmatrix} x \\ y \end{pmatrix} \rightarrow W(x + jy).$$

The matrix $T(\varphi)$ is the Nth root of the identity matrix I, when $\varphi = 2\pi/N$.

We have constructed the roots of the identity matrix, which are $C(\varphi)$, and then the transformation $C^k(\varphi) \rightarrow T^k(\varphi)$, $k = 0 : (N-1)$. The transformation with the block-matrix $X(C)$ composed by the powers of the matrix C is the N-block elliptic Fourier transformation. It should be mentioned that the mul-tiplication by the matrix $C(\varphi)$ does not reduce to the complex multiplication, as the matrix $T(\varphi)$ does. Since the EFT is the linear transformation, we can propose that the multiplication by $C(\varphi)$ assumes the following multiplication $C(\varphi)z = \alpha(\varphi)z + \beta(\varphi)z^*$, where $\alpha(\varphi)$ and $\beta(\varphi)$ are some complex functions to be found. In other words, to the complex matrix operation

$$C(\varphi)z = [-x\cos(\varphi) + y(\cos(\varphi) - 1)] + j[-x(\cos(\varphi) + 1) - y\cos(\varphi)],$$

we want to put in correspondence the function $\xi^\varphi(z) = \alpha(\varphi)z + \beta(\varphi)z^*$.

The calculation shows that $\alpha(\varphi) = -\cos(\varphi) + j$ and $\beta(\varphi) = j\cos(\varphi)$. Thus $\xi^\varphi(z) = [-\cos(\varphi) + j]z + j\cos(\varphi)z^*$. The correspondence $C(\varphi) \rightarrow \xi^\varphi$ leads to the following property of the homomorphism: $C^n(\varphi) \rightarrow \xi^{n\varphi}$, $n = \pm 1, \pm 2, \ldots$, and $\xi^{n\varphi} = \xi^{(n-m)\varphi} \circ \xi^{m\varphi}$, $z \in Z$, where the operation $\circ$ denotes the composition of the function. Since the matrix $C^n(\varphi)$ is the $2N$ root of the identity matrix, then $\xi^{2N\varphi}(z) = z$. If N is an even integer, then $\xi^{N\varphi}(z) = z$, for any complex z.

2.6   Codes for the block DFT

Below are examples of MATLAB-based codes for computing the N-block disc-rete Fourier transform. These codes have been used to compute the traditional and elliptic DFTs of the signal as shown in Figure 2.40.

```
%  ----------------------------------------------------------------
%  demo_vdft.m file of programs (library of codes of Grigoryans)
%  List of codes for processing signals of the length N>1:
%    1-D N-block discrete Fourier transform   - 'fft_real_mat.m'
%    1-D N-block elliptic Fourier transform   - 'fft_real_matC.m'
```

```
%   The main program for the 128-block DFT and elliptic DFT (N=128)
%   1. Define the real and imaginary parts of the signal of length 2N
    fid=fopen('Boli.sig','rb');
    X=fread(fid,'float');  fclose(fid); clear fid;
    N=128;
    x_boli=X(120:120+N-1)';
    xe2=X(120+N:120+2*N-1)';
%   The composition of the matrix C:
    f=2*pi/N; c=cos(f);
    U=[-1  1; 1 -1];   V=[ 0 -1; 1  0];
    C=c*U+V;
%   2N-point signal composition:
    step_t=2*pi/(N-1);
    t=0:step_t:2*pi;
    x_signal=[];
    for n=1:N
        x_signal=[x_signal x_boli(n) xe2(n)];
    end;
%   The N-block DFT of the signal:
    XF=fft_real_mat(N);
    yf=XF*x_signal';
    yf=yf';
    real_y=yf(1:2:2*N-1);
      im_y=yf(2:2:2*N);
%   The N-block C-generated Elliptic DFT of the signal:
    XC=fft_real_matC(C,N);
    ye=XC*x_signal';
    ye=ye';
    ye_r=ye(1:2:2*N);
    ye_i=ye(2:2:2*N);
    yx=[x_boli,xe2];
    step_t2=2*pi/(2*N-1);
    t2=0:step_t2:2*pi;
%   Plot the composed signal
    figure;
    subplot(4,1,1);
    h_0=plot(t2,yx);
    axis([-0.1,2*pi+0.1,-2,35]);
    h_x=text(6.5,15,'(a)');
    set(h_x,'FontName','Times','FontSize',12);
    h_1=text(.35,29,'"real" part');
    set(h_1,'FontName','Times','FontSize',11);
    h_2=text(4.5,29,'"imaginary" part');
    set(h_2,'FontName','Times','FontSize',11);
    hold on;
    X=[pi,pi]; Y=[-1,34];
    h_l=line(X,Y);
    set(h_l,'LineStyle','-');
%   The permutation of the EFT:
```

```
        n=0:N-1;
        T=mod(63*n,N);
        % The permutation of the "real part" of the EFT:
        y_rp=ye_r(T+1);
        subplot(4,1,2); hold on;
        h_2=plot(t,fftshift(y_rp)/N,'b');
        axis([-0.1,2*pi+0.1,-5,15]);
        h_b=text(6.5,5,'(b)');
        set(h_b,'FontName','Times','FontSize',12);
%       The permutation of the "imaginary part" of the EFT:
        y_ip=ye_i(T+1);
        subplot(4,1,3);
        h_2=plot(t,fftshift(y_ip)/N,'b');
        axis([-0.1,2*pi+0.1,-100,100]);
        h_c=text(6.5,0,'(c)');
        set(h_c,'FontName','Times','FontSize',12);
%       Calculation of the amplitude spectrum of the EFT:
        y_asp=sqrt(y_rp.^2+(y_ip/64).^2);
        subplot(4,1,4);
        plot(t,fftshift(y_asp)/N,'b');
        axis([-0.1,2*pi+0.1,0,15]);
        h_d=text(6.5,8,'(d)');
        set(h_d,'FontName','Times','FontSize',12);
%       end of the main program.
%       --------------------------------------------
        function F=fft_real_mat(N)
          w=2*pi/N;
          c=cos(w); s=sin(w);
          T=[c s; -s c];     % or use T=[c -s; s c];
          for i1=1:N
              for j1=1:N
                  k=(i1-1)*(j1-1);
                  F(:,i1,:,j1)=T^k;
              end;
          end;
          N2=N*2;
          F=reshape(F,[N2 N2]);
%       --------------------------------------------
        function F=fft_real_matC(C,N)
          for i1=1:N
              for j1=1:N
                  k=(i1-1)*(j1-1);
                  F(:,i1,:,j1)=C^k;
              end;
          end;
          N2=N*2;
          F=reshape(F,[N2 N2]);
%       --------------------------------------------
```

2.7   General elliptic Fourier transforms

The above described Nth roots of the identity matrix (2×2), which are defined
by the transformations of type $C(\varphi)$, can be generalized. In this section, we
consider a new class of such transformations which are parameterized by two
angles and isoparametric. We also introduce a new class of the elliptic-type
Fourier transformations and describe their properties.

We first note that for a given integer $N > 1$ and angle $\varphi = \varphi_N = 2\pi/N$,
the matrix

$$C = C(\varphi) = \begin{bmatrix} -\cos\varphi & \cos\varphi - 1 \\ \cos\varphi + 1 & -\cos\varphi \end{bmatrix} = \cos\varphi \cdot U + V$$

can be written as $C = C(\varphi) = -\cos\varphi \cdot I + \sin\varphi \cdot R$, where

$$R = R(\varphi) = \begin{bmatrix} 0 & -\tan(\varphi/2) \\ \cot(\varphi/2) & 0 \end{bmatrix}, \quad \text{and} \quad \det R = 1.$$

The matrix C can be also changed as $C = \cos\varphi \cdot I + \sin\varphi \cdot R$. The matrix
U in this case consists only of plus 1. This definition leads to the equality
$C^N(\varphi) = I$ regardless of the fact that N is even or odd. For example, when
$N = 5$ and 8, we obtain the following matrices, respectively:

$$C(\varphi_5) = \begin{bmatrix} 0.3090 & -0.6910 \\ 1.3090 & 0.3090 \end{bmatrix}, \quad C(\varphi_8) = \begin{bmatrix} 0.7071 & -0.2929 \\ 1.7071 & 0.7071 \end{bmatrix},$$

where $\varphi_5 = 1.2566$ and $\varphi_8 = 0.7854$. We obtain $C^5(\varphi_5) = I$ and $C^8(\varphi_8) = I$.
However, the desired property $C(2\varphi) = C^2(\varphi)$ does not hold for both cases
$\varphi = \varphi_5$ and φ_8. One can notice that $R^2(\varphi) = -I$, and the equality sign holds
for any angle φ. Therefore, we introduce the following matrix.

DEFINITION 2.1   Given an integer $N > 1$ and angle ϕ, the matrix
$C_\phi = C_\phi(\varphi) = \cos\varphi \cdot I + \sin\varphi \cdot R(\phi)$ is called *the generalized elliptic matrix*
(GEM).

The following property holds for the GEM:

$$C_\phi(\varphi_1)C_\phi(\varphi_2) = C_\phi(\varphi_1 + \varphi_2), \tag{2.31}$$

for any angles φ_1 and φ_2. To show that, we perform simple calculations as

$$\begin{aligned} C_\phi(\varphi_1)C_\phi(\varphi_2) &= \cos(\varphi_1)\cos(\varphi_2)I + \sin(\varphi_1)\sin(\varphi_2)R^2(\phi) \\ &+ [\cos(\varphi_1)\sin(\varphi_2) + \sin(\varphi_1)\cos(\varphi_2)]R(\phi) \\ &= \cos(\varphi_1 + \varphi_2)I + \sin(\varphi_1 + \varphi_2)R(\phi) = C_\phi(\varphi_1 + \varphi_2). \end{aligned}$$

It should be noted that if we define the matrix by $C_\phi(\varphi) = -\cos\varphi \cdot I + \sin\varphi \cdot R(\phi)$, then such an isomorphism does not hold, but $C_\phi(\varphi_1)C_\phi(\varphi_2) = -C_\phi(\varphi_1 + \varphi_2)$.

According to (2.31), $C_\phi(2\varphi) = C_\phi^2(\varphi)$, and $C_\phi(m\varphi) = C_\phi^m(\varphi)$, for any integer m. Therefore, if $\varphi = \varphi_N = 2\pi/N$, then $C_\phi^N(\varphi) = C_\phi(N\varphi) = C_\phi(2\pi) = I$, and $C_\phi^{-1}(\varphi) = C_\phi(-\varphi)$, $(C_\phi^k(\varphi))^{-1} = C_\phi^{-k}(\varphi)$, $k = 1 : (N-1)$. The matrix C_ϕ defines a one-parametric group with period N. This group moves consequently the point $\mathbf{x} = (1,0)'$ along an ellipse, $\mathbf{x} \to \mathbf{y}_1 = C\mathbf{x} \to \mathbf{y}_2 = C\mathbf{y}_1 \to \mathbf{y}_3 = C\mathbf{y}_2 \to \cdots \to \mathbf{y}_{N-1} = C\mathbf{y}_{N-2} \to \mathbf{x} = C\mathbf{y}_{N-1}$ where $\mathbf{y}_k = C^k\mathbf{x}$, for $k = 1 : (N-1)$.

For the $N = 9$ example, Figure 2.42 in part a shows the locations of 9 points $\mathbf{y}_k$, $k = 0 : 8$, and the drawings of the full path of the point $\mathbf{x}$ during its movement for the cases $\phi = \pi/9$ and $\pi/18$. The large ellipse corresponds to the $\phi = \pi/18$ case. Similar drawings of the full path of the point $\mathbf{x}$ during

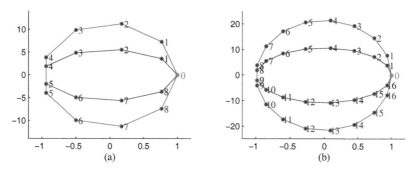

FIGURE 2.42
Locations of all points $\mathbf{y}_k$ and scheme of the movement of the point $(1,0)$, when (a) $N = 9$ and (b) $N = 17$.

its movement for the $N = 17$ case, when $\phi = \pi/17$ and $\pi/34$, are shown in part b. The large ellipse corresponds to the $\phi = \pi/34$ case.

The ellipse around which the point $\mathbf{x}$ is moving is defined by $x^2 + y^2/b^2 = 1$, $b = \cot(\phi/2)$. Indeed, if the point $C\mathbf{x}$ lies on the ellipse, then

$$C\mathbf{x} = \begin{bmatrix} \cos(\varphi) \\ \sin(\varphi)\cot(\frac{\phi}{2}) \end{bmatrix} \Rightarrow \cos^2(\varphi) + \sin^2(\varphi)\frac{\cot^2(\frac{\phi}{2})}{b^2} = 1 \Rightarrow b^2 = \cot^2(\frac{\phi}{2}).$$

This ellipse does not depend on the angle φ, but ϕ. The point $\mathbf{x}$ is moving around the same ellipse for $N = 9, 17$, and any other integer N. The ellipse height, b, rises as the angle ϕ becomes small, as shown in Figure 2.42.

According to Definition 2.1, the matrix R satisfies the condition $R^2(\phi) = -I$. In the general $R^2 = -I$ case, for a given integer $N > 1$ and angle φ, we also call the matrix $C = C(\varphi) = \cos \varphi \cdot I + \sin \varphi \cdot R$ the *generalized elliptic matrix (GEM)*.

We now consider a few examples of the matrix R. Since

$$R = \begin{bmatrix} a & b \\ c & d \end{bmatrix} \quad \rightarrow \quad R^2 = \begin{bmatrix} a^2 + cb & b(a+d) \\ c(a+d) & d^2 + bc \end{bmatrix},$$

then for the coefficients of this matrix we have the following equations: $a+d = 0$, $a^2 + bc = 1$. Therefore, we consider the matrix R in the form

$$R = \begin{bmatrix} a & b \\ c & -a \end{bmatrix} \quad \rightarrow \quad a^2 + bc = 1.$$

As an example, consider the matrix

$$R = \begin{bmatrix} -1 & 2 \\ -1 & 1 \end{bmatrix}.$$

This matrix can be expressed as GEM, $C = \cos(\varphi) \cdot I + \sin(\varphi) \cdot R(\phi)$, when $\cos(\varphi) = 0$, i.e., $\varphi = \pi/2$. The angle $\varphi = \pi/2$ corresponds to the $N = 4$ case, i.e., when $\varphi_4 = 2\pi/N$ and $C^4 = I$. Thus,

$$C = \cos(\varphi)I + \sin(\varphi) \begin{bmatrix} -1 & 2 \\ -1 & 1 \end{bmatrix}.$$

Similarly, for the case $N = 4$, we can consider matrices

$$R = \frac{1}{\sqrt{5}} \begin{bmatrix} 1 & -2 \\ 1 & -1 \end{bmatrix}, \quad \text{and} \quad R = \begin{bmatrix} 0 & -1 \\ 1 & 0 \end{bmatrix}.$$

2.7.1 *N*-block GEFT

The above described generalized elliptic matrices, $C_\phi = C_\phi(\varphi)$, where $\varphi = \varphi_N = 2\pi/N$, can be used for constructing the N-block EFT. The matrix of this transformation equals

$$X(C_\phi) = \begin{bmatrix} I & I & I & I & \cdots & I \\ I & C_\phi^1 & C_\phi^2 & C_\phi^3 & \cdots & C_\phi^{N-1} \\ I & C_\phi^2 & C_\phi^4 & C_\phi^6 & \cdots & C_\phi^{2N-2} \\ I & \cdots & \cdots & \cdots & \cdots & \cdots \\ I & C_\phi^{N-1} & C_\phi^{2N-2} & C_\phi^{3N-3} & \cdots & C_\phi^1 \end{bmatrix}, \qquad (2.32)$$

where we consider the property $C^{kN+m} = C^m$, when $k, m = 0 : (N - 1)$. We call this transformation *the N-block generalized elliptic Fourier transformation*, or *the N-block GEFT*. The matrix of the N-block GEFT is parameterized by two angles, φ_N and ϕ.

The following property holds for the N-block GEFT, as for the DFT:

$$R(\mathbf{x})_{N-k} = R(\mathbf{x})_k, \quad I(\mathbf{x})_{N-k} = -I(\mathbf{x})_k, \quad k = 1 : (N/2 - 1),$$

where we denote by $R(\mathbf{x})$ and $I(\mathbf{x})$ the "real" and "imaginary" parts of the N-block EFT of a real input $\mathbf{x}$, respectively.

Example 2.15
For the $N = 5$ case, the 5-block GEFT has the matrix

$$X(C) = \begin{bmatrix} I & I & I & I & I \\ I & C^1 & C^2 & C^3 & C^4 \\ I & C^2 & C^4 & C^1 & C^3 \\ I & C^3 & C^1 & C^4 & C^2 \\ I & C^4 & C^3 & C^2 & C^1 \end{bmatrix}, \tag{2.33}$$

where $C = C_\phi(\varphi)$, $\varphi = 2\pi/5$. Let $\phi = \pi/5$, then the elliptic matrix C defines the one-parametric group with period 5,

$$C = C^1 = \begin{bmatrix} 0.3090 & -0.3090 \\ 2.9271 & 0.3090 \end{bmatrix}, \quad C^2 = \begin{bmatrix} -0.8090 & -0.1910 \\ 1.8090 & -0.8090 \end{bmatrix},$$

$$C^3 = \begin{bmatrix} -0.8090 & 0.1910 \\ -1.8090 & -0.8090 \end{bmatrix}, \quad C^4 = \begin{bmatrix} 0.3090 & 0.3090 \\ -2.9271 & 0.3090 \end{bmatrix}, \quad C^5 = \begin{bmatrix} 1 & 0 \\ 0 & 1 \end{bmatrix}.$$

The matrix $X(C)$ of the 5-point GEFT equals

$$\begin{bmatrix} 1\,0 & 1 & 0 & 1 & 0 & 1 & 0 & 1 & 0 \\ 0\,1 & 0 & 1 & 0 & 1 & 0 & 1 & 0 & 1 \\ 1\,0 & 0.3090 & -0.3090 & -0.8090 & -0.1910 & -0.8090 & 0.1910 & 0.3090 & 0.3090 \\ 0\,1 & 2.9271 & 0.3090 & 1.8090 & -0.8090 & -1.8090 & -0.8090 & -2.9271 & 0.3090 \\ 1\,0 & -0.8090 & -0.1910 & 0.3090 & 0.3090 & 0.3090 & -0.3090 & -0.8090 & 0.1910 \\ 0\,1 & 1.8090 & -0.8090 & -2.9271 & 0.3090 & 2.9271 & 0.3090 & -1.8090 & -0.8090 \\ 1\,0 & -0.8090 & 0.1910 & 0.3090 & -0.3090 & 0.3090 & 0.3090 & -0.8090 & -0.1910 \\ 0\,1 & -1.8090 & -0.8090 & 2.9271 & 0.3090 & -2.9271 & 0.3090 & 1.8090 & -0.8090 \\ 1\,0 & 0.3090 & 0.3090 & -0.8090 & 0.1910 & -0.8090 & -0.1910 & 0.3090 & -0.3090 \\ 0\,1 & -2.9271 & 0.3090 & -1.8090 & -0.8090 & 1.8090 & -0.8090 & 2.9271 & 0.3090 \end{bmatrix}.$$

This matrix is orthogonal, the determinant is $\det(X(C)) = 3125$, and the inverse matrix equals

$$X^{-1}(C) = \frac{1}{5} X(C^{-1}) = \frac{1}{5} \begin{bmatrix} I & I & I & I & I \\ I & C^{-1} & C^{-2} & C^{-3} & C^{-4} \\ I & C^{-2} & C^{-4} & C^{-1} & C^{-3} \\ I & C^{-3} & C^{-1} & C^{-4} & C^{-2} \\ I & C^{-4} & C^{-3} & C^{-2} & C^{-1} \end{bmatrix} = \frac{1}{5} \begin{bmatrix} I & I & I & I & I \\ I & C^4 & C^3 & C^2 & C^1 \\ I & C^3 & C^1 & C^4 & C^2 \\ I & C^2 & C^4 & C^1 & C^3 \\ I & C^1 & C^2 & C^3 & C^4 \end{bmatrix}.$$

☐

We consider the example of the 128-block GEFT of the "complex" discrete-time signal $\bar{y}$ of length 128, which is shown in Figure 2.38(a). The angle $\varphi = 2 \cdot \pi/128$ and ϕ is considered to be equal $\pi/6$. The elliptic matrix equals

$$C_{\pi/6}(\varphi) = \begin{bmatrix} 0.9988 & -0.0131 \\ 0.1831 & 0.9988 \end{bmatrix}.$$

Figure 2.43 shows the real part of the 128-block GEFT of $\bar{y}$ in part a, along with the imaginary part in b. In parts c and d, the real and imaginary parts of the 128-block GEF of the same signal are shown for the case when $\phi = \pi/9$. The elliptic matrix in this case equals

$$C_{\pi/9}(\varphi) = \begin{bmatrix} 0.9988 & -0.0087 \\ 0.2783 & 0.9988 \end{bmatrix}.$$

One can see that the real part of the transform changes slightly, when angle

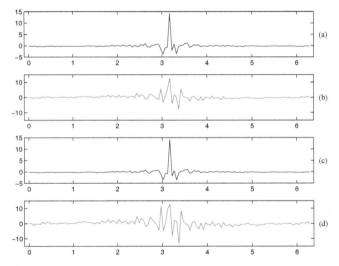

FIGURE 2.43

(a) The real and (b) imaginary parts the 128-block GEFT of the signal $\bar{y}$, when $\phi = \pi/6$. (c) The real and (d) imaginary parts the 128-block GEFT of the signal $\bar{y}$, when $\phi = \pi/9$. (The transforms have been shifted cyclicly to the center.)

ϕ varies. The imaginary parts differ much.

Figure 2.44 shows the gray-level images of the matrices ($2N \times 2N$) of the N-block GEFT for the cases $N = 17, 32$, and 50, respectively, in parts a, b, and c. For comparison, the images of the matrices of the N-block DFT of the same orders are shown in Figure 2.45.

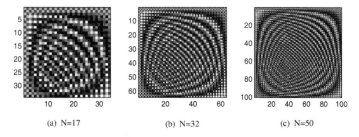

(a) N=17 (b) N=32 (c) N=50

FIGURE 2.44

The image of the matrix of the N-block GEFT, when (a) $N = 17$, (b) $N = 32$, and (c) $N = 50$.

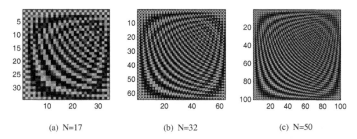

(a) N=17 (b) N=32 (c) N=50

FIGURE 2.45

The image of the matrix of the N-block DFT, when (a) $N = 17$, (b) $N = 32$, and (c) $N = 50$.

Problems

Problem 2.1 A. Calculate the eight integer DFT of the signal

$$\mathbf{f} = \{1, 4, 2, 3, 7, 2, 1, 8\},$$

by using the method of the three-step lifting scheme.

B. Show the full signal-flow graph of the 8-point integer DFT with data of all intermediate calculations, when using lifting schemes.

C. Calculate the error of integer approximation of the DFT of this signal.

Problem 2.2 Calculate the one-point integer transforms $\mathcal{A}_a[x]$ of the numbers $x = 2, 5, 8, 17$, and 22, for the $a = 0.7071$ case.

Problem 2.3 A. Calculate the reversible integer transforms $\mathcal{B}[x]$ of the numbers $x = 6$ and 9 that approximate the complex multiplications wx, when $w = 0.3827 - j0.9239$.

B. Calculate the reversible integer transforms $\mathcal{B}[x]$ of the numbers $x = 5$ and 11 that approximate the complex multiplications wx, when $w = -0.9239 - j0.3827$.

Problem 2.4 * Define the reversible integer transforms $\mathcal{A}[x]$ and $\mathcal{B}[x]$ for approximating the complex multiplication ax and wx, where $a = 0.7071$, w is a twiddle coefficient, and x is a complex number.

Problem 2.5 Calculate the eight-point integer DFT of the signal

$$\mathbf{f} = \{1, 4, 2, 3, 7, 2, 1, 8\}$$

by using the method of control bits. Calculate the error of integer approximation of the DFT of this signal and make a comparison with the method of lifting schemes.

Problem 2.6 Consider the cyclic shift of the signal $\mathbf{f} = \{1, 4, 2, 5, 7, 2, 1, 8\}$ as

$$\mathbf{g} = \{1, 8, 1, 4, 2, 5, 7, 2\}.$$

A. Calculate the eight-point integer DFT of the signal $\mathbf{g}$ by using the method of control bits.
B. Calculate the inverse integer DFT.
C. Calculate the error of integer approximation of the DFT.
D. Analyze if the integer DFT of $\mathbf{g}$ can be expressed by the integer DFT of $\mathbf{f}$.

Problem 2.7 Consider two signals $\mathbf{f} = \{1, 4, 2, 3, 7, 2, 1, 8\}$ and $\mathbf{g} = \{1, 8, 1, 4, 2, 3, 7, 2\}$.
A. Calculate the eight-point integer DFTs of the signals $\mathbf{f}$ and $\mathbf{g}$, by using the method of lifting schemes.
B. Analyze if the integer DFT of $\mathbf{g}$ can be expressed by the integer DFT of $\mathbf{f}$.

Problem 2.8 * Consider the signal $\mathbf{f} = \{1, 4, 7, 2, 1, 8\}$.
A. Calculate the six-point DFT of this signal by the paired method.
B. Calculate the six-point integer DFT of the signal $\mathbf{f}$ by using the method of control bits.
C. Calculate the error of integer approximation of the DFT.

Problem 2.9 Calculate the five-point DFT of the signal $\mathbf{f} = \{3, 2, 5, 1, 4\}$ by using the vector form of the transform, i.e., calculate the five-block DFT. Calculate the matrix of the five-block DFT.

Problem 2.10 Consider the matrix

$$T = \begin{bmatrix} 1 & 1 \\ -1 & 0 \end{bmatrix}.$$

Calculate the matrix of the 6-block T-GDT and the 6-block T-GDT of the signal $\mathbf{f} = \{2, 5, 1, 4, 1, 3\}$.

Problem 2.11 Given integer $N = 21$ and angle $\varphi = \varphi_N = 2\pi/N$, consider the matrix

$$C = C(\varphi) = \begin{bmatrix} -\cos\varphi & \cos\varphi - 1 \\ \cos\varphi + 1 & -\cos\varphi \end{bmatrix}. \tag{2.34}$$

A. Draw the scheme of movement of the point $(1, 0)$ by the one-parametric group defined by C.

B. Calculate the N-block C-GDT of the discrete-time signal which is uniformly sampled in the interval $[0, 2\pi]$ from the following wave:

$$x(t) = 2\cos(5t) - 3\sin(4t), \quad t \in [0, 2\pi].$$

Plot the sampled signal and the "real" and "imaginary" parts of the transform.

Problem 2.12 Consider the matrix

$$V = \frac{1}{2} \begin{bmatrix} 1 & 1 & 1 & 1 \\ 1 & -1 & 1 & -1 \\ 1 & -1 & -1 & 1 \\ 1 & 1 & -1 & -1 \end{bmatrix}. \tag{2.35}$$

Show that $V^6 = I$ and $VV' = I$. Find the characteristic equation for the matrix V.

Problem 2.13 Verify if we can construct the vector 4-block Fourier transform by using the Hadamard matrix (4×4).

Problem 2.14 Consider the matrix

$$T = \begin{bmatrix} 0 & 1 & 0 & 0 \\ 0 & 0 & j & 0 \\ 0 & 0 & 0 & -1 \\ -j & 0 & 0 & 0 \end{bmatrix}$$

with $\det(T) = 1$, and for which $\det(T - I) = 2$. Construct the T-generated block discrete transform, T-GDT. Show the code for calculating this transform.

Problem 2.15 Consider the matrix

$$T = \begin{bmatrix} 0 & 1 \\ j & 0 \end{bmatrix}.$$

A. Show that this matrix can be used to define the block T-GDT.
B. Define the order N of this T-GDT, and calculate the direct and inverse matrices of the transform.
C. Represent this N-block T-GDT in the complex plane.

Problem 2.16 Consider the matrix

$$T = \begin{bmatrix} j & 1 \\ 1 & 0 \end{bmatrix}.$$

Show that this matrix can be used to define the T-GDT.

Problem 2.17 Given integer $N > 1$, consider the angle $\varphi = \varphi_N = 2\pi/N$ and the matrix

$$T = T(\varphi) = \begin{bmatrix} -\cos\varphi & j\sin\varphi \\ j\sin\varphi & -\cos\varphi \end{bmatrix}. \tag{2.36}$$

A. Prove that $T(n\varphi)T(\varphi) = T((n+1)\varphi)$ and $T^{2N}(\varphi) = I$.
B. Construct the N-block T-GDT.

3

Cosine Transform

The discrete cosine transform (DCT) is one of the widely used tools in digital signal and image processing, especially in image compression and transform-based coding in telecommunication [45]-[49]. This transform approaches the Karhuner-Loeve transform for first-order Markov stationary random data and can be used in linear filtration and pattern recognition [50, 51]. Since introducing the DCT [2], many fast algorithms have been proposed to reduce the computational complexity of the DCT. We note the fast 1-D DCT proposed in [56], radix-2 (including cases with even sequence of lengths) and vector-radix [61, 62], fast approximate computation of the DCT [63], as well as different recursive algorithms for the DCT with effective VLSI hardware implementation [52, 57, 58], algorithms based on Clenshaw's recurrent formula [59], and recursive order reduction of the Tchebycheff polynomial [53, 55]. We consider effective algorithms for splitting the 2^r-point DCT by short 2^k-point DCTs, $k = 1 : (r - 1)$. The splitting is defined by the paired representation of signals with respect to the cosine transform. Two recursive forms of splitting the N-point type-II DCT are described. We also consider a method of calculating the DCT which is based on using the Coxeter type matrices. The cases $N = 4$ and 8 are illustrated in detail. The integer approximations of the DCT are also considered, and methods of control bits, nonlinear equations, and lifting scheme are described.

3.1 Partitioning the DCT

Let f be a real even sequence f_n of the length $2N$, which is determined by the following relation: $f_{2N-n-1} = f_n$, $n = 0 : (N - 1)$. It is not difficult to see that the $2N$-point discrete shifted Fourier transform over the sequence f

$$F_p = \sum_{n=0}^{2N-1} f_n W^{(n+\frac{1}{2})p}, \quad p = 0 : (N - 1),$$

where $W = W_{2N} = \exp(-2\pi j/(2N))$.

The transform is real and can be expressed in the following form:

$$F_p = 2 \sum_{n=0}^{N-1} f_n \cos\left(\frac{\pi}{N}\left[(n+\frac{1}{2})p\right]\right), \quad p = 0 : (2N-1). \tag{3.1}$$

The following relations hold between components of this transform: $F_{2N-p} = \bar{F}_p$, $p = 1 : (N-1)$, and $F_N = 0$. Therefore, the calculation in (3.1) can be performed only for the half of frequency-points p of the set of integers $X_{2N} = \{0, 1, 2, ..., (2N-1)\}$. The factor of 2 can also be omitted from the definition. The N-point discrete cosine transform of f_n is defined by

$$C_p = \sum_{n=0}^{N-1} f_n \cos\left(\frac{\pi}{2N}[2n+1]\,p\right) = \sum_{n=0}^{N-1} f_n \cos_N\left(\left[n + \frac{1}{2}\right]p\right), \tag{3.2}$$

where $\cos_N(t) = \cos(\pi t/N)$. The transformation $f_n \to C_p$ is also called the N-point DCT of type II. The normalization factors $k_0 = 1/\sqrt{2}$ and $k_p = 1$, for $p = 1 : (N-1)$, can also be considered in the definition of the transform. For instance, the matrix of the four-point DCT with normalization factors equals

$$[C_4] = \begin{Vmatrix} 0.7071 & 0.7071 & 0.7071 & 0.7071 \\ 0.9239 & 0.3827 & -0.3827 & -0.9239 \\ 0.7071 & -0.7071 & -0.7071 & 0.7071 \\ 0.3827 & -0.9239 & 0.9239 & -0.3827 \end{Vmatrix}, \quad \det[C_4] = 4.$$

This matrix is orthogonal, and the inverse matrix equals the transpose matrix up to the factor of $1/2$, i.e., $2[C_4]^{-1} = [C_4]^T$. For simplicity of further calculations, the normalization factors k_p will be omitted.

Given a pair of numbers p and t, where p is an integer of the interval $X_N = \{0, 1, 2, ..., N-1\}$ and t is from the set $Y = \{0, 1/2, 1, 3/2, 2, ..., 2N - 1, 2N - 1/2\}$, we denote by $V_{p,t}$ the set

$$V_{p,t} = \left\{n; \ \left(n + \frac{1}{2}\right)p = t \bmod 2N, \ n = 0 : (N-1)\right\} \tag{3.3}$$

and define the component

$$\bar{f}_{p,t} = \sum_{n \in V_{p,t}} f_n, \quad t = 0 : 1/2 : (2N - 1/2). \tag{3.4}$$

We use the notation $t = 0 : \alpha : M$, to denote the numbers that run from 0 to M with step α, and the notation $t = 0 : M$ if the step is 1. Because of the property $\cos_N(t+N) = \cos(\pi(t+N)/N) = \cos(\pi t/N + \pi) = -\cos_N(t)$, $t \in [0, N)$, the discrete cosine transform can be written as

$$C_p = \sum_{t=0:0.5:(N-0.5)} \bar{f}_{p,t} \cos_N(t), \quad p = 0 : (N-1). \tag{3.5}$$

The components $f_{p,t}$ are defined as

$$f_{p,t} = \bar{f}_{p,t} - \bar{f}_{p,t+N} = \sum_{n \in V_{p,t}} f_n - \sum_{n \in V_{p,t+N}} f_n. \qquad (3.6)$$

Note that $f_{p,N/2} \cos_N(N/2) = f_{p,N/2} \cos(\pi/2) = 0$.

Example 3.1

Let $N = 8$ and $p = 3$. The variable $t = [(n + 1/2)p] \bmod 16 = k + 1/2$ takes values from the following table:

n	0	1	2	3	4	5	6	7
t	1.5	4.5	7.5	10.5	13.5	0.5	3.5	6.5

Therefore $f_{3,n+0.5} = \bar{f}_{3,n+0.5} - \bar{f}_{3,n+8.5} = \pm f_{3n+5 \bmod 8}$, $n = 0 : 7$, and $f_{3,t} = 0$ for other numbers t of Y. For instance, $f_{3,0.5} = f_5$, $f_{3,1.5} = f_0$, $f_{3,2.5} = -f_3$, and $f_{3,1} = f_{3,2} = 0$. The spectral component at frequency-point $p = 3$ can be written as

$$C_3 = \sum_{t=0.5:1:7.5} f_{3,t} \cos_N(t) = \sum_{n=0}^{7} f_{3,n+1/2} \cos_N(n + 1/2).$$

For $p = 2$, the variable $t = [(n + 1/2)p] \bmod 16 = 2n + 1$ takes values from the table

n	0	1	2	3	4	5	6	7
t	1	3	5	7	9	11	13	15

and the components $f_{2,t}$ are calculated by

$$f_{2,1} = \bar{f}_{2,1} - \bar{f}_{2,9} = f_0 - f_4, \quad f_{2,3} = \bar{f}_{2,3} - \bar{f}_{2,11} = f_1 - f_5$$
$$f_{2,5} = \bar{f}_{2,5} - \bar{f}_{2,13} = f_2 - f_6, \quad f_{2,7} = \bar{f}_{2,7} - \bar{f}_{2,15} = f_3 - f_7.$$

All other components $f_{2,t} = 0$ for $t \neq 2n + 1$, where $n = 0 : 3$. The spectral component at frequency-point $p = 2$ can be written as

$$C_2 = \sum_{t=1:2:7} f_{2,t} \cos_N(t) = \sum_{n=0}^{3} f_{2,2n+1} \cos_N(2n+1) = \sum_{n=0}^{3} f_{2,2n+1} \cos_{N/2}\left(n + \frac{1}{2}\right).$$

▯

Property 3.1 For a given p integer,

$$C_{\overline{kp}} = C_{kp \bmod 2N} = \sum_{t=0:0.5:(N-0.5)} f_{p,t} \cos_N(kt), \quad k = 1 : 2 : (N - 1), \quad (3.7)$$

where the summation is taken only for t such that $f_{p,t} \neq 0$.

Indeed, for a given p, the sets $V_{p,t}$ are manually disjoint when $t = 0 : 0.5 : (2N - 0.5)$. Therefore, the following calculations hold for odd k:

$$\sum_{t=0:0.5:(N-0.5)} f_{p,t} \cos_N(kt) = \sum_{t=0:0.5:(N-0.5)} \left[\left(\sum_{n \in V_{p,t}} f_n - \sum_{n \in V_{p,t+N}} f_n \right) \cos_N(kt) \right]$$

$$= \sum_{t=0:0.5:(N-0.5)} \left[\left(\sum_{n \in V_{p,t}} f_n \cos_N(kt) - \sum_{n \in V_{p,t+N}} f_n \cos_N(kt) \right) \right]$$

$$= \sum_{t=0:0.5:(N-0.5)} \left[\left(\sum_{n \in V_{p,t}} f_n \cos_N(kt) + \sum_{n \in V_{p,t+N}} f_n \cos_N(k[t + N]) \right) \right]$$

$$= \sum_{t=0:0.5:(2N-0.5)} \left[\sum_{n \in V_{p,t}} f_n \cos_N(kt) \right] = \sum_{n=0}^{N-1} f_n \cos_N \left(k \left[n + \frac{1}{2} \right] p \right)$$

$$= \sum_{n=0}^{N-1} f_n \cos_N \left(\left[n + \frac{1}{2} \right] kp \right) = C_{\overline{kp}} = C_{kp \bmod 2N}.$$

Because of the property $\cos_N(k[N - t]) = \cos(k\pi(N - t)/N) = \cos(k\pi - \pi kt/N) = -\cos_N(kt)$, $t \in [0, N]$, we introduce the following components, for odd numbers k,

$$f'_{p,t} = f_{p,t} - f_{p,N-t}, \quad t = 0.5 : 0.5 : (N/2 - 0.5),$$
$$f'_{p,0} = f_{p,0} = f_{p,0} - f_{p,N},$$

where, for simplicity of calculation, we consider $f_{p,N} = 0$.

Then, for odd $k = 2m + 1$, equation (3.7) can be written as follows:

$$C_{\overline{kp}} = \sum_{t=0:0.5:(N/2-0.5)} f_{p,t} \cos_N(kt) + \sum_{t=N/2:0.5:(N-0.5)} f_{p,t} \cos_N(kt)$$

$$= \sum_{t=0:0.5:(N/2-0.5)} f_{p,t} \cos_N(kt) + \sum_{t=0.5:0.5:N/2} f_{p,N-t} \cos_N(k[N - t])$$

$$= \sum_{t=0:0.5:(N/2-0.5)} [f_{p,t} - f_{p,N-t}] \cos_N(kt) = \sum_{t=0:0.5:(N/2-0.5)} f'_{p,t} \cos_N(kt).$$

Thus

$$C_{\overline{(2m+1)p}} = \sum_{t=0:0.5:(N/2-0.5)} f'_{p,t} \cos_{N/2} \left(\left[m + \frac{1}{2} \right] t \right), \quad m = 0 : (N/2 - 1).$$

(3.8)

The right side of the last equality represents a $N/2$-point DCT of type III or IV. The type of the DCT depends on the value of p, when t takes values of odd integers, mixed numbers, or even integers. We will apply equation (3.8) for calculating the spectral components $C_{\overline{(2m+1)p}}$, when $p = 2^n$, $n = 0 : (r - 1)$. It is not difficult to note that for these p, the sets $U_{p,t} = V_{p,t} \cup V_{p,t+N}$ are not empty only for $L = N/p$ values of t. We denote those values by $t_0, t_1, t_2, ..., t_{L-1}$. For instance, when $N = 8$ and $p = 1$, then $L = 8$, and $\{t_0, t_1, t_2, ..., t_7\} = \{0.5, 1.5, 2.5, ..., 7.5\}$. If $p = 2$, then $L = 4$ and $\{t_0, t_1, t_2, t_3\} = \{1, 3, 5, 7\}$.

We now consider the partition $\sigma' = (T')$ composed by the subsets of X of the form

$$T'_p = \{\, \overline{(2m+1)p};\ m = 0 : (N/2 - 1)\}. \tag{3.9}$$

According to (3.8), at frequency-points of set T'_p, the N-point DCT of the signal f represents a $N/2$-point DCT of type III or IV of the sequence $f_{T'_p} = \{f'_{p,t_0}, f'_{p,t_1}, f'_{p,t_2}, ..., f'_{p,t_{L-1}}\}$.

Example 3.2

Consider the eight-point DCT

$$C_p = \sum_{n=0}^{7} f_n \cos_8\left(\left[n + \frac{1}{2}\right]p\right), \quad p = 0 : 7,$$

with the matrix $[C_8]$ equal to

$$
\begin{Vmatrix}
1.0000 & 1.0000 & 1.0000 & 1.0000 & 1.0000 & 1.0000 & 1.0000 & 1.0000 \\
0.9808 & 0.8315 & 0.5556 & 0.1951 & -0.1951 & -0.5556 & -0.8315 & -0.9808 \\
0.9239 & 0.3827 & -0.3827 & -0.9239 & -0.9239 & -0.3827 & 0.3827 & 0.9239 \\
0.8315 & -0.1951 & -0.9808 & -0.5556 & 0.5556 & 0.9808 & 0.1951 & -0.8315 \\
0.7071 & -0.7071 & -0.7071 & 0.7071 & 0.7071 & -0.7071 & -0.7071 & 0.7071 \\
0.5556 & -0.9808 & 0.1951 & 0.8315 & -0.8315 & -0.1951 & 0.9808 & -0.5556 \\
0.3827 & -0.9239 & 0.9239 & -0.3827 & -0.3827 & 0.9239 & -0.9239 & 0.3827 \\
0.1951 & -0.5556 & 0.8315 & -0.9808 & 0.9808 & -0.8315 & 0.5556 & -0.1951
\end{Vmatrix}.
$$

The set of frequency-points $X = \{0, 1, 2, 3, 4, 5, 6, 7\}$ is covered by the partition $\sigma' = (T'_1, T'_2, T'_4, T'_0)$ with sets $T'_1 = \{1, 3, 5, 7\}$, $T'_2 = \{2, 6\}$, $T'_4 = \{4\}$, $T'_0 = \{0\}$. For the considered generators $p = 1, 2, 4$, and 0 of the sets $T'_p \in \sigma'$, we compose the matrix 8×4 of values of $t = \left(n + \frac{1}{2}\right)p \bmod 16$,

$$
||t||_{n=0:7, p=1,2,4,0} =
\begin{Vmatrix}
0.5 & 1.5 & 2.5 & 3.5 & 4.5 & 5.5 & 6.5 & 7.5 \\
1 & 3 & 5 & 7 & 9 & 11 & 13 & 15 \\
2 & 6 & 10 & 14 & 2 & 6 & 10 & 14 \\
0 & 0 & 0 & 0 & 0 & 0 & 0 & 0
\end{Vmatrix}.
$$

For $p = 1$, the variable t takes values of $n + 0.5$, where $n = 0 : 7$. Moreover $f_{1,n+0.5} = \bar{f}_{1,n+0.5} = f_n$, $n = 0 : 7$, and the components $f'_{1,t} = f_{1,t} - f_{1,8-t}$ are calculated as $f'_{1,0.5} = f_0 - f_7$, $f'_{1,1.5} = f_1 - f_6$, $f'_{1,2.5} = f_2 - f_5$, $f'_{1,3.5} = f_3 - f_4$.

For $p = 2$, the variable t takes values $1, 3, 5, ..., 15$, and

$$f_{2,1} = \bar{f}_{2,1} - \bar{f}_{2,9} = f_0 - f_4, \quad f_{2,3} = \bar{f}_{2,3} - \bar{f}_{2,11} = f_1 - f_5$$
$$f_{2,5} = \bar{f}_{2,5} - \bar{f}_{2,13} = f_2 - f_6, \quad f_{2,7} = \bar{f}_{2,7} - \bar{f}_{2,15} = f_3 - f_7.$$

The components $f'_{2,t}$ are calculated as

$$f'_{2,1} = f_{2,1} - f_{2,7} = (f_0 - f_4) - (f_3 - f_7)$$
$$f'_{2,3} = f_{2,3} - f_{2,5} = (f_1 - f_5) - (f_2 - f_6).$$

For $p = 4$, the variable t takes values $2, 6, 10$, and 14. Therefore

$$f_{4,2} = \bar{f}_{4,2} - \bar{f}_{4,10} = (f_0 + f_4) - (f_2 + f_6)$$
$$f_{4,6} = \bar{f}_{4,6} - \bar{f}_{4,14} = (f_1 + f_5) - (f_3 + f_7)$$

and the component $f'_{4,2}$ is calculated as

$$f'_{4,2} = f_{4,2} - f_{4,6} = (f_0 + f_4) - (f_2 + f_6) - (f_1 + f_5) + (f_3 + f_7).$$

Finally, for $p = 0$, the variable t takes only value 0. Therefore

$$f'_{0,0} = f_{0,0} = f_0 + f_1 + f_2 + f_3 + f_4 + f_5 + f_6 + f_7.$$

Thus, the sequence f is represented as four short sequences $f_{T'_1} = \{f'_{1,0.5},$ $f'_{1,1.5}, f'_{1,2.5}, f'_{1,3.5}\}$, $f_{T'_2} = \{f'_{2,1}, f'_{2,3}\}$, $f_{T'_4} = \{f'_{4,2}\}$, $f_{T'_0} = \{f'_{0,0}\}$, and this representation is performed by the transformation χ'_8 whose matrix is

$$[\chi'_8] = \begin{bmatrix} 1 & 0 & 0 & 0 & 0 & 0 & 0 & -1 \\ 0 & 1 & 0 & 0 & 0 & 0 & -1 & 0 \\ 0 & 0 & 1 & 0 & 0 & -1 & 0 & 0 \\ 0 & 0 & 0 & 1 & -1 & 0 & 0 & 0 \\ 1 & 0 & 0 & -1 & -1 & 0 & 0 & 1 \\ 0 & 1 & -1 & 0 & 0 & -1 & 1 & 0 \\ 1 & -1 & -1 & 1 & 1 & -1 & -1 & 1 \\ 1 & 1 & 1 & 1 & 1 & 1 & 1 & 1 \end{bmatrix} = \begin{bmatrix} [\chi'_{1,0.5}] \\ [\chi'_{1,1.5}] \\ [\chi'_{1,2.5}] \\ [\chi'_{1,3.5}] \\ [\chi'_{2,1}] \\ [\chi'_{2,3}] \\ [\chi'_{4,2}] \\ [\chi'_{0,0}] \end{bmatrix}. \qquad (3.10)$$

According to (3.8), when $p = 1$, the spectral components at frequency-points $(2m + 1)p = 2m + 1$, $m = 0 : 3$, are calculated as

$$C_{2m+1} = \sum_{n=0}^{3} f'_{1,n+1/2} \cos_{N/2} \left(\left[n + \frac{1}{2} \right] \left[m + \frac{1}{2} \right] \right).$$

The spectral components at frequency-points 2 and 6 are calculated as

$$C_{(2m+1)2} = \sum_{n=0,1} f'_{2,2n+1} \cos_{N/2} \left([2n + 1] \left[m + \frac{1}{2} \right] \right)$$

$$= \sum_{n=0,1} f'_{2,2n+1} \cos_{N/4} \left(\left[n + \frac{1}{2} \right] \left[m + \frac{1}{2} \right] \right)$$

where $m = 0, 1$. The spectral component at point $(2m + 1)4 = 4$, $(m = 0)$, is calculated as

$$C_4 = f'_{4,2} \cos_{N/2}\left(\left[m + \frac{1}{2}\right]2\right) = f'_{4,2} \cos_{N/2}(1) = f'_{4,2} \cos(\pi/4).$$

At point $p = 0$, $C_0 = f'_{0,0}$.

Thus, by means of the transformation χ'_8, the 8-point DCT is decomposed by 4-, 2-, and 1-point DCTs of type IV, and one 1-point DCT which is an identity transformation and will be denoted by I. In matrix form, the factorization can be written as

$$[C_8]=
\begin{bmatrix}
[C_{4|IV}] & & & \\
& [C_{2|IV}] & & \\
& & [C_{1|IV}] & \\
& & & 1
\end{bmatrix}
[\chi'_8]$$

$$=
\begin{bmatrix}
0.9808 & 0.8315 & 0.5556 & 0.1951 & 0 & 0 & 0 & 0 \\
0.8315 & -0.1951 & -0.9808 & -0.5556 & 0 & 0 & 0 & 0 \\
0.5556 & -0.9808 & 0.1951 & 0.8315 & 0 & 0 & 0 & 0 \\
0.1951 & -0.5556 & 0.8315 & -0.9808 & 0 & 0 & 0 & 0 \\
0 & 0 & 0 & 0 & 0.9239 & 0.3827 & 0 & 0 \\
0 & 0 & 0 & 0 & 0.3827 & -0.9239 & 0 & 0 \\
0 & 0 & 0 & 0 & 0 & 0 & 0.7071 & 0 \\
0 & 0 & 0 & 0 & 0 & 0 & 0 & 1
\end{bmatrix}
[\chi'_8]$$

which results in the following matrix:

$$
\begin{bmatrix}
0.9808 & 0.8315 & 0.5556 & 0.1951 & -0.1951 & -0.5556 & -0.8315 & -0.9808 \\
0.8315 & -0.1951 & -0.9808 & -0.5556 & 0.5556 & 0.9808 & 0.1951 & -0.8315 \\
0.5556 & -0.9808 & 0.1951 & 0.8315 & -0.8315 & -0.1951 & 0.9808 & -0.5556 \\
0.1951 & -0.5556 & 0.8315 & -0.9808 & 0.9808 & -0.8315 & 0.5556 & -0.1951 \\
0.9239 & 0.3827 & -0.3827 & -0.9239 & -0.9239 & -0.3827 & 0.3827 & 0.9239 \\
0.3827 & -0.9239 & 0.9239 & -0.3827 & -0.3827 & 0.9239 & -0.9239 & 0.3827 \\
0.7071 & -0.7071 & -0.7071 & 0.7071 & 0.7071 & -0.7071 & -0.7071 & 0.7071 \\
1.0000 & 1.0000 & 1.0000 & 1.0000 & 1.0000 & 1.0000 & 1.0000 & 1.0000
\end{bmatrix}
$$

The 8-point DCT is thus calculated by the paired transform in the order $C_1, C_3, C_5, C_7, C_2, C_6, C_4$, and C_0.

Figure 3.1 shows the block diagram of calculation of the 8-point DCT by the paired transform χ'_8, which is reduced to 4-, 2-, and 1-point DCTs of type IV. The 8-point paired transform requires $2 \cdot 8 - 2 = 14$ additions. The 4- and 2-point DCTs of type IV use respectively 8 and 3 multiplications and 12 and 3 additions [54]. One multiplication is needed for the 1-point DCTs of type IV. Therefore, the method of paired transform requires $\mu(C_8) = 8+3+1 = 12$ multiplications and $\alpha(C_8) = 14 + 12 + 3 = 29$ additions. The arithmetical complexity (μ, α) of each transform is shown in the corresponding block in

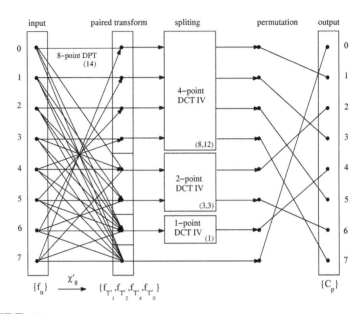

FIGURE 3.1
Block diagram of splitting the 8-point DCT of type II into 4-, 2-, and 1-point DCTs of type IV. The decomposition requires 12 multiplications and 29 additions.

the figure. For instance, the complexity $(8, 12)$ is shown for the 4-point DCT of type IV. ☐

Example 3.3
Consider the case when $N = 4$,

$$C_p = \sum_{n=0}^{3} f_n \cos_4\left(\left[n + \frac{1}{2}\right]p\right), \quad p = 0 : 3.$$

The matrix of the DCT equals

$$[C_4] = \begin{Vmatrix} 1.0000 & 1.0000 & 1.0000 & 1.0000 \\ 0.9239 & 0.3827 & -0.3827 & -0.9239 \\ 0.7071 & -0.7071 & -0.7071 & 0.7071 \\ 0.3827 & -0.9239 & 0.9239 & -0.3827 \end{Vmatrix}.$$

The set $X = \{0, 1, 2, 3\}$ is covered by the partition $\sigma' = (T_1', T_2', T_0')$ with sets $T_1' = \{1, 3\}$, $T_2' = \{2\}$, and $T_0' = \{0\}$. For the generators $p = 1, 2$, and 0, we compose the following matrix 4×3 with values of $t = (n + 1/2)p \mod 8$, when $n = 0 : 3$

$$||t||_{n=0:3, p=1,2,0} = \begin{Vmatrix} 0.5 & 1.5 & 2.5 & 3.5 \\ 1 & 3 & 5 & 7 \\ 0 & 0 & 0 & 0 \end{Vmatrix}.$$

Therefore, the sequence f is represented by three short sequences $f_{T_1'} = \{f_{1,0.5}', f_{1,1.5}'\}$, $f_{T_2'} = \{f_{2,1}'\}$, $f_{T_0'} = \{f_{0,0}'\}$, and this representation is performed by the paired transformation χ_4' with the matrix

$$[\chi_4'] = \begin{bmatrix} 1 & 0 & 0 & -1 \\ 0 & 1 & -1 & 0 \\ 1 & -1 & -1 & 1 \\ 1 & 1 & 1 & 1 \end{bmatrix} = \begin{bmatrix} [\chi_{1,0.5}'] \\ [\chi_{1,1.5}'] \\ [\chi_{2,1}'] \\ [\chi_{0,0}'] \end{bmatrix}.$$

Calculations similar to ones given in Example 2 show that, by means of the transformation χ_4', the 4-point DCT can be decomposed by 2-, 1-point DCTs of type IV, and one 1-point DCT. In the matrix form, the decomposition has the form

$$[C_4] = \begin{bmatrix} [C_{2|IV}] & \\ & [C_{1|IV}] \\ & & 1 \end{bmatrix} [\chi_4'] = \begin{bmatrix} 0.9239 & 0.3827 & 0 & 0 \\ 0.3827 & -0.9239 & 0 & 0 \\ 0 & 0 & 0.7071 & 0 \\ 0 & 0 & 0 & 1 \end{bmatrix} [\chi_4']$$

$$= \begin{bmatrix} 0.9239 & 0.3827 & -0.3827 & -0.9239 \\ 0.3827 & -0.9239 & 0.9239 & -0.3827 \\ 0.7071 & -0.7071 & -0.7071 & 0.7071 \\ 1.0000 & 1.0000 & 1.0000 & 1.0000 \end{bmatrix}.$$

The 4-point DCT is calculated by the paired transform in the order C_1, C_3, C_2, and C_0. Figure 3.2 shows the block diagram of calculation of the 4-point DCT by the paired transformation χ_4'. The 4-point discrete paired transform (DPT) requires $2 \cdot 4 - 2 = 6$ additions. Therefore, the method of paired

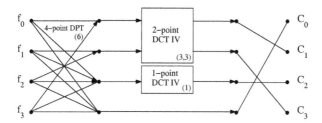

FIGURE 3.2
Block diagram of splitting the 4-point DCT of type II into 2- and 1-point DCTs of type IV. The decomposition requires 4 multiplications and 9 additions.

transform requires $\mu(C_4) = 3 + 1 = 4$ multiplications and $\alpha(C_4) = 6 + 3 = 9$ additions. ▯

Analyzing the results of the two examples considered, one can notice that the following recursive relations hold for the first three transformations χ_2', χ_4', and χ_8' :

$$[\chi_4'] = \begin{bmatrix} I_2 & -J_2 \\ [\chi_2'] & D_2[\chi_2'] \end{bmatrix}, \quad [\chi_2'] = \begin{bmatrix} 1 & -1 \\ 1 & 1 \end{bmatrix} \quad (3.11)$$

$$[\chi_8'] = \begin{bmatrix} I_4 & -J_4 \\ [\chi_4'] & D_4[\chi_4'] \end{bmatrix} = \begin{bmatrix} I_4 & & -J_4 \\ I_2 & -J_2 & -I_2 & J_2 \\ [\chi_2'] & D_2[\chi_2'] & [\chi_2'] & D_2[\chi_2'] \end{bmatrix},$$

where J_M denotes the anti-diagonal matrix $M \times M$. The matrices D_4 and D_2 are diagonal and are defined respectively as

$$D_4 = \begin{bmatrix} -1 & 0 & 0 & 0 \\ 0 & -1 & 0 & 0 \\ 0 & 0 & 1 & 0 \\ 0 & 0 & 0 & 1 \end{bmatrix}, \quad D_2 = \begin{bmatrix} -1 & 0 \\ 0 & 1 \end{bmatrix}.$$

Transformations χ_8', χ_4', and χ_2' with matrices $[\chi_8']$, $[\chi_4']$, and $[\chi_2']$ we name *paired transformations* with respect to the cosine transform. The representation of 8- and 4-point signals respectively in forms of $\{f_{T_p'}\}$ relates to the *paired representation* of signals f with respect to the cosine transform. Note that $D_M[\chi_M'] = [\chi_M']J_M$, for $M = 4$ and 2. Therefore, the recursive relation in (3.11) leads to the following matrix representation of the 8-point DCT:

$$[C_8] = \begin{bmatrix} [C_{4|IV}] & & & \\ & [C_{2|IV}] & & \\ & & [C_{1|IV}] & \\ & & & 1 \end{bmatrix} \begin{bmatrix} I_4 & -J_4 \\ [\chi_4'] & [\chi_4']J_4 \end{bmatrix}$$

$$= \begin{bmatrix} [C_{4|IV}] & & & \\ & [C_{2|IV}] & & \\ & & [C_{1|IV}] & \\ & & & 1 \end{bmatrix} \begin{bmatrix} I_4 & -J_4 \\ [\chi_4'] \end{bmatrix} \begin{bmatrix} I_4 & -J_4 \\ I_4 & J_4 \end{bmatrix}$$

$$= \begin{bmatrix} [C_{4|IV}] & \\ & [C_4] \end{bmatrix} \begin{bmatrix} I_4 & -J_4 \\ I_4 & J_4 \end{bmatrix} =$$

$$\begin{bmatrix} 0.9808 & 0.8315 & 0.5556 & 0.1951 & -0.1951 & -0.5556 & -0.8315 & -0.9808 \\ 0.8315 & -0.1951 & -0.9808 & -0.5556 & 0.5556 & 0.9808 & 0.1951 & -0.8315 \\ 0.5556 & -0.9808 & 0.1951 & 0.8315 & -0.8315 & -0.1951 & 0.9808 & -0.5556 \\ 0.1951 & -0.5556 & 0.8315 & -0.9808 & 0.9808 & -0.8315 & 0.5556 & -0.1951 \\ 1.0000 & 1.0000 & 1.0000 & 1.0000 & 1.0000 & 1.0000 & 1.0000 & 1.0000 \\ 0.9239 & 0.3827 & -0.3827 & -0.9239 & -0.9239 & -0.3827 & 0.3827 & 0.9239 \\ 0.7071 & -0.7071 & -0.7071 & 0.7071 & 0.7071 & -0.7071 & -0.7071 & 0.7071 \\ 0.3827 & -0.9239 & 0.9239 & -0.3827 & -0.3827 & 0.9239 & -0.9239 & 0.3827 \end{bmatrix}.$$

In this decomposition, the rows of the matrix are rearranged in the order $1, 3, 5, 7, 0, 2, 4$, and 6.

Figure 3.3 shows the block diagram of decomposition of the 8-point DCT by the 4-point DCT and 4-point DCT of type IV. The complexity of this decomposition is $(12, 29)$. Indeed, 8 operations of addition are required for the discrete transform (DT) with the matrix $[I_4, -J_4; I_4, D_4]$, in order to compose the inputs for both the 4-point DCTs. The 4-point DCT of type IV requires 8 multiplications and 12 additions. The 4-point DCT requires 4 multiplications and 9 additions. Therefore, $\mu(C_8) = 8 + 4 = 12$ and $\alpha(C_8) = 8 + 12 + 9 = 29$.

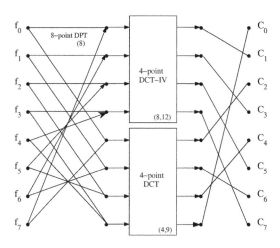

FIGURE 3.3

Block diagram of decomposition of the 8-point DCT of type II into 4-point DCT and 4-point DCT of type IV.

Figure 3.4 shows the block diagram of decomposition of the 4-point DCT by the 2-point DCTs of types IV and II,

$$
[C_4] = \begin{bmatrix} 0.9239 & 0.3827 & 0 & 0 \\ 0.3827 & -0.9239 & 0 & 0 \\ 0 & 0 & 1 & 1 \\ 0 & 0 & 0.7071 & -0.7071 \end{bmatrix} \begin{bmatrix} 1 & 0 & 0 & -1 \\ 0 & 1 & -1 & 0 \\ 1 & 0 & 0 & 1 \\ 0 & 1 & 1 & 0 \end{bmatrix}
$$

$$
= \begin{bmatrix} 0.9239 & 0.3827 & -0.3827 & -0.9239 \\ 0.3827 & -0.9239 & 0.9239 & -0.3827 \\ 1.0000 & 1.0000 & 1.0000 & 1.0000 \\ 0.7071 & -0.7071 & -0.7071 & 0.7071 \end{bmatrix}.
$$

$$(3.12)$$

The rows of the matrix of the 4-point DCT are rearranged in the order $1, 3, 0,$ and 2. The arithmetical complexity of splitting the 4-point DCT by this method equals $(\mu(C_4), \alpha(C_4)) = (3 + 1, 6 + 3) = (4, 9)$.

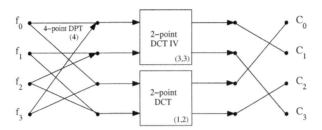

FIGURE 3.4
Block diagram of decomposition of the 4-point DCT of type II by the 2-point DCT and 2-point DCT of type IV.

3.1.1   4-point DCT of type IV

We now analyze the four-point DCT of type IV, which is used in the decomposition of the eight-point DCT,

$$C_p = \sum_{n=0}^{3} f_n \cos\left(\frac{\pi}{4}(n + 0.5)(p + 0.5)\right), \quad p = 0:3.$$

For that, we consider the matrix of this transformation

$$[C_{4;IV}] = \begin{bmatrix} 0.9808 & 0.8315 & 0.5556 & 0.1951 \\ 0.8315 & -0.1951 & -0.9808 & -0.5556 \\ 0.5556 & -0.9808 & 0.1951 & 0.8315 \\ 0.1951 & -0.5556 & 0.8315 & -0.9808 \end{bmatrix} \tag{3.13}$$

in the form

$$\mathbf{Q} = \mathbf{P}[C_{4;IV}]\mathbf{P}' = \begin{bmatrix} -0.1951 & -0.9808 & -0.5556 & 0.8315 \\ -0.9808 & 0.1951 & 0.8315 & 0.5556 \\ -0.5556 & 0.8315 & -0.9808 & 0.1951 \\ 0.8315 & 0.5556 & 0.1951 & 0.9808 \end{bmatrix} \tag{3.14}$$

where the permutation matrix $\mathbf{P}$ equals

$$\mathbf{P} = \begin{bmatrix} & & & 1 \\ & & 1 & \\ & 1 & & \\ 1 & & & \end{bmatrix}.$$

For the matrix $\mathbf{Q}$, the following factorization is valid:

$$\mathbf{Q} = \begin{bmatrix} \mathbf{A} & \\ & \mathbf{A} \end{bmatrix} \begin{bmatrix} -\mathbf{T}_{\frac{\pi}{4}} & \mathbf{I} \\ \mathbf{I} & \mathbf{T}_{-\frac{\pi}{4}} \end{bmatrix}$$

$$= \begin{bmatrix} -0.5556 & 0.8315 & & \\ 0.8315 & 0.5556 & & \\ & & -0.5556 & 0.8315 \\ & & 0.8315 & 0.5556 \end{bmatrix} \begin{bmatrix} -0.7071 & +0.7071 & 1 & 0 \\ -0.7071 & -0.7071 & 0 & 1 \\ 1 & 0 & 0.7071 & 0.7071 \\ 0 & 1 & -0.7071 & 0.7071 \end{bmatrix}$$

where the matrix

$$\mathbf{A} = \begin{bmatrix} -0.5556 & 0.8315 \\ 0.8315 & 0.5556 \end{bmatrix} = \begin{bmatrix} -1 & 0 \\ 0 & 1 \end{bmatrix} \begin{bmatrix} 0.5556 & -0.8315 \\ 0.8315 & 0.5556 \end{bmatrix}.$$

Using the matrix of rotation by the angle $5/16\pi$, we can write

$$\mathbf{A} = \begin{bmatrix} -1 & 0 \\ 0 & 1 \end{bmatrix} \mathbf{T}_{\frac{5\pi}{16}} = \mathbf{D}\mathbf{T}_{\frac{5\pi}{16}}.$$

Thus we obtain the following decomposition of the matrix of type IV DCT:

$$[C_{4;IV}] = \mathbf{P'QP}$$

$$= \begin{bmatrix} & & & 1 \\ & -1 & & \\ & & 1 & \\ & & & -1 \end{bmatrix} \begin{bmatrix} \mathbf{T}_{\frac{5\pi}{16}} & \\ & \mathbf{T}_{\frac{5\pi}{16}} \end{bmatrix} \begin{bmatrix} -\mathbf{T}_{\frac{\pi}{4}} & \mathbf{I} \\ \mathbf{I} & \mathbf{T}_{-\frac{\pi}{4}} \end{bmatrix} \begin{bmatrix} & & & 1 \\ & 1 & & \\ & & 1 & \\ 1 & & & \end{bmatrix}.$$

Four rotations are used in the decomposition. Two rotations by $\pm\pi/4$ require two multiplications by a factor of 0.7071 and two additions each. The rotation by $5/16\pi$ requires four multiplications and two additions, or three multiplications and four additions if the lifting scheme is used. The total number of operations for the four-point DCT of type IV by the above decomposition equals $\mu(C_{4;IV}) = 2(2 + \mu(T)) = 2(2+3) = 10, \alpha(C_{4;IV}) = 2(2 + \alpha(T)) + 4 = 2(2 + 4) + 4 = 16$, where we consider the number of multiplications for one rotation $\mu(T) = 3$ and additions $\alpha(T) = 4$.

We can also derive the following decomposition of the matrix of the four-point DCT of type IV:

$$[C_{4;IV}] = \mathbf{P'} \begin{bmatrix} \mathbf{T}_{-\frac{\pi}{16}} & \\ & \mathbf{T}_{-\frac{\pi}{16}} \end{bmatrix} \left(\mathbf{M} + \frac{1}{\sqrt{2}} \begin{bmatrix} -1 & 1 & & \\ 1 & 1 & & \\ & & -1 & 1 \\ & & 1 & 1 \end{bmatrix} \right) \mathbf{P'},$$

where

$$\mathbf{M} = \begin{bmatrix} & & & -1 \\ & -1 & & \\ -1 & & & \\ & & 1 & \end{bmatrix}, \quad \mathbf{T}_{-\frac{\pi}{16}} = \begin{bmatrix} 0.9808 & 0.1951 \\ -0.1951 & 0.9808 \end{bmatrix}.$$

The same number of multiplications and additions is used.

3.1.2   Fast four-point type IV DCT

We here describe the effective algorithm of the four-point DCT with eight multiplications. For that we first note that the matrix of the transform

$$[C_{4;IV}] = \begin{bmatrix} 0.9808 & 0.8315 & 0.5556 & 0.1951 \\ 0.8315 & -0.1951 & -0.9808 & -0.5556 \\ 0.5556 & -0.9808 & 0.1951 & 0.8315 \\ 0.1951 & -0.5556 & 0.8315 & -0.9808 \end{bmatrix}$$

is composed only by four different coefficients $c_n = \cos(2n+1)\pi/16$, $n = 0 : 3$. These coefficients are $c_0 = 0.9808$, $c_1 = 0.8315$, $c_2 = 0.5556$, and $c_3 = 0.1951$. Each coefficient is located on four different rows and columns of the matrix.

We therefore consider the following four Coxeter type matrices corresponding to these coefficients:

$$\mathbf{A}_0 = \mathbf{A}(c_0) = \begin{bmatrix} 1 & & & \\ & & -1 & \\ & -1 & & \\ & & & -1 \end{bmatrix}, \quad \mathbf{A}_1 = \mathbf{A}(c_1) = \begin{bmatrix} & & & 1 \\ & 1 & & \\ & & 1 & \\ & & & 1 \end{bmatrix}$$

$$\mathbf{A}_2 = \mathbf{A}(c_2) = \begin{bmatrix} & & 1 & \\ & & & -1 \\ 1 & & & \\ & -1 & & \end{bmatrix}, \quad \mathbf{A}_3 = \mathbf{A}(c_3) = \begin{bmatrix} & & & 1 \\ & -1 & & \\ & & 1 & \\ 1 & & & \end{bmatrix}$$

and the decomposition of the DCT matrix by these four matrices

$$[C_{4;IV}] = \sum_{n=0}^{3} c_n \mathbf{A}_n.$$

For the Coxeter matrices, we have $\mathbf{A}_n^2 = \mathbf{I}$, and $\mathbf{A}_3\mathbf{A}_0 = -\mathbf{J}$, $\mathbf{A}_2\mathbf{A}_1 = \mathbf{J}$, where $\mathbf{I}$ is the identity matrix and $\mathbf{J}$ is the diagonal matrix

$$\mathbf{J} = \begin{pmatrix} & & & 1 \\ & & -1 & \\ & 1 & & \\ -1 & & & \end{pmatrix}.$$

The following equalities hold for the trigonometric coefficients: $c_2 = \cos 5\pi/16 = s_1 = \sin 3\pi/16$, $c_3 = \cos 7\pi/16 = s_0 = \sin \pi/16$. Therefore we can write the matrix of the DCT as follows:

$$[C_{4;IV}] = c_0\mathbf{A}_0 + c_1\mathbf{A}_1 + s_1\mathbf{A}_2 + s_0\mathbf{A}_3 = (c_0\mathbf{I} + s_0\mathbf{A}_3\mathbf{A}_0)\mathbf{A}_0$$
$$+(c_1\mathbf{I} + s_1\mathbf{A}_2\mathbf{A}_1)\mathbf{A}_1 = (c_0\mathbf{I} - s_0\mathbf{J})\mathbf{A}_0 + (c_1\mathbf{I} + s_1\mathbf{J})\mathbf{A}_1.$$

Note that the matrix

$$c_0\mathbf{I} - s_0\mathbf{J} = \begin{pmatrix} c_0 & & & -s_0 \\ & c_0 & s_0 & \\ & -s_0 & c_0 & \\ s_0 & & & c_0 \end{pmatrix} = \begin{pmatrix} c_0 & & & -s_0 \\ & 1 & & \\ & & 1 & \\ s_0 & & & c_0 \end{pmatrix} \begin{pmatrix} 1 & & & \\ & c_0 & s_0 & \\ & -s_0 & c_0 & \\ & & & 1 \end{pmatrix}$$

describes two rotations by angles $\pm\pi/16$ in the subspaces spanned on the bases e_0, e_3 and e_1, e_2, respectively. We denote these rotations by $R_{0,3}$ and $R_{1,2}$ respectively with matrices

$$\mathbf{R}_{0,3}\left(\frac{\pi}{16}\right) = \begin{pmatrix} c_0 & & & -s_0 \\ & 1 & & \\ & & 1 & \\ s_0 & & & c_0 \end{pmatrix}, \quad \mathbf{R}_{1,2}\left(-\frac{\pi}{16}\right) = \begin{pmatrix} 1 & & & \\ & c_0 & s_0 & \\ & -s_0 & c_0 & \\ & & & 1 \end{pmatrix}.$$

Similarly, we can consider another pair of rotations $R_{0,3}$ and $R_{1,2}$ by angles $\pm 3\pi/16$ respectively with matrices

$$\mathbf{R}_{0,3}\left(-\frac{3\pi}{16}\right) = \begin{pmatrix} c_1 & & & s_1 \\ & 1 & & \\ & & 1 & \\ -s_1 & & & c_1 \end{pmatrix}, \quad \mathbf{R}_{1,2}\left(\frac{3\pi}{16}\right) = \begin{pmatrix} 1 & & & \\ & c_1 & -s_1 & \\ & s_1 & c_1 & \\ & & & 1 \end{pmatrix}$$

and write the following:

$$c_1\mathbf{I} + s_1\mathbf{J} = \begin{pmatrix} c_1 & & & s_1 \\ & c_1 & -s_1 & \\ & s_1 & c_1 & \\ -s_1 & & & c_1 \end{pmatrix} = \mathbf{R}_{0,3}\left(-\frac{3\pi}{16}\right)\mathbf{R}_{1,2}\left(\frac{3\pi}{16}\right).$$

We thus obtain the following decomposition of the matrix of the four-point DCT of type IV:

$$[C_{4;IV}] = \left[\mathbf{R}_{0,3}\left(\frac{\pi}{16}\right)\mathbf{R}_{1,2}\left(-\frac{\pi}{16}\right)\right]\mathbf{A}_0 + \left[\mathbf{R}_{0,3}\left(-\frac{3\pi}{16}\right)\mathbf{R}_{1,2}\left(\frac{3\pi}{16}\right)\right]\mathbf{A}_1. \quad (3.15)$$

This formula can be modified in two different ways: nonsymmetric and symmetric. Indeed, we first can write the following:

$$[C_{4;IV}] = \left[\mathbf{R}_{0,3}\left(\frac{\pi}{16}\right)\mathbf{R}_{1,2}\left(-\frac{\pi}{16}\right)\right]\left[\mathbf{A}_0 + \mathbf{R}_{0,3}\left(-\frac{\pi}{4}\right)\mathbf{R}_{1,2}\left(\frac{\pi}{4}\right)\mathbf{A}_1\right]$$

and, then,

$$[C_{4;IV}] = \mathbf{R}_{0,3}\left(\frac{\pi}{16}\right)\mathbf{R}_{1,2}\left(\frac{3\pi}{16}\right)\left[\mathbf{R}_{1,2}\left(-\frac{\pi}{4}\right)\mathbf{A}_0 + \mathbf{R}_{0,3}\left(-\frac{\pi}{4}\right)\mathbf{A}_1\right]$$

$$= \mathbf{R}_{0,3}\left(\frac{\pi}{16}\right)\mathbf{R}_{1,2}\left(\frac{3\pi}{16}\right)\mathbf{Q} \quad (3.16)$$

$$= \mathbf{R}_{0,3}\left(\frac{\pi}{16}\right)\mathbf{R}_{1,2}\left(-\frac{\pi}{16}\right)\mathbf{A}_0 + \mathbf{R}_{0,3}\left(-\frac{3\pi}{16}\right)\mathbf{R}_{1,2}\left(\frac{3\pi}{16}\right)\mathbf{A}_1.$$

We now analyze the matrix $\mathbf{Q}$ and consider the coefficients $c = s = \cos(\pi/4)$. The simple mathematical calculations show that this matrix can be represented as follows:

$$\mathbf{Q} = \begin{pmatrix} 1 & c & c & 0 \\ 1 & -c & -c & 0 \\ 0 & -c & c & 1 \\ 0 & -c & c & -1 \end{pmatrix} = \mathbf{Q}_1\mathbf{Q}_2 = \begin{pmatrix} 1 & 1 & 1 & 0 \\ 1 & -1 & -1 & 0 \\ 0 & -1 & 1 & 1 \\ 0 & -1 & 1 & -1 \end{pmatrix}\begin{pmatrix} 1 & & & \\ & c & & \\ & & c & \\ & & & 1 \end{pmatrix}.$$

Thus the matrix of the four-point DCT of type IV can be decomposed as

$$[C_{4;IV}] = \mathbf{R}_{0,3}(\frac{\pi}{16})\mathbf{R}_{1,2}(\frac{3\pi}{16})\mathbf{Q}_1\mathbf{Q}_2 \qquad (3.17)$$

$$= \begin{pmatrix} c_0 & & & -s_0 \\ & 1 & & \\ & & 1 & \\ s_0 & & & c_0 \end{pmatrix} \begin{pmatrix} 1 & & & \\ & c_1 & -s_1 & \\ & s_1 & c_1 & \\ & & & 1 \end{pmatrix} \begin{pmatrix} 1 & 1 & 1 & 0 \\ 1 & -1 & -1 & 0 \\ 0 & -1 & 1 & 1 \\ 0 & -1 & 1 & -1 \end{pmatrix} \begin{pmatrix} 1 & & & \\ & c & & \\ & & c & \\ & & & 1 \end{pmatrix}.$$

The signal-flow graph of the four-point DCT of type IV by this algorithm is given in Figure 3.5.

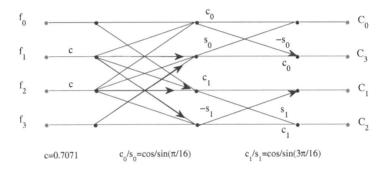

FIGURE 3.5
Signal-flow graph of the four-point DCT of type IV.

Two rotations and the diagonal matrix with two nontrivial coefficients are used in this decomposition. Eight multiplications are required for calculation of the DCT by this decomposition, $\mu(C_{4;IV}) = \mu(T_{\frac{\pi}{16}}) + \mu(T_{\frac{3\pi}{16}}) + 2 = 2 \cdot 3 + 2 = 8$. Note that the multiplication by the matrix

$$\begin{pmatrix} 1 & 1 & 1 & 0 \\ 1 & -1 & -1 & 0 \\ 0 & -1 & 1 & 1 \\ 0 & -1 & 1 & -1 \end{pmatrix} = \begin{pmatrix} 0 & 1 & 1 & 0 \\ 0 & -1 & -1 & 0 \\ 0 & -1 & 1 & 0 \\ 0 & -1 & 1 & 0 \end{pmatrix} + \begin{pmatrix} 1 & 0 & 0 & 0 \\ 1 & 0 & 0 & 0 \\ 0 & 0 & 0 & 1 \\ 0 & 0 & 0 & -1 \end{pmatrix}$$

requires $2 + 4 = 6$ additions. Therefore the number of operations of addition for the four-point DCT of type IV equals $\alpha(C_{4;IV}) = \alpha(T_{\frac{\pi}{16}}) + \alpha(T_{-\frac{3\pi}{16}}) + 6 = 2 \cdot 4 + 6 = 14$. For comparison, we consider the well-known algorithm of decomposition of the type IV DCT matrix [33]:

$$\begin{bmatrix} 1 & & & \\ & 1 & & \\ & & 1 & \\ & & & 1 \end{bmatrix} \begin{bmatrix} 1 & 1 & & \\ & 1 & 1 & \\ 1 & -1 & & \\ 1 & & -1 & \end{bmatrix} \begin{bmatrix} 1 & & & \\ & 1 & & \\ & & 1 & \\ & & & -1 \end{bmatrix} \begin{bmatrix} c_1 & s_1 & & \\ s_1 & -c_1 & & \\ & & c_1 & s_1 \\ & & s_1 & -c_1 \end{bmatrix}$$

$$\times \begin{bmatrix} 1 & & & \\ & 1 & & \\ & & 1 & \\ & & & 1 \end{bmatrix} \begin{bmatrix} 1 & & & \\ & 1 & & \\ & & 1 & \\ & & & 1 \end{bmatrix} \begin{bmatrix} c_2 & -s_2 & & \\ s_2 & c_2 & & \\ & & c_3 & -s_3 \\ & & s_3 & c_3 \end{bmatrix} \begin{bmatrix} 1 & & & \\ & 1 & & \\ & 1 & & \\ & & & 1 \end{bmatrix}$$

where $c_1 = \cos(\pi/8) = 0.9239$, $s_1 = \sin(\pi/8) = 0.3827$, $c_2 = \cos(\pi/16) = 0.9808$, $s_2 = \sin(\pi/16) = 0.1951$, $c_3 = \cos(3\pi/16) = 0.8315$, $s_3 = \sin(3\pi/16) = 0.5556$. This decomposition uses four rotations and arithmetical complexity equals $\mu(C_{4;IV}) = 4\mu(T) = 12$, $\alpha(C_{4;IV}) = 4\alpha(T) + 4 = 16 + 4 = 20$.

Below is the MATLAB-based code for calculating the four-point DCT of type IV by the proposed method in (3.29).

```
% ----------------------------------------------------------
%  demo_dct4type4.m file (library of codes of Grigoryans)
%  Calculation of the 4-point DCT IV with 8 multiplications
      c=1/sqrt(2);
      Q2=diag([ 1 c c 1]);
      Q1=[1  1  1  0
          1 -1 -1  0
          0 -1  1  1
          0 -1  1 -1];
      w0=pi/16;
      c0=cos(w0); s0=sin(w0);
      R03=[c0  0  0 -s0
            0  1  0  0
            0  0  1  0
           s0  0  0  c0];
      w1=3*w0;
      c1=cos(w1); s1=sin(w1);
      R12=[1  0   0  0
           0  c1 -s1 0
           0  s1  c1 0
           0  0   0  1 ];
      DCTtype4=R03*R12*Q1*Q2;
% ----------------------------------------------------------
```

3.1.3 8-point DCT of type IV

The above described method of decomposition of the four-point DCT by the Coxeter-type matrices can be used for other orders of the transform. To show that, we consider the $N = 8$ case, when the calculation of the 8-point DCT of type IV can be performed in parallel by the pair of four rotations.

The matrix $[C_{8;IV}]$ of this transform

$$
\begin{bmatrix}
0.9952 & 0.9569 & 0.8819 & 0.7730 & 0.6344 & 0.4714 & 0.2903 & 0.0980 \\
0.9569 & 0.6344 & 0.0980 & -0.4714 & -0.8819 & -0.9952 & -0.7730 & 0.2903 \\
0.8819 & 0.0980 & -0.7730 & -0.9569 & 0.2903 & 0.6344 & 0.9952 & 0.4714 \\
0.7730 & -0.4714 & -0.9569 & 0.0980 & 0.9952 & 0.2903 & -0.8819 & -0.6344 \\
0.6344 & -0.8819 & -0.2903 & 0.9952 & -0.0980 & -0.9569 & 0.4714 & 0.7730 \\
0.4714 & -0.9952 & 0.6344 & 0.2903 & -0.9569 & 0.7730 & 0.0980 & -0.8819 \\
0.2903 & -0.7730 & 0.9952 & -0.8819 & 0.4714 & 0.0980 & -0.6344 & 0.9569 \\
0.0980 & -0.2903 & 0.4714 & -0.6344 & 0.7730 & -0.8819 & 0.9569 & -0.9952
\end{bmatrix}
$$

is composed by eight cosine coefficients: $c_n = \cos(2n+1)\pi/32$, $n = 0 : 7$. The following eight Coxeter-type matrices correspond to these coefficients:

$$
\mathbf{A}_0 = \begin{bmatrix}
1 & & & & & & & \\
 & & & -1 & & & & \\
 & & & & 1 & & & \\
 & & 1 & & & & & \\
 & 1 & & & & & & \\
 & & & & & -1 & & \\
 & & & & & & 1 & \\
 & & & & & & & -1
\end{bmatrix}, \quad
\mathbf{A}_1 = \begin{bmatrix}
 & 1 & & & & & & \\
1 & & & & & & & \\
 & & & -1 & & & & \\
 & & -1 & & & & & \\
 & & & & & -1 & & \\
 & & & & -1 & & & \\
 & & & & & & & 1 \\
 & & & & & & 1 &
\end{bmatrix}
$$

$$
\mathbf{A}_2 = \begin{bmatrix}
 & 1 & & & & & & \\
 & & & -1 & & & & \\
1 & & & & & & & \\
 & & & & & 1 & & \\
 & & -1 & & & & & \\
 & & & & & & & -1 \\
 & & & & -1 & & & \\
 & & & & & & -1 &
\end{bmatrix}, \quad
\mathbf{A}_3 = \begin{bmatrix}
 & & 1 & & & & & \\
 & & & & & & -1 & \\
 & & & & -1 & & & \\
1 & & & & & & & \\
 & & & & & & & 1 \\
 & & & & & 1 & & \\
 & -1 & & & & & & \\
 & & & 1 & & & &
\end{bmatrix}
$$

$$
\mathbf{A}_4 = \begin{bmatrix}
 & & & 1 & & & & \\
 & & 1 & & & & & \\
 & & & & 1 & & & \\
 & & & & & & -1 & \\
1 & & & & & & & \\
 & 1 & & & & & & \\
 & & & & & & & -1 \\
 & & & & & -1 & &
\end{bmatrix}, \quad
\mathbf{A}_5 = \begin{bmatrix}
 & & & & 1 & & & \\
 & & & -1 & & & & \\
 & & & & & & 1 & \\
 & -1 & & & & & & \\
 & & & & & & & 1 \\
1 & & & & & & & \\
 & & & & & 1 & & \\
 & & 1 & & & & &
\end{bmatrix}
$$

$$
\mathbf{A}_6 = \begin{bmatrix} & & & & 1 & \\ & & & -1 & & \\ & & -1 & & & \\ & & 1 & & & \\ & -1 & & & & \\ & 1 & & & & \\ 1 & & & & & \\ -1 & & & & & \end{bmatrix}, \quad \mathbf{A}_7 = \begin{bmatrix} & & & & & & 1 \\ & & & & & 1 & \\ & & & & 1 & & \\ & & & 1 & & & \\ & & -1 & & & & \\ & 1 & & & & & \\ 1 & & & & & & \end{bmatrix}.
$$

For these Coxeter matrices the following equations hold: $\mathbf{J} = \mathbf{A}_0\mathbf{A}_7 = -\mathbf{A}_7\mathbf{A}_0 = -\mathbf{A}_1\mathbf{A}_6 = \mathbf{A}_6\mathbf{A}_1 = \mathbf{A}_2\mathbf{A}_5 = -\mathbf{A}_5\mathbf{A}_2 = -\mathbf{A}_3\mathbf{A}_4 = \mathbf{A}_4\mathbf{A}_3$, and $\mathbf{A}_n\mathbf{A}_n = \mathbf{I}$, where $\mathbf{J}$ is the diagonal matrix

$$
\mathbf{J} = \begin{pmatrix} & & & & & & & 1 \\ & & & & & & -1 & \\ & & & & & 1 & & \\ & & & & -1 & & & \\ & & & 1 & & & & \\ & & -1 & & & & & \\ & 1 & & & & & & \\ -1 & & & & & & & \end{pmatrix}.
$$

The matrix of the DCT can be written as follows:

$$
\begin{aligned}
[C_{8;IV}] &= c_0\mathbf{A}_0 + c_1\mathbf{A}_1 + c_2\mathbf{A}_2 + c_3\mathbf{A}_3 + c_4\mathbf{A}_4 + c_5\mathbf{A}_5 + c_6\mathbf{A}_6 + c_7\mathbf{A}_7 \\
&= \mathbf{A}_0(c_0\mathbf{I} + c_7\mathbf{J}) + \mathbf{A}_1(c_1\mathbf{I} - c_6\mathbf{J}) + \mathbf{A}_2(c_2\mathbf{I} + c_5\mathbf{J}) + \mathbf{A}_3(c_3\mathbf{I} - c_4\mathbf{J}) \\
&= (c_0\mathbf{I} - s_0\mathbf{J})\mathbf{A}_0 + (c_1\mathbf{I} + s_1\mathbf{J})\mathbf{A}_1 + (c_2\mathbf{I} - s_2\mathbf{J})\mathbf{A}_2 + (c_3\mathbf{I} + s_3\mathbf{J})\mathbf{A}_3
\end{aligned}
\tag{3.18}
$$

since $c_n = s_{7-n} = \sin\left(2(7-n)+1\right)\pi/32$, $n = 0:7$. The X-type matrix

$$
c_0\mathbf{I} - s_0\mathbf{J} = \begin{pmatrix} c_0 & & & & & & & -s_0 \\ & c_0 & & & & & s_0 & \\ & & c_0 & & & -s_0 & & \\ & & & c_0 & s_0 & & & \\ & & & -s_0 & c_0 & & & \\ & & s_0 & & & c_0 & & \\ & -s_0 & & & & & c_0 & \\ s_0 & & & & & & & c_0 \end{pmatrix}
$$

is the combination of four rotations by angles $\pm(\varphi_0 = \pi/32)$ in four subspaces spanned on bases $\{e_0, e_7\}$, $\{e_1, e_6\}$, $\{e_2, e_5\}$, and $\{e_3, e_4\}$, which we write as

$$
c_0\mathbf{I} - s_0\mathbf{J} = \mathbf{R}(-\varphi_0, \varphi_0, -\varphi_0, \varphi_0) = \mathbf{R}_{3,4}\left(-\varphi_0\right)\mathbf{R}_{2,5}\left(\varphi_0\right)\mathbf{R}_{1,6}\left(-\varphi_0\right)\mathbf{R}_{0,7}\left(\varphi_0\right)
$$

or shortly as $c_0\mathbf{I} - s_0\mathbf{J} = \widehat{\mathbf{R}}(\varphi_0)$.

Similar representations have other such X-matrices in (3.18) $c_1\mathbf{I} + s_1\mathbf{J} = \widehat{\mathbf{R}}(\varphi_1) = \mathbf{R}(\varphi_1, -\varphi_1, \varphi_1, -\varphi_1)$, $c_2\mathbf{I} - s_2\mathbf{J} = \widehat{\mathbf{R}}(-\varphi_2) = \mathbf{R}(-\varphi_2, \varphi_2, -\varphi_2, \varphi_2)$,

$c_3\mathbf{I}+s_3\mathbf{J} = \widehat{\mathbf{R}}(\varphi_3) = \mathbf{R}(\varphi_3, -\varphi_3, \varphi_3, -\varphi_3)$, where $\varphi_k = (2k+1)\pi/32, k = 1:3$. Therefore, we can write the matrix of the DCT as

$$[C_{8;IV}] = \widehat{\mathbf{R}}(-\varphi_0)\mathbf{A}_0 + \widehat{\mathbf{R}}(\varphi_1)\mathbf{A}_1 + \widehat{\mathbf{R}}(-\varphi_2)\mathbf{A}_2 + \widehat{\mathbf{R}}(\varphi_3)\mathbf{A}_3. \qquad (3.19)$$

Given any angles a_1, b_1, c_1, d_1 and a_2, b_2, c_2, d_2, the following holds for the rotation matrices:

$$\mathbf{R}(a_1, b_1, c_1, d_1)\mathbf{R}(a_2, b_2, c_2, d_2) = \mathbf{R}(a_1 + a_2, b_1 + b_2, c_1 + c_2, d_1 + d_2).$$

Since, $\varphi_0 + \varphi_3 = \varphi_1 + \varphi_2 = \pi/4$, the matrix of the DCT can be written as

$$[C_{8;IV}] = \widehat{\mathbf{R}}(-\varphi_0)[\mathbf{A}_0 + \widehat{\mathbf{R}}(\varphi_3 + \varphi_0)\mathbf{A}_3] + \widehat{\mathbf{R}}(\varphi_1)[\mathbf{A}_1 + \widehat{\mathbf{R}}(-\varphi_2 - \varphi_1)\mathbf{A}_2]$$
$$= \widehat{\mathbf{R}}(-\varphi_0)[\mathbf{A}_0 + \widehat{\mathbf{R}}(\pi/4)\mathbf{A}_3] + \widehat{\mathbf{R}}(\varphi_1)[\mathbf{A}_1 + \widehat{\mathbf{R}}(-\pi/4)\mathbf{A}_2].$$

Note that $\widehat{\mathbf{R}}(\pi/4) = \frac{1}{\sqrt{2}}(\mathbf{I} + \mathbf{J})$, $\widehat{\mathbf{R}}(-\pi/4) = \frac{1}{\sqrt{2}}(\mathbf{I} - \mathbf{J})$. Therefore, we obtain the following decomposition of the matrix of the eight-point DCT of type IV:

$$[C_{8;IV}] = \widehat{\mathbf{R}}(-\varphi_0)[\mathbf{A}_0 + \frac{1}{\sqrt{2}}(\mathbf{I} + \mathbf{J})\mathbf{A}_3] + \widehat{\mathbf{R}}(\varphi_1)[\mathbf{A}_1 + \frac{1}{\sqrt{2}}(\mathbf{I} - \mathbf{J})\mathbf{A}_2]. \quad (3.20)$$

Below is the MATLAB-based code for calculating the eight-point DCT of type IV by the proposed method.

```
% ------------------------------------------------------------
%  demo_dct4type8.m file (library of codes of Grigoryans)
%  Calculation of the 8-point DCT IV with 40 multiplications
    I=diag([1 1 1 1 1 1 1 1]);
    N=diag([1 -1 1 -1 1 -1 1 -1]);
    N=fliplr(N);
    AO=[1  0  0  0  0  0  0  0
        0  0  0  0  0  0 -1  0  0
        0  0  0  0  0  0  1  0
        0  0  0  0  1  0  0  0
        0  0  0  1  0  0  0  0
        0 -1  0  0  0  0  0  0
        0  0  1  0  0  0  0  0
        0  0  0  0  0  0  0 -1];
    A1=[0  1  0  0  0  0  0  0
        1  0  0  0  0  0  0  0
        0  0  0 -1  0  0  0  0
        0  0 -1  0  0  0  0  0
        0  0  0  0  0 -1  0  0
        0  0  0  0 -1  0  0  0
        0  0  0  0  0  0  0  1
        0  0  0  0  0  0  1  0];
    A2=[0  0  1  0  0  0  0  0
        0  0  0  0 -1  0  0  0
```

```
      1  0  0  0  0  0  0  0
      0  0  0  0  0  0 -1  0
      0 -1  0  0  0  0  0  0
      0  0  0  0  0  0  0 -1
      0  0  0 -1  0  0  0  0
      0  0  0  0  0 -1  0  0];
  A3=[0  0  0  1  0  0  0  0
      0  0  0  0  0  0 -1  0
      0  0 -1  0  0  0  0  0
      1  0  0  0  0  0  0  0
      0  0  0  0  0  0  0  1
      0  0  0  0  0  1  0  0
      0 -1  0  0  0  0  0  0
      0  0  0  0  1  0  0  0];
  w0=pi/32; c0=cos(w0); s0=sin(w0);
  w1=3*w0;  c1=cos(w1); s1=sin(w1);
  S0=c0*I-s0*N;
  S1=c1*I+s1*N;
  d2=1/sqrt(2);
  C=S0*(A0+d2*(I+N)*A3)+S1*(A1+d2*(I-N)*A2);
% modification with 32 multiplications:
  K=A0/A3;
  D=I+d2*(I-N)*K;
  C=S0*A0*D+S1*A1*(2*I-D);
%------------------------------------------------
```

Each of two summands in this sum is composed by four rotations by angle φ_0 or φ_1, which requires four multiplications, or three multiplications when using the lifting schemes. Eight multiplications are also required for multiplication of each of the X-type matrices $(\mathbf{I} + \mathbf{J})$ and $(\mathbf{I} - \mathbf{J})$ with coefficients $1/\sqrt{2}$. Therefore, the required number of multiplications and additions for calculating the eight-point DCT of type IV by (3.20) equal, respectively, $\mu(C_{8;IV}) = 2[4\mu(T) + 8] = 2[4 \cdot 3 + 8] = 40$, $\alpha(C_{8;IV}) = 2[4\alpha(T) + 16] + 8 = 2[4 \cdot 4 + 16] + 8 = 72$.

It is possible to reduce the number of multiplications to 32, by noticing that

$$\mathbf{K} = \mathbf{A}_0\mathbf{A}_3^{-1} = -\mathbf{A}_2\mathbf{A}_1^{-1} = \begin{pmatrix} 0 & 0 & 0 & 1 & 0 & 0 & 0 & 0 \\ 0 & 0 & 0 & 0 & 0 & -1 & 0 & 0 \\ 0 & -1 & 0 & 0 & 0 & 0 & 0 & 0 \\ 0 & 0 & 0 & 0 & 0 & 0 & 0 & 1 \\ 1 & 0 & 0 & 0 & 0 & 0 & 0 & 0 \\ 0 & 0 & 0 & 0 & 0 & 0 & 1 & 0 \\ 0 & 0 & -1 & 0 & 0 & 0 & 0 & 0 \\ 0 & 0 & 0 & 0 & -1 & 0 & 0 & 0 \end{pmatrix}.$$

Therefore, the following decomposition of the matrix of the eight-point DCT of type IV can be derived from (3.20):

$$[C_{8;IV}] = \widehat{\mathbf{R}}(-\varphi_0)\mathbf{A}_0\mathbf{D} + \widehat{\mathbf{R}}(-\varphi_1)\mathbf{A}_1(2\mathbf{I} - \mathbf{D}), \tag{3.21}$$

where the matrix $\mathbf{D} = \mathbf{I} + \frac{1}{\sqrt{2}}(\mathbf{I}-\mathbf{J})\mathbf{K}$. The required numbers of multiplications and additions for calculating the eight-point DCT of type IV by this decomposition equal, respectively, $\mu(C_{8;IV}) = 2[4\mu(T)] + \mu(D) = 2[4 \cdot 3] + 8 = 32$, $\alpha(C_{8;IV}) = 2[4\alpha(T)] + [\alpha(D) + 8] + 8 = 2[4 \cdot 4] + [16 + 8] + 8 = 64$. If we implement the parallel processing of two summands in (3.21), the calculation of the eight-point DCT can be done for the time close to the time of $4\mu(T) + \mu(D) = 20$ multiplications.

The number $\mu(C_{8;IV})$ is greater than 22 multiplications in the known Wang method [60], where six rotations by angles $5\pi/8, 9\pi/8, (4k+1)\pi/32, (k = 0:3)$, are used plus two rotations by $\pi/4$. The factorization of the matrix in this method is the following:

$$[C_{8;IV}] = \mathbf{P}_1 \begin{bmatrix} \mathbf{M} \cdot \mathbf{T}_{\frac{\pi}{32}} & & & \\ & \mathbf{M} \cdot \mathbf{T}_{\frac{5\pi}{32}} & & \\ & & \mathbf{M} \cdot \mathbf{T}_{\frac{9\pi}{32}} & \\ & & & \mathbf{M} \cdot \mathbf{T}_{\frac{13\pi}{32}} \end{bmatrix} \cdot$$

$$\cdot \mathbf{P}_2 \begin{bmatrix} \mathbf{I}_2 & & \\ & \mathbf{I}_2 & \\ & & \mathbf{M} \cdot \mathbf{T}_{\frac{5\pi}{8}} \\ & & & \mathbf{M} \cdot \mathbf{T}_{\frac{9\pi}{8}} \end{bmatrix} \mathbf{P}_3 \begin{bmatrix} \mathbf{I}_2 & & \\ & \mathbf{M} \cdot \mathbf{T}_{\frac{\pi}{4}} & \\ & & \mathbf{I}_2 \\ & & & \mathbf{M} \cdot \mathbf{T}_{\frac{\pi}{4}} \end{bmatrix} \mathbf{D},$$

where the binary matrices $\mathbf{P}_1, \mathbf{P}_2, \mathbf{P}_3$, and $\mathbf{D}$ are defined as

$$\mathbf{P}_1 = \begin{pmatrix} 1&0&0&0&0&0&0&0 \\ 0&0&0&0&0&0&0&1 \\ 0&0&1&0&0&0&0&0 \\ 0&0&0&0&0&1&0&0 \\ 0&0&0&0&1&0&0&0 \\ 0&0&0&1&0&0&0&0 \\ 0&0&0&0&0&0&1&0 \\ 0&1&0&0&0&0&0&0 \end{pmatrix}, \quad \mathbf{P}_2 = \begin{pmatrix} 1&0&0&0&1&0&0&0 \\ 0&1&0&0&0&1&0&0 \\ 0&0&1&0&0&0&1&0 \\ 0&0&0&1&0&0&0&1 \\ 1&0&0&0&-1&0&0&0 \\ 0&1&0&0&0&-1&0&0 \\ 0&0&1&0&0&0&-1&0 \\ 0&0&0&1&0&0&0&-1 \end{pmatrix}$$

$$\mathbf{P}_3 = \begin{pmatrix} 1&0&1&0&0&0&0&0 \\ 0&1&0&1&0&0&0&0 \\ 1&0&-1&0&0&0&0&0 \\ 0&1&0&-1&0&0&0&0 \\ 0&0&0&0&1&0&1&0 \\ 0&0&0&0&0&1&0&1 \\ 0&0&0&0&1&0&-1&0 \\ 0&0&0&0&0&1&0&-1 \end{pmatrix}, \quad \mathbf{D} = \begin{pmatrix} 1&0&0&0&0&0&0&0 \\ 0&0&0&0&0&0&0&1 \\ 0&0&0&1&0&0&0&0 \\ 0&0&0&0&1&0&0&0 \\ 0&1&0&0&0&0&0&0 \\ 0&0&0&0&0&0&1&0 \\ 0&0&1&0&0&0&0&0 \\ 0&0&0&0&0&1&0&0 \end{pmatrix}$$

and the diagonal matrix $\mathbf{M} = \text{diag}\{1, -1\}$. The arithmetical complexity of the eight-point DCT of type IV by this decomposition is estimated as

$$\mu(C_{8;IV}) = 6\mu(T) + 2\mu(T_{\pi/4}) = 6 \cdot 3 + 2 \cdot 2 = 22,$$
$$\alpha(C_{8;IV}) = 6\alpha(T) + 2\alpha(T_{\pi/4}) + 16 = 6 \cdot 4 + 2 \cdot 2 + 16 = 44.$$

3.2   Paired algorithm for the N-point DCT

In this section, the paired representation of the signals is described in the general case $N \geq 8$. The representation is with respect to the cosine transformation. We first consider the N-point DCT in form (3.8), which yields the representation of the transform at points of sets T'_p by the $N/2^k$-point DCT of type IV

$$C_{\overline{(2m+1)p}} = \sum_{t=0:0.5:(N/2-1)} f'_{p,t} \cos_{N/2}\left(\left[m+\frac{1}{2}\right]t\right), \quad m = 0 : (N/2 - 1), \quad (3.22)$$

where integer $k \in [1, r]$ and $r = \log_2 N$. Indeed, for $p = 1$, the variable $t = \left(n + \frac{1}{2}\right) p \bmod 2N = n + \frac{1}{2}$ takes N different values, when $n = 0 : (N-1)$. Then

$$C_{\overline{(2m+1)}} = \sum_{n=0}^{N/2-1} f'_{1,n+1/2} \cos_{N/2}\left(\left[m+\frac{1}{2}\right]\left[n+\frac{1}{2}\right]\right), \quad m = 0 : (N/2 - 1).$$

When $p = 2^k$, $k = 1 : (r - 1)$, the variable

$$t = t_n = \left(n + \frac{1}{2}\right) p \bmod 2N = \left(2^k n + 2^{k-1}\right) \bmod 2N, \quad (n = 0 : (N - 1))$$

takes 2^{r+1-k} different values. Therefore, the basis functions of the cosine transformation in (3.22) can be written as follows:

$$\cos_{N/2}\left(\left[m+\frac{1}{2}\right]t\right) = \cos_{N/2}\left(\left[m+\frac{1}{2}\right]\left(2^k n + 2^{k-1}\right)\right) =$$
$$= \cos_{N/2}\left(\left[m+\frac{1}{2}\right]2^k\left[n+\frac{1}{2}\right]\right) = \cos_{N/2^{k+1}}\left(\left[m+\frac{1}{2}\right]\left[n+\frac{1}{2}\right]\right).$$

The right part of (3.22) can be considered as the $N/2^{k+1}$-point DCT of type IV

$$C_{\overline{(2m+1)2^k}} = \sum_{n=0}^{N/2^{k+1}-1} f'_{2^k,t_n} \cos_{N/2^{k+1}}\left(\left[m+\frac{1}{2}\right]\left[n+\frac{1}{2}\right]\right), \qquad (3.23)$$

$m = 0 : (N/2^{k+1} - 1)$, of the sequence $f'_{T'_{2^k}} = \left\{ f'_{2^k,t_0}, f'_{2^k,t_1}, ..., f'_{2^k,t_{N/2^{k+1}-1}} \right\}$.
The set of such sequences

$$\left\{ f'_{T'_1}, f'_{T'_2}, f'_{T'_4}, f'_{T'_8}, \ldots, f'_{T'_{N/2}}, f'_{T'_0} \right\} \qquad (3.24)$$

is called the *paired representation* of the signal f_n with respect to the cosine transform. The transformation performing this representation

$$\chi'_N : f \rightarrow \left\{ f'_{T'_1}, f'_{T'_2}, f'_{T'_4}, f'_{T'_8}, ..., f'_{T'_{N/2}}, f'_{T'_0} \right\}$$

is called a *paired transformation* with respect to the cosine transformation.

The matrix of the paired transformation can be defined by using the following recursive procedure:

$$[\chi'_N] = \begin{bmatrix} I_{N/2} & -J_{N/2} \\ \begin{bmatrix} \chi'_{N/2} \end{bmatrix} \begin{bmatrix} \chi'_{N/2} \end{bmatrix} J_{N/2} \end{bmatrix}.$$

By means of the paired transformation χ'_N, the N-point DCT of type II is decomposed by one $N/2$-, $N/4$-,..., and 1-point DCTs of type IV, and one 1-point DCT. It should be noted that similar decompositions can be applied to $N/2^n$-point DCTs of type IV, where $n = 1 : (r-2)$. Indeed, it was shown in [58] that the $N/2^n$-point DCT of type IV can be represented by two $N/2^{n+1}$-point type-II cosine and sine transformations. And for the sine transformation, the paired representation is similar to the above described paired representation with respect to the cosine transformation.

In matrix form, the decomposition of the N-point DCT is described as

$$[\mathcal{C}_N] = \begin{bmatrix} \begin{bmatrix} \mathcal{C}_{N/2|IV} \end{bmatrix} & & & & \\ & \begin{bmatrix} \mathcal{C}_{N/4|IV} \end{bmatrix} & & & \\ & & \begin{bmatrix} \mathcal{C}_{N/8|IV} \end{bmatrix} & & \\ & & & \ddots & \\ & & & & \begin{bmatrix} \mathcal{C}_{1|IV} \end{bmatrix} \\ & & & & & 1 \end{bmatrix} [\chi'_N]$$

$$= \begin{bmatrix} \begin{bmatrix} \mathcal{C}_{N/2|IV} \end{bmatrix} & \\ & \begin{bmatrix} \mathcal{C}_{N/2} \end{bmatrix} \end{bmatrix} \begin{bmatrix} I_{N/2} & -J_{N/2} \\ I_{N/2} & J_{N/2} \end{bmatrix}.$$

The second equality shows that the N-point DCT is also decomposed by the $N/2$-point DCT and the $N/2$-point DCT of type IV.

3.2.1   Paired functions

The basic paired functions of the N-point discrete paired transformation with respect to the cosine transformation are defined in the following way. For given $p \in X$ and $t \in Y$, let $\chi_{p,t}(n)$ be the binary function

$$\chi_{p,t}(n) = \begin{cases} 1; & \left(n + \frac{1}{2} \right) p = t \bmod 2N, \\ 0; & \text{otherwise}, \end{cases} \quad n = 0 : (N - 1). \quad (3.25)$$

In other words, $\chi_{p,t}$ is the characteristic function of the set $V_{p,t}$.

DEFINITION 3.1   The function

$$\chi'_{p,t}(n) = \chi_{p,t}(n) - \chi_{p,t+N}(n) - \chi_{p,N-t}(n) + \chi_{p,2N-t}(n) \qquad (3.26)$$

$$= \begin{cases} 1; \ \left(n + \frac{1}{2}\right)p = t \text{ or } (2N - t) \bmod 2N \\ -1; \ \left(n + \frac{1}{2}\right)p = (N - t) \text{ or } (t + N) \bmod 2N \\ 0; \text{ otherwise} \end{cases} \qquad (3.27)$$

where $n = 0 : (N - 1)$, is called *the paired function* with number (p, t).

To compose a complete set of paired functions, we consider the following partition of X: $\sigma' = \left(T'_1, T'_2, T'_4, \ldots, T'_{N/2}, T'_0\right)$. As was mentioned above, the variable $t = (n + 1/2)p \bmod 2N$ takes N values of $n + 1/2$ for $p = 0$, when $n = 0 : (N - 1)$, and $2N/p$ values of $\left(2^k n + 2^{k-1}\right)$ for $n = 0 : (2^{r-k+1} - 1)$, when $p = 2^k$, $1 \le k \le r - 1$. Therefore, the complete set of paired functions is defined as

$$\left\{ \{\chi'_{2^k, \tau + 2^{k-1}}; \ \tau = 0 : 2^k : (N/2 - 1), \ \ k = 0 : (r - 1)\}, 1 \right\}. \qquad (3.28)$$

For instance, when $N = 8, 4$, and 2, the corresponding sets of basic paired functions are $\{\chi'_{1,0.5}, \chi'_{1,1.5}, \chi'_{1,2.5}, \chi'_{1,3.5}, \chi'_{2,1}, \chi'_{2,3}, \chi'_{4,2}, \chi'_{0,0}\}$, $\{\chi'_{1,0.5}, \chi'_{1,1.5}, \chi'_{2,1}, \chi'_{0,0}\}$, and $\{\chi'_{1,0.5}, \chi'_{0,0}\}$, where $\chi'_{0,0}(n) \equiv 1$. Figure 3.6 illustrates these three sets of basic paired functions. These complete sets of functions are given in the time-intervals $[0, 1]$, $[0, 3]$, and $[0, 7]$, respectively. Similar to the Fourier transform case, the second part of functions for the 8-point paired transformation represents the system of functions for the 4-point transformation, which are periodically extended in the second period $[4, 7]$. The second part of functions for the 4-point paired transformation equals the paired functions of the 2-point paired transformation, which are extended periodically in the second period $[2, 3]$.

3.2.2   Complexity of the calculation

The number of operations of multiplication and addition required for calculating the N-point DCT by the paired transform can be estimated by the following recursive formulas:

$$\mu(C_N) = \mu(C_{N/2|IV}) + \mu(C_{N/4|IV}) + \ldots + \mu(C_{2|IV}) + \mu(C_{1|IV})$$
$$\alpha(C_N) = \alpha(C_{N/2|IV}) + \alpha(C_{N/4|IV}) + \ldots + \alpha(C_{2|IV}) + \alpha(C_{1|IV}) + 2N - 2$$

or $\mu(C_N) = \mu(C_{N/2|IV}) + \mu(C_{N/2})$, $\alpha(C_N) = \alpha(C_{N/2|IV}) + \alpha(C_{N/2}) + N$. Here $\mu(C_{M|IV})$, $\mu(C_M)$ and $\alpha(C_M)$, $\alpha(C_{M|IV})$ denote, respectively, the numbers of multiplications and additions required for calculating the M-point DCT and DCT of type IV. To derive the estimates for the 2^k-point DCT, we use the following estimates given in [54]: $\mu(C_{2^k|IV}) = 2^{k-1}(k + 2)$, $\alpha(C_{2^k|IV}) = 2^{k-1}3k$. For instance, $\mu(C_4) = \mu(C_{2|IV}) + \mu(C_{1|IV}) = 3 + 1 = 4$, and if

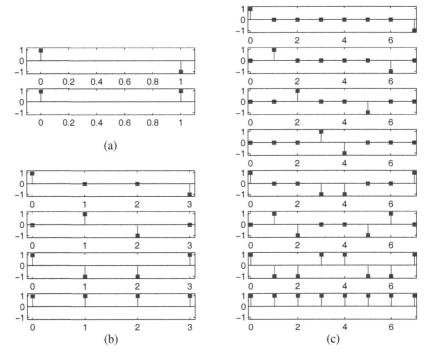

FIGURE 3.6
The complete sets of the discrete paired functions for (a) $N = 2$, (b) $N = 4$, and (b) $N = 8$.

$N = 8$, then $\mu(C_8) = \mu(C_{4|IV}) + \mu(C_4) = 8 + 4 = 12$. If $N = 16$, then $\mu(C_{16}) = \mu(C_{8|IV}) + \mu(C_8) = 20 + 12 = 32$.

In the general $N \geq 2$ case, the following estimations hold: $\mu(C_N) = \mu(C_{N/2}) + 2^{r-2}(r + 1)$ and $\alpha(C_N) = \alpha(C_{N/2}) + 2^{r-2}3(r - 1) + N$. Therefore, $\mu(C_N) = Nr/2$ and $\alpha(C_N) = 3/2Nr - N + 1$, and $\mu(C_2) = 1$, $\alpha(C_2) = 2$. The radix-2 algorithm for the N-point DCT uses the same number of operations of multiplication and addition [55, 62]. These estimations for $N = 2^r$, $r = 2 : 11$, as well as the number of additions required for calculating the paired transform, are given in Table 3.1.

TABLE 3.1
Number of operations of multiplication and addition

N	4	8	16	32	64	128	256	512	1024	2048
$\mu(C_N)$	4	12	32	80	192	448	1024	2304	5120	11264
$\alpha(C_N)$	9	29	81	209	513	1217	2817	6401	14337	31745
$\alpha(\chi'_N)$	6	14	30	62	126	254	510	1022	2046	4094

3.3 Codes for the paired transform

Below are examples of MATLAB-based codes for computing the discrete paired transform (and its matrix) with respect to the cosine transform.

```
% ------------------------------------------------------------------
%  demo_pdct.m file of programs (library of codes of Grigoryans)
%  List of codes for processing signals of length N=2^r, r>1:
%     1-D fast direct paired transform   -  'dctpaired_1d.m'
%     matrix of the paired transformation - 'dctpaired_matrix.m'
%     call: dctpaired.m
      function y=dctpaired(x)
      N=length(x);
      a=1;
      % a=1/sqrt(2);
      if N==1
         y=x;
      else
         N2=bitshift(N,-1);
         x1=x(1:N2);
         x2=x(N:-1:N2+1);
         y1=x1+x2;
         y2=x1-x2;
         y=[y2 dctpaired(y1)]*a;
      end;

%     call: dctpaired_matrix.m
      function T=dctpaired_matrix(N)
      T=zeros(N);
      for k=1:N
          y=zeros(1,N);
          y(k)=1;
          T(:,k)=dctpaired(y);
      end;
```

3.4 Reversible integer DCT

The fast DCT of large order requires a great number of float-point multiplications even in a case most important in practice, when the input data are integer numbers. The integer-to-integer discrete cosine transform without float-point multiplication has become popular in recent years [34]-[38]. The study of integer transforms shows that they can be used in lossless image coding, mobile computing, filter banks, and other areas. One can mention the integer wavelet transforms based on the lifting scheme [39, 40], the reversible transform coding [41, 42], the decomposition of the DCT by the Walsh-Hadamard transform [41, 43, 44]. These methods use effective decompositions by transforms of small orders and then the diagonal matrices in the decomposition with trigonometric coefficients are substituted by products of lifting matrices, floor functions, or simple 2-point integer transforms.

We briefly discuss main methods of integer-to-integer transforms, which are based on the lifting scheme with flooring function. We discuss a general parameterized integer-to-integer discrete cosine transform when the parameter relates to the operation of floor function. Different examples for the 2-, 4-, and 8-point integer reversible and inverse DCTs of types II and IV are analyzed in detail, and optimal parameters for estimation of these transforms by integer DCTs are given.

3.4.1   Integer four-point DCTs

The lifting schemes can be used for calculating the integer approximation of DCTs of types II and IV, when using the factorization of the transforms and all rotations in the factorization are substituted by integer lifting operations.

As an example, we first analyze the application of the lifting scheme for the four-point DCT of type IV, the signal-flow graph of which is given in Figure 3.5. The factorization of the matrix of this transform is described as

$$[C_{4;IV}] = \mathbf{R}_{0,3}(\frac{\pi}{16})\mathbf{R}_{1,2}(\frac{3\pi}{16})\mathbf{Q}_1\mathbf{Q}_2. \tag{3.29}$$

Two lifting schemes can be used for approximation of the Givens rotations by the angles $\varphi_0 = \pi/16$ and $\varphi_1 = 3\pi/16$. The diagonal matrix $\mathcal{Q}_2$ has two coefficients $c = 1/\sqrt{2}$. We therefore need two schemes for multiplications of inputs f_1 and f_2 by the coefficient c. The one-to-two integer transform $\mathcal{A}$, namely, the integer transform with one control bit (TOCB) can be used for such a purpose. For the factor $c = 1/\sqrt{2}$ and integer $x > 0$, this integer transformation with one control bit is defined as

$$\mathcal{A} = \mathcal{A}_c : x \rightarrow \begin{cases} \vartheta_0 = [cx], \\ \vartheta_1 = \dfrac{1 + \text{sign}(cx - \vartheta_0)}{2}, \end{cases}$$

where ϑ_0 is the integer approximation of the multiplication cx, and ϑ_1 is the control bit.

The signal-flow graph of the integer approximation of the four-point DCT of type IV with two lifting schemes and two integer multiplications with control bits α_1 and α_2 is shown in Figure 3.7. The calculation uses the operations of multiplication and addition in numbers $\mu(C_{4;IV}) = 2 + 2(3) = 8$ and $\alpha(C_{4;IV}) = 6 + 2(4) = 14$, respectively.

The inverse integer transform performs first two rotations by angles $-\pi/16$ and $-3\pi/16$. Then the outputs of these rotations are processed in accordance

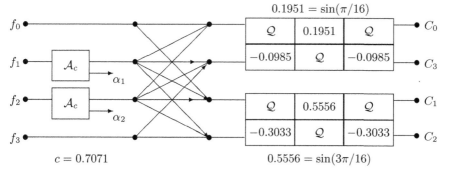

FIGURE 3.7

Signal-flow graph of the integer approximation 4-point DCT of type IV.

with the inverse matrix

$$
\begin{pmatrix} 1 & 1 & 1 & 0 \\ 0 & -1 & 1 & -1 \\ 1 & -1 & -1 & 0 \\ 0 & -1 & 1 & 1 \end{pmatrix}^{-1} = \begin{pmatrix} \frac{1}{2} & 0 & 0 & 0 \\ 0 & \frac{1}{4} & 0 & 0 \\ 0 & 0 & \frac{1}{4} & 0 \\ 0 & 0 & 0 & \frac{1}{2} \end{pmatrix} \begin{pmatrix} 1 & 0 & 1 & 0 \\ 1 & -1 & -1 & -1 \\ 1 & 1 & -1 & 1 \\ 0 & -1 & 0 & 1 \end{pmatrix}.
$$

Two inverse integer operations are also needed for approximating the inverse transform $cx \rightarrow x$, for which the control bit is used. The signal-flow graph of the integer approximation of the inverse four-point DCT of type IV with two lifting schemes is shown in Figure 3.8.

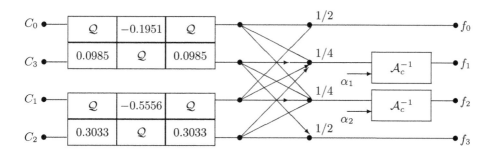

FIGURE 3.8

Signal-flow graph of the integer approximation of the inverse four-point DCT of type IV.

We also consider possible applications of the three-step lifting schemes for calculating the four-point DCT of type II. The factorization of the matrix $[\mathcal{C}_4]$

by the paired transform is given in (3.12) and can be written as

$$
[\mathcal{C}_4] = \mathbf{diag}
\begin{bmatrix} 1 \\ -1 \\ 1 \\ 1 \end{bmatrix}
\begin{bmatrix}
0.9239 & 0.3827 & 0 & 0 \\
-0.3827 & 0.9239 & 0 & 0 \\
0 & 0 & 1 & 1 \\
0 & 0 & 0.7071 & -0.7071
\end{bmatrix}
\begin{bmatrix}
1 & 0 & 0 & -1 \\
0 & 1 & -1 & 0 \\
1 & 0 & 0 & 1 \\
0 & 1 & 1 & 0
\end{bmatrix}
\tag{3.30}
$$

The matrix (2×2) in the left corner of the 2nd matrix represents the Givens rotation by the angle $\varphi = \pi/8$. For this rotation, the lifting scheme can be used as shown in Figure 3.9. The integer approximation of the multiplication

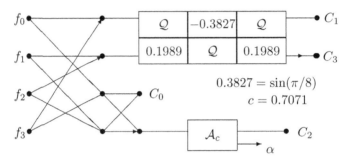

FIGURE 3.9
Block diagram of the four-point integer DCT of type II.

by the coefficient $c = 0.7071$ is also required. Four operations of multiplication and $(4 + 2) + 4 = 10$ additions are used in this calculation; $\mu(C_{4;II}) = 4$ and $\alpha(C_{4;II}) = 10$.

Note that in the definition of the cosine transform, the first component, C_0, was not normalized; this is why the third row of the second matrix in (3.30) consists of 1s. The necessity in using the one-point integer transform can be dropped if we use the normalized factors $k(p)$ in the definition (3.2) of the DCT. In this case, the matrix of the DCT is decomposed by the paired transform as follows:

$$
[\mathcal{C}_4] =
\begin{bmatrix}
0.9239 & 0.3827 & -0.3827 & -0.9239 \\
0.3827 & -0.9239 & 0.9239 & -0.3827 \\
0.7071 & 0.7071 & 0.7071 & 0.7071 \\
0.7071 & -0.7071 & -0.7071 & 0.7071
\end{bmatrix}
$$

$$
=
\begin{bmatrix}
0.9239 & 0.3827 & 0 & 0 \\
0.3827 & -0.9239 & 0 & 0 \\
0 & 0 & 0.7071 & 0.7071 \\
0 & 0 & 0.7071 & -0.7071
\end{bmatrix}
\begin{bmatrix}
1 & 0 & 0 & -1 \\
0 & 1 & -1 & 0 \\
1 & 0 & 0 & 1 \\
0 & 1 & 1 & 0
\end{bmatrix}
$$

$$
\mathbf{diag}\begin{bmatrix} 1 \\ -1 \\ 1 \\ -1 \end{bmatrix}\begin{bmatrix} 0.9239 & 0.3827 & 0 & 0 \\ -0.3827 & 0.9239 & 0 & 0 \\ 0 & 0 & 0.7071 & 0.7071 \\ 0 & 0 & -0.7071 & 0.7071 \end{bmatrix}\begin{bmatrix} 1 & 0 & 0 & -1 \\ 0 & 1 & -1 & 0 \\ 1 & 0 & 0 & 1 \\ 0 & 1 & 1 & 0 \end{bmatrix} \tag{3.31}
$$

Two rotations are required for this decomposition, one rotation by angle $-\pi/8$ and another by $-\pi/4$. The use of lifting schemes for these rotations leads to the integer approximation of the four-point DCT, as shown in Figure 3.10. The

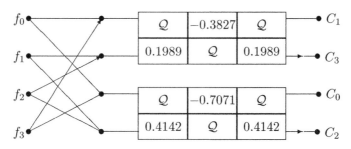

FIGURE 3.10
Signal-flow graph of the integer approximation of the 4-point DCT of type II.

arithmetical complexity of this approximation equals $\mu(C_{4,Int}) = \mu(R_{-\frac{\pi}{8}}) + \mu(R_{-\frac{\pi}{4}}) = 3 + 2 = 5$, $\alpha(C_{4,Int}) = 4 + \alpha(R_{-\frac{\pi}{8}}) + \alpha(R_{-\frac{\pi}{4}}) = 4 + 4 + 2 = 10$, plus six operations of rounding.

3.4.2 Integer eight-point DCT

Using the above described integer approximations of the four-point DCT of types II and IV, we can propose the integer eight-point DCT. For that we use the following decomposition of the DCT by the paired transform:

$$
[C_8] = \begin{bmatrix} [C_{4|IV}] & \\ & [C_4] \end{bmatrix}\begin{bmatrix} I_4 & -J_4 \\ I_4 & J_4 \end{bmatrix}.
$$

The integer eight-point transform thus can be defined as the transform calculated through the following decomposition by the integer four-point DCTs:

$$
[C_{8,Int}] = \begin{bmatrix} [C_{4,Int|IV}] & \\ & [C_{4,Int}] \end{bmatrix}\begin{bmatrix} I_4 & -J_4 \\ I_4 & J_4 \end{bmatrix}.
$$

The signal-flow graph of the integer eight-point DCT which is based on this decomposition is given in Figure 3.11. Four rotations are used and three TOCBs with one control bit each. Therefore, the arithmetical complexity of the proposed integer approximation of the eight-point DCT equals $\mu(C_{8,Int}) = \mu(C_{4|IV}) + \mu(C_{4|II}) = 8 + 4 = 12$, $\alpha(C_{8,Int}) = 8 + \alpha(C_{4|IV}) + \alpha(C_{4|II}) =$

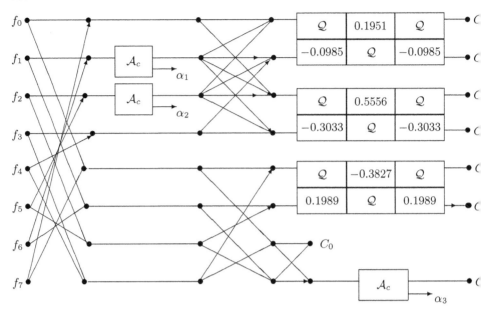

FIGURE 3.11
Signal-flow graph of the eight-point integer DCT of type II.

$8 + 14 + 10 = 32$, plus nine operations of rounding for three rotations and three roundings for the TOCBs.

For comparison, we consider the known method [41] of decomposition of the eight-point DCT by two-point DCTs and one four-point DCT. The diagram of this algorithm is shown in Figure 3.12. Eight two-point DCTs are used in this decomposition, namely, seven two-point DCTs of type II each and one two-point DCT of type IV, and one four-point DCT of type IV. This diagram can be used for integer approximation of the eight-point DCT, if we consider all short DCTs to be integer transforms. The implementation of lifting schemes for such decomposition leads to the following arithmetical complexity of the method: $\mu(C_{8,Int}) = 7\mu(C_{2|II}) + \mu(C_{2|IV}) + \mu(C_{4|IV}) = 7 \cdot 2 + 3 + 8 = 25$, $\alpha(C_{8,Int}) = 7\alpha(C_{2|II}) + \alpha(C_{2|IV}) + \alpha(C_{4|IV}) = 7 \cdot 2 + 4 + 14 = 32$, plus operations of rounding in number $8 \cdot 3 + 8 = 32$.

3.5 Method of nonlinear equations

We consider the well-known concept of the integer transform, which is used to approximate the discrete cosine transforms by solving a system of nonlinear

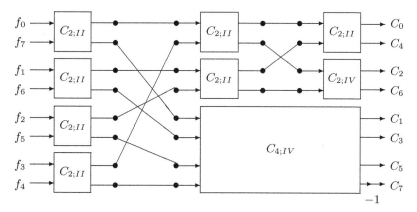

FIGURE 3.12
Block scheme of the 8-point integer DCT of type II.

equations with the floor function. This approach can also be applied for other unitary transforms.

Let f_n be an input sequence of length N, i.e., $n = 0 : (N-1)$. We denote by ϑ_p the components of the N-point discrete cosine transforms of types II and IV, which respectively are calculated by

$$\vartheta_{p,II} = \sum_{n=0}^{N-1} f_n \cos\left(\frac{\pi(n+1/2)p}{N}\right) = \sum_{n=0}^{N-1} f_n \cos_N\left(\left[n + \frac{1}{2}\right]p\right), \quad (3.32)$$

$$\vartheta_{p,IV} = \sum_{n=0}^{N-1} f_n \cos\left(\frac{\pi(n+1/2)(p+1/2)}{N}\right) = \sum_{n=0}^{N-1} f_n \cos_N\left(\left[n + \frac{1}{2}\right]\left[p + \frac{1}{2}\right]\right),$$
$$p = 0 : (N-1),$$
$$(3.33)$$

where $\cos_N(t)$ denotes the basis function $\cos(\pi t/N)$. The N-point reversible integer transform of the input f_n is the set of coefficients $[\vartheta_1, \vartheta_2, \ldots, \vartheta_N]$ calculated by the following system of nonlinear equations:

$$\vartheta_j = f_j + \left\lfloor \sum_{i=0}^{j-1} c_{i,j}\vartheta_i + \sum_{i=j+1}^{N-1} c_{i,j}f_i + 0.5 \right\rfloor, \quad j = 0 : (N-1),$$
$$\vartheta_N = \vartheta_0 + \left\lfloor \sum_{i=1}^{N-1} c_{i,N}\vartheta_i + 0.5 \right\rfloor,$$
$$(3.34)$$

where $\lfloor \cdot \rfloor$ denotes the floor function. The coefficients $c_{i,j}$ are defined by comparing equations (3.32) and (3.34), or (3.33) and (3.34) for the DCT of type II or IV, respectively.

The algorithm of calculation of the inverse N-point transform is reduced to the solution of the same system (3.34) of nonlinear equations but considered with respect to unknown variables x_j, $j = 0 : (N - 1)$. In other words, the following system of equations is considered

$$\vartheta_0 = \vartheta_N - \left\lfloor \sum_{i=1}^{N-1} c_{i,N}\vartheta_i + 0.5 \right\rfloor,$$

$$f_j = \vartheta_j - \left\lfloor \sum_{i=0}^{j-1} c_{i,j}\vartheta_i + \sum_{i=j+1}^{N-1} c_{i,j}f_i + 0.5 \right\rfloor, \tag{3.35}$$

where $j = (N - 1), (N - 2), ..., 2, 1, 0$ and $[f_j]$ are the output and $[\vartheta_j]$ are the input of the inverse transform. The coefficients $c_{i,j}$ play an important role and should be carefully found. The described $N = 2$ case generalizes the concept of the three-step lifting scheme. In the general case $N \geq 2$, the calculation of those coefficients is not a simple task and requires much handwritten work. We now show a way of calculating these coefficients.

3.5.1   Calculation of coefficients

The coefficients $c_{i,j}$ are calculated by comparing the matrices of the N-point reversible DCT and original N-point DCT. Each element of the matrix of the discrete cosine transform is compared with the corresponding element of the matrix of the reversible integer cosine transform. That leads in general to a large system of equations to be solved with respect to coefficients $c_{i,j}$. Then the obtained values of coefficients are substituted in equations (3.34) and (3.35) for both integer reversible and inverse discrete cosine transforms. We now describe briefly the cases of the most interest when $N = 2, 4$, and 8.

For the $N = 2$ case, the matrix of the DCT equals

$$[C_{2;II}] = \begin{bmatrix} a_{0,0} & a_{0,1} \\ a_{1,0} & a_{1,1} \end{bmatrix} = \begin{bmatrix} a & a \\ a & -a \end{bmatrix}, \tag{3.36}$$

where $a = 1/\sqrt{2}$. The system of equations (3.34) is described as

$$\vartheta_0 = f_0 + \lfloor c_{1,0}f_1 + 0.5 \rfloor,$$
$$\vartheta_1 = f_1 + \lfloor c_{0,1}\vartheta_0 + 0.5 \rfloor = f_1 + \lfloor c_{0,1}(f_0 + \lfloor c_{1,0}f_1 + 0.5 \rfloor) + 0.5 \rfloor$$
$$\vartheta_2 = \vartheta_0 + \lfloor c_{1,2}\vartheta_1 + 0.5 \rfloor = \vartheta_0 + \lfloor c_{1,2}(f_1 + \lfloor c_{0,1}(f_0 + \lfloor c_{1,0}f_1 + 0.5 \rfloor) + 0.5 \rfloor) + 0.5 \rfloor$$

Omitting the rounding parameter 0.5 and floor function, we obtain the following system:

$$\hat{\vartheta}_0 = f_0 + c_{1,0}f_1,$$
$$\hat{\vartheta}_1 = f_1 + c_{0,1}(f_0 + c_{1,0}f_1) = c_{0,1}f_0 + (1 + c_{0,1}c_{1,0})f_1,$$
$$\hat{\vartheta}_2 = \hat{\vartheta}_0 + c_{1,2}(f_1 + c_{0,1}(f_0 + c_{1,0}f_1)) = (1 + c_{1,2}c_{0,1})f_0$$
$$+ (c_{1,0} + c_{1,2}(1 + c_{0,1}c_{1,0}))f_1$$

which in matrix form equals

$$\begin{bmatrix} \hat{\vartheta}_1 \\ \hat{\vartheta}_2 \end{bmatrix} = \begin{bmatrix} c_{0,1} & 1 + c_{0,1}c_{1,0} \\ 1 + c_{1,2}c_{0,1} & c_{1,0} + c_{1,2}(1 + c_{0,1}c_{1,0}) \end{bmatrix} \begin{bmatrix} f_0 \\ f_1 \end{bmatrix}. \tag{3.37}$$

Equating elements of the matrix of (3.37) with the corresponding elements of the matrix of (3.36), we obtain the following equations in terms of $c_{1,0}$, $c_{0,1}$, and $c_{1,2}$:

$$c_{0,1} = a, \ 1 + c_{0,1}c_{1,0} = a, \ 1 + c_{1,2}c_{0,1} = a, \ c_{1,0} + c_{1,2}a = -a. \quad (3.38)$$

The solution of this system equals $c_{0,1} = 0.7071$, $c_{1,0} = c_{1,2} = -0.4142$. Substituting the obtained coefficients in (3.34), we obtain the reversible integer discrete cosine transform of type II algorithm

$$\begin{cases} \vartheta_0 = f_0 + \lfloor -0.4142f_1 + 0.5 \rfloor, & \vartheta_1 = f_1 + \lfloor 0.7071\vartheta_0 + 0.5 \rfloor, \\ \vartheta_2 = \vartheta_0 + \lfloor -0.4142\vartheta_1 + 0.5 \rfloor. \end{cases}$$

One can notice that this system describes the three-step lifting scheme, with quantization operation $\mathcal{Q}(x) = \lfloor x + 0.5 \rfloor$ (see for comparison the two-step lifting scheme in Figure 3.10).

In a similar way, equating all elements of the matrix of (3.37) with the corresponding elements of the matrix of the two-point integer DCT of type IV, we obtain the following coefficients: $c_{0,1} = 0.9238$, $c_{1,0} = c_{1,2} = -0.6682$. The reversible integer DCT of type IV is thus calculated by

$$\begin{cases} \vartheta_1 = f_1 + \lfloor 0.9238\vartheta_0 + 0.5 \rfloor, \\ \vartheta_2 = \vartheta_0 + \lfloor -0.6682\vartheta_1 + 0.5 \rfloor, \end{cases} \quad \text{where} \quad \vartheta_0 = f_0 + \lfloor -0.6682f_1 + 0.5 \rfloor.$$

For the 4-point integer DCT of type II, the coefficients $c_{i,j}$ are defined by

$$\begin{array}{lll} c_{0,1} = 0.5, & c_{1,0} = -1, & c_{0,2} = 0.19135, \\ c_{1,2} = 0.9239, & c_{2,0} = -9.0543, & c_{2,1} = 5.0272, \\ c_{0,3} = -0.2706, & c_{1,3} = -3.7207, & c_{2,3} = 4.0272, \\ c_{3,0} = -14.7497 & c_{3,1} = 7.8748, & c_{3,2} = 1.7071, \\ c_{1,4} = 11.9021, & c_{2,4} = -12.4682, & c_{3,4} = 1.8477, \end{array} \quad (3.39)$$

and for the 4-point integer DCT of type IV we obtain the following coefficients:

$$\begin{array}{ll} c_{1,0} = -c_{3,4} = -0.5942, & c_{2,0} = c_{2,4} - 4.0250, \\ c_{3,0} = c_{1,4} = -1.435, & c_{0,1} = c_{0,3} = 0.6935, \\ c_{0,2} = 0.4413, & c_{1,2} = c_{3,2} = 0.2114, \\ c_{1,3} = -c_{3,1} = -1.133, & c_{2,1} = c_{2,3} = 3.185. \end{array} \quad (3.40)$$

It should be noted that when substituting the values of coefficients $c_{i,j}$ of (3.39) or (3.40) respectively in the system of equations (3.34) or (3.35), for calculating the corresponding 4-point integer DCT of type II or IV, the sign of the last output of the transform should be changed, $\vartheta_4 = -\vartheta_4$. Otherwise, the last row of the matrix of the transform will be opposite in sign to the last row of the matrix of the 4-point DCT. If the 4-point DCT is used in the decomposition of the 8-point DCT as shown in Figure 3.11, then it should also be considered that $\vartheta_8 = -\vartheta_8$ (or $C_7 = -C_7$).

3.5.2   Error of approximation

It can be seen from the above described construction of the reversible integer DCT by (3.34) and (3.35) that the number 0.5 is added and then the floor function is used. We consider the general case of an integer transform, by introducing a parameter α which is not necessarily 0.5 [14]. The parameterized integer transform is defined by

$$
\vartheta_j = f_j + \left\lfloor \sum_{i=0}^{j-1} c_{i,j}\vartheta_i + \sum_{i=j+1}^{N-1} c_{i,j}f_i + \alpha \right\rfloor, \quad j = 0:(N-1)
$$

$$
\vartheta_N = \vartheta_0 + \left\lfloor \sum_{i=1}^{N-1} c_{i,N}\vartheta_i + \alpha \right\rfloor.
$$

(3.41)

The values of parameter α are considered in the interval $[0,1]$. The parameterized inverse N-point integer transform algorithm is defined by

$$
\vartheta_0 = \vartheta_N - \left\lfloor \sum_{i=1}^{N-1} c_{i,N}\vartheta_i + \alpha \right\rfloor
$$

$$
f_j = \vartheta_j - \left\lfloor \sum_{i=0}^{j-1} c_{i,j}\vartheta_i + \sum_{i=j+1}^{N-1} c_{i,j}f_i + \alpha \right\rfloor,
$$

(3.42)

when $j = (N-1), (N-2), ..., 2, 1, 0$. The goal of introducing α parameter is to investigate the values of the parameter that yield an integer reversible transform, as well as to find the best or optimal values of the parameter providing the best approximation of the integer transform to the original transform. We would like also to know if the value 0.5 is a good choice for this parameter in the general case of inputs. Here we dwell on the discrete cosine transforms of types II and IV. As a criterion of approximation we consider the root-mean-square error which as the function of α is analyzed for different inputs and orders of the DCT. We now present a few results when applying the parameterized integer transform for $N = 2, 4$, and 8.

The root-mean-square (RMS) error for different N-point samples $f = \{f_n\}$ is defined by

$$
RMSE(f) = \sqrt{\frac{1}{N} \sum_{j=0}^{N-1} (\theta_j - \vartheta_j)^2},
$$

(3.43)

where θ_j is the original DCT and ϑ_j is an integer parameterized DCT of f. RMS error for $N = 2, 4$, and 8 can be calculated easily for different values of α. By varying the value of α, the RMS error function can be plotted and analyzed. We now consider a few examples of the error function, which has been calculated for different inputs for the $N = 2, 4$, and 8 cases. From the figures given below, one can see that there are values of α that differ from 0.5

and give minimum errors between the DCT and reversible integer DCT. Such optimum values of α can be found for each of the considered cases of inputs. Figure 3.13 shows examples of approximation of the 2-point DCT of type II, for given inputs. It can be seen from Figure 3.13a that the value $\alpha = 0.5$ gives a minimum RMS error, when approximating the 2-point DCT of type II by the integer DCT, which equals 0.1632. Figure 3.13b shows that the value of α equal to 0.6 gives the minimum RMS error, which equals 0.1589. One can see that in parts c and d, the values of α equal to 0.3 and 0.4 yield the minimums of the RMS error, which are equal to 0.3668, and 0.2381.

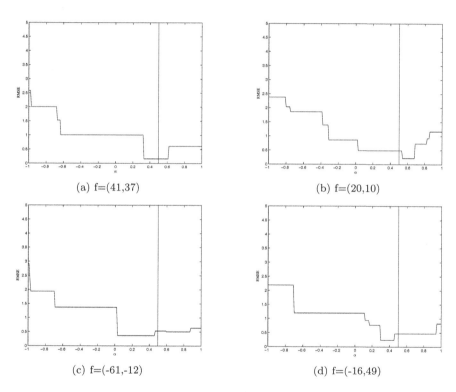

(a) f=(41,37)   (b) f=(20,10)

(c) f=(-61,-12)   (d) f=(-16,49)

FIGURE 3.13
Graphs of the error function for the two-point DCT of type II.

Examples of the invertibility for the 2-point parameterized integer DCT of type II are given in Table 3.2. The first column of the table shows the input, the second column shows the corresponding values of the 2-point DCT. The third column shows the values of α for which the integer DCT has been calculated, and results are given in column 4. For all these examples, the results of the inverse integer DCT coincide with inputs.

TABLE 3.2
Data of the two-point DCT of type II

Inputs	DCT	α	IDCT
41,37	55.1543, 2.8284	0.5	55,3
20,10	21.2132, 7.0711	0.5	21,7
-61,-12	-51.6188, -34.6482	0.1	-52,-35
-61,-12	-51.6188, -34.6482	0.2	-52,-35
-16,49	23.3345, -45.9619	0.4	23,-46

Figure 3.14 shows a few examples of approximation of the 2-point DCT of type IV by the integer transform. The value $\alpha = 0.4$ gives the minimum RMS error equal to 0.3090 and 0.2975, in parts a and b, respectively. Table 3.3

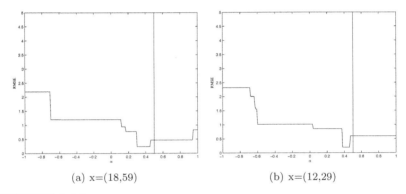

(a) x=(18,59)   (b) x=(12,29)

FIGURE 3.14
Error functions for the 2-point DCT of type IV.

shows data used for verifying the invertibility for the 2-point parameterized integer DCT of type IV. The parameterized integer DCT is invertible. It

TABLE 3.3
Data of the 2-point DCT of type IV

Inputs	DCT	α	IDCT
18,59	39.2077, -47.6156	0.3	38,-48
18,59	39.2077, -47.6156	0.4	39,-48
12,29	22.1839, -22.1978	0.4	22,-22

means that if the outputs and values of α are substituted in the system of nonlinear equations of the inverse integer DCT in (3.35), then we should get the exact input values.

When approximating the 4-point DCT of type IV for the inputs $f = (19, 45, 74, 34)$ and $(42, 13, 46, 35)$, the value $\alpha = 0.2$ gives the minimum RMS errors in both cases, which are equal to 0.3366 and 0.4904, respectively. In Table 3.4, data of inputs and outputs of the 4-point parameterized integer DCT of type IV are given, for which the property of invertibility has been verified.

TABLE 3.4
Data of the 4-point DCT of type IV

Inputs	DCT	α	IDCT
19,45,74,34	73.3878,-59.7096,6.4489, 4.8697	0.2	73,-60,6,5
42,13,46,35	59.6650,-22.7499,34.4020,3.4563	0.2	59,-23,34,4

A few examples of the 8-point parameterized integer DCT of type II are given in Table 3.5. The integer approximation of the eight-point DCT is calculated by the diagram which is given in Figure 3.12. The values of the

TABLE 3.5
Data of the 8-point DCT of type II

Inputs	DCT	α	IDCT
10,20,30,40,50,60,70,80	C1	0.5	127,0,-63,-5,-13,4,-4,2
32,19,46,38,61,13,41,51	C2	0.5	106,-6,-17,-12,22,-16,1,24
81,47,59,23,18,40,62,21	C3	0.4	124,25,25,-3,29,10,6,-37
10,20,30,40,35,26,37,48	C4	0.6	88,-5,-23,-15,-7,-4,-2,10
14,27,43,14,35,46,71,18	C5	0.6	94,-2,-27,2,8,-18,2,-41
14,27,43,14,35,46,71,18	C5	0.65	95,-2,-26,2,9,-18,4,40
36,60,29,45,32,65,59,78	C6	0.6	143,22,-23,3,-29,13,-9,-2

original 8-point DCT calculated for this table are defined as follows:

$C1 = [127.2756, 0, -63.0808, -4.4859, -12.8134, 4.4993, -3.0059, 2.5491]$
$C2 = [106.4165, -5.3031, -17.0241, -12.4641, 22.0573, -16.1364, 0.2111, 23.9032]$
$C3 = [124.0937, 25.1016, 25.3774, -2.4748, 29.5255, 10.0241, 6.6887, -35.9345]$
$C4 = [86.9716, -5.6567, -23.6819, -14.6811, -7.6336, -4.5401, -1.5689, 10.9378]$
$C5 = [94.7496, -2.8283, -26.7631, 1.9008, 7.7933, -17.5585, 3.1281, -39.8849]$
$C6 = [142.8315, 21.9197, -23.3385, 2.7783, -29.4386, 13.0428, -9.9362, -2.2974].$

3.6   Canonical representation of the integer DCT

In this section, we present matrix representations of the reversible integer discrete cosine transforms (IDCT) that are based on the canonical representation
and floor function. The kernel integer discrete cosine transform is introduced
that allows for reducing the calculation of the IDCT of type II to the kernel
IDCT with fewer operations of multiplication and floor function. The application of the kernel IDCT for calculating the eight-point IDCT of type II saves
seven multiplications and seven floor functions. The parameterized two-point
DCT of type IV and its particular case require two operations of multiplication, four additions, and two floor functions. We also consider the golden
two-point DCT that minimizes the error of the transform approximation by
the IDCT.

We first consider the well-known two-point integer-to-integer parameterized
transformation

$$\mathcal{A}_2 = \mathcal{A}_{2,\alpha} : \begin{bmatrix} x_0 \\ x_1 \end{bmatrix} \rightarrow \begin{bmatrix} \vartheta_0 \\ \vartheta_1 \end{bmatrix} = \begin{bmatrix} x_0 + \lfloor c_0 x_1 + \alpha \rfloor \\ \lfloor c_1 x_0 + \alpha \rfloor + x_1 \end{bmatrix} \qquad (3.44)$$

where the operation $\lfloor \cdot \rfloor$ denotes the floor function and c_0 and c_1 are coefficients
such that $c_0 c_1 \leq 0$, when the inverse transform $\mathcal{A}_2^{-1}$ exists [41]. The value
of α is considered to be equal to 0.5, although other values can also be used
in (3.44) to achieve an integer transform being the best approximation of the
desired unitary transform, for instance, the discrete cosine transform. We
define the operation $\circ$ over the set of integers by $c \circ x = \lfloor cx + 0.5 \rfloor$. One can
see that $(c_0 + c_1) \circ x = c_0 \circ x + c_1 \circ x$, when c_0 or c_1 is an integer. We call
this operation *a truncation multiplication*. The truncation multiplication $\circ$
is not a linear operation, i.e., for a given coefficient c and integers x and y,
$c \circ (x+y) = c \circ x + c \circ y + \alpha$, where, depending on values of x, y, and c, the function
$\alpha = \alpha_c(x, y)$ is not equal to zero and takes values of 0 and ± 1, in general. For
instance, $0.2 \circ (1 + 17) = 0.2 \circ 18$, but $0.2 \circ 1 + 0.2 \circ 17 = 0 + 3 \neq 4 = 0.2 \circ 18$.

3.6.1   Reversible two-point transforms

We consider the transformation which is described by the following two-step
lifting scheme:

$$\mathcal{A}_2 : \begin{bmatrix} x_0 \\ x_1 \end{bmatrix} \rightarrow \begin{bmatrix} \vartheta_0 \\ \vartheta_1 \end{bmatrix}$$

where coefficients ϑ_0 and ϑ_1 are defined in matrix form by

$$\begin{bmatrix} \vartheta_0 \\ \vartheta_1 \end{bmatrix} = \begin{bmatrix} 1 & c_0 & 0 \\ 0 & 1 & c_1 \end{bmatrix} \circ \begin{bmatrix} x_0 \\ x_1 \\ \vartheta_0 \end{bmatrix},$$

for given coefficients c_0 and c_1. We name this form *a canonical representation* of transformation $\mathcal{A}_2$. This transformation is reversible for any coefficients c_0 and c_1. The direct calculation results in the following:

$$
\begin{bmatrix} \vartheta_0 \\ \vartheta_1 \end{bmatrix} = \begin{bmatrix} x_0 + c_0 \circ x_1 \\ c_1 \circ \vartheta_0 + x_1 \end{bmatrix} = \begin{bmatrix} x_0 + c_0 \circ x_1 \\ c_1 \circ (x_0 + c_0 \circ x_1) + x_1 \end{bmatrix}
$$
$$
= \begin{bmatrix} x_0 + c_0 \circ x_1 \\ c_1 \circ x_0 + c_1 c_0 \circ x_1 + x_1 + \alpha_1 \end{bmatrix},
$$

where number $\alpha_1 = \alpha_1(x_0, x_1)$ appears because of nonlinearity of the truncation multiplication, i.e.,

$$
\alpha_1 = c_1 \circ (x_0 + c_0 \circ x_1) - (c_1 \circ x_0 + c_1 c_0 \circ x_1) \tag{3.45}
$$

and α_1 takes values of 0 and ± 1. The matrix of this transformation has one trivial coefficient equal to 1, and thus this transformation cannot be used for integer approximation of the two-point DCTs.

We now consider a more general and parameterized two-point reversible transformation which describes the three-step lifting scheme:

$$
\mathcal{B}_2 : \begin{bmatrix} x_0 \\ x_1 \end{bmatrix} \rightarrow \begin{bmatrix} \vartheta_1 \\ \vartheta_2 \end{bmatrix},
$$

where coefficients ϑ_1 and ϑ_2 of the transform are defined as follows

$$
\begin{bmatrix} \vartheta_0 \\ \vartheta_1 \\ \vartheta_2 \end{bmatrix} = \begin{bmatrix} 1 & c_0 & 0 & 0 \\ 0 & 1 & c_1 & 0 \\ 0 & 0 & 1 & c_2 \end{bmatrix} \circ \begin{bmatrix} x_0 \\ x_1 \\ \vartheta_0 \\ \vartheta_1 \end{bmatrix} = \tag{3.46}
$$

$$
= \begin{bmatrix} x_0 + c_0 \circ x_1 \\ x_1 + c_1 \circ \vartheta_0 \\ \vartheta_0 + c_2 \circ \vartheta_1 \end{bmatrix} = \begin{bmatrix} x_0 + c_0 \circ x_1 \\ x_1 + c_1 \circ (x_0 + c_0 \circ x_1) \\ x_0 + c_0 \circ x_1 + c_2 \circ (x_1 + c_1 \circ (x_0 + c_0 \circ x_1)) \end{bmatrix} \tag{3.47}
$$

$$
= \begin{bmatrix} 1 & c_0 \\ c_1 & 1 + c_1 c_0 \\ 1 + c_2 c_1 & c_0 + c_2(1 + c_1 c_0) \end{bmatrix} \circ \begin{bmatrix} x_0 \\ x_1 \end{bmatrix} + \begin{bmatrix} 0 \\ \alpha_1 \\ \alpha_2 \end{bmatrix}.
$$

Additional coefficients α_1 and α_2 occur because of nonlinearity of the truncation operation $\circ$ in the last two rows of the vector in (3.47). For given coefficients c_0, c_1, and c_2, the coefficient-function $\alpha_1 = \alpha_1(x_0, x_1)$ is calculated by (3.45). The next coefficient $\alpha_2 = \alpha_2(x_0, x_1)$ is calculated by

$$
\alpha_2 = c_0 \circ x_1 + c_2 \circ (x_1 + c_1 \circ (x_0 + c_0 \circ x_1)) - c_1 c_2 \circ x_0 - (c_0 + c_2(1 + c_1 c_0)) \circ x_1. \tag{3.48}
$$

For all cases of c_0, c_1, and c_2 of interest, the coefficients α_1 and α_2 take only values of 0 and ± 1. We call the representation in (3.46) *the canonical representation* of $\mathcal{B}_2$ and the transformation defined by

$$
\begin{bmatrix} \vartheta_0 \\ \vartheta_1 \\ \vartheta_2 \end{bmatrix} = \mathbf{K}(\mathcal{B}_2) \circ \begin{bmatrix} x_0 \\ x_1 \end{bmatrix} = \begin{bmatrix} 1 & c_0 \\ c_1 & 1 + c_1 c_0 \\ 1 + c_2 c_1 & c_0 + c_2(1 + c_1 c_0) \end{bmatrix} \circ \begin{bmatrix} x_0 \\ x_1 \end{bmatrix}
$$

to be a *kernel* of transformation $\mathcal{B}_2$ or *kernel transformation* which is denoted by $\mathcal{K}_2$. We next consider the kernel transformation $\mathcal{K}_2$ defined by

$$\mathcal{K}_2 : \begin{bmatrix} x_0 \\ x_1 \end{bmatrix} \rightarrow \begin{bmatrix} \vartheta_1 \\ \vartheta_2 \end{bmatrix} = \begin{bmatrix} c_1 & 1 + c_1 c_0 \\ 1 + c_2 c_1 & c_0 + c_2(1 + c_1 c_0) \end{bmatrix} \circ \begin{bmatrix} x_0 \\ x_1 \end{bmatrix}. \tag{3.49}$$

The determinant of the transform matrix is -1, and the inverse transform is defined as

$$\begin{bmatrix} x_0 \\ x_1 \end{bmatrix} = \begin{bmatrix} -c_0 - c_2(1 + c_1 c_0) & 1 + c_1 c_0 \\ 1 + c_1 c_2 & -c_1 \end{bmatrix} \circ \begin{bmatrix} \vartheta_1 \\ \vartheta_2 \end{bmatrix}.$$

The kernel transform can be used for integer approximation of the two-point discrete cosine transforms of types II and IV.

3.6.2   Reversible two-point DCT of type II

In this section, the following two-point discrete cosine transformation is considered:

$$\mathcal{C}_2 : \begin{bmatrix} x_0 \\ x_1 \end{bmatrix} \rightarrow \begin{bmatrix} \vartheta'_1 \\ \vartheta'_2 \end{bmatrix} = \begin{bmatrix} a & a \\ a & -a \end{bmatrix} \begin{bmatrix} x_0 \\ x_1 \end{bmatrix} \tag{3.50}$$

where $a = \sqrt{2}/2 = 0.7071$. Instead of kernel transformation $\mathcal{K}_2$ with matrix representation of (3.49), we define the two-point transformation $\mathcal{K}'_2$ with the following matrix representation

$$\begin{bmatrix} \vartheta'_1 \\ \vartheta'_2 \end{bmatrix} = \begin{bmatrix} c_1 & 1 + c_1 c_0 \\ 1 + c_1 c_2 & c_0 + c_2(1 + c_1 c_0) \end{bmatrix} \begin{bmatrix} x_0 \\ x_1 \end{bmatrix}. \tag{3.51}$$

In other words, it is assumed that the traditional operation of matrix multiplication holds in (3.49) instead of the truncation multiplication. Coefficients of this transform can be found by assuming that $\mathcal{K}'_2 = \mathcal{C}_2$. These coefficients c_0, c_1, and c_2 will then be used for construction of the matrix of transformation $\mathcal{B}_2$ in (3.46).

By comparing two matrix equations (3.50) and (3.51), we obtain

$$\begin{bmatrix} c_1 & 1 + c_1 c_0 \\ 1 + c_1 c_2 & c_0 + c_2(1 + c_1 c_0) \end{bmatrix} = \begin{bmatrix} a & a \\ a & -a \end{bmatrix}$$

and the following equations to be therefore solved (as in (3.38)) $c_1 = a$, $1 + c_1 c_0 = a$, $1 + c_1 c_2 = a$, $c_0 + c_2 a = -a$. From the first three equations, we obtain $c_1 = a$, $c_0 = c_2 = 1 - a^{-1}$, and the last condition $(1 - a^{-1}) + (1 - a^{-1})a = -a$ holds only for $a = \sqrt{2}/2$. The canonical representation of the two-point transform in (3.46) takes the form

$$\begin{bmatrix} \vartheta_0 \\ \vartheta_1 \\ \vartheta_2 \end{bmatrix} = \begin{bmatrix} 1 & 1 - a^{-1} & 0 & 0 \\ 0 & 1 & a & 0 \\ 0 & 0 & 1 & 1 - a^{-1} \end{bmatrix} \circ \begin{bmatrix} x_0 \\ x_1 \\ \vartheta_0 \\ \vartheta_1 \end{bmatrix}.$$

The inverse IDCT is represented in the canonical form as

$$
\begin{bmatrix} \vartheta_0 \\ x_1 \\ x_0 \end{bmatrix} = \begin{bmatrix} 1 & -c_2 & 0 & 0 \\ 0 & 1 & -c_1 & 0 \\ 0 & 0 & 1 & -c_0 \end{bmatrix} \circ \begin{bmatrix} \vartheta_2 \\ \vartheta_1 \\ \vartheta_0 \\ x_1 \end{bmatrix}.
$$

To simplify future calculations, we write the above canonical form in the form similar to (3.46), namely, as

$$
\begin{bmatrix} \vartheta_0 \\ x_1 \\ x_0 \end{bmatrix} = \begin{bmatrix} 1 & \bar{c}_0 & 0 & 0 \\ 0 & 1 & \bar{c}_1 & 0 \\ 0 & 0 & 1 & \bar{c}_2 \end{bmatrix} \circ \begin{bmatrix} \vartheta_2 \\ \vartheta_1 \\ \vartheta_0 \\ x_1 \end{bmatrix},
$$

where $\bar{c}_0 = -c_2$, $\bar{c}_1 = -c_1$, and $\bar{c}_2 = -c_0$. Then, the following calculations hold

$$
\begin{bmatrix} \vartheta_0 \\ x_1 \\ x_0 \end{bmatrix} = \begin{bmatrix} 1 & \bar{c}_0 \\ \bar{c}_1 & 1 + \bar{c}_1\bar{c}_0 \\ 1 + \bar{c}_2\bar{c}_1 & \bar{c}_0 + \bar{c}_2(1 + \bar{c}_1\bar{c}_0) \end{bmatrix} \circ \begin{bmatrix} \vartheta_2 \\ \vartheta_1 \end{bmatrix} + \begin{bmatrix} 0 \\ \beta_1 \\ \beta_2 \end{bmatrix}
$$

$$
= \begin{bmatrix} 1 & -c_2 \\ -c_1 & 1 + c_1c_2 \\ 1 + c_0c_1 & -c_2 - c_0(1 + c_1c_2) \end{bmatrix} \circ \begin{bmatrix} \vartheta_2 \\ \vartheta_1 \end{bmatrix} + \begin{bmatrix} 0 \\ \beta_1 \\ \beta_2 \end{bmatrix}
$$

and the kernel transform is thus written as

$$
\begin{bmatrix} x_1 \\ x_0 \end{bmatrix} = \begin{bmatrix} -c_1 & 1 + c_1c_2 \\ 1 + c_0c_1 & -c_2 - c_0(1 + c_1c_2) \end{bmatrix} \circ \begin{bmatrix} \vartheta_2 \\ \vartheta_1 \end{bmatrix} + \begin{bmatrix} \beta_1 \\ \beta_2 \end{bmatrix}.
$$

Additional coefficients β_1 and β_2 as functions of (x_0, x_1) can be derived from formulas similar to the formulas given in (3.45) and (3.48) for coefficients α_1 and α_2, respectively. Indeed, by substituting $\bar{c}_k$ instead of c_k as well as x_0 by ϑ_2 and x_1 by ϑ_1, we obtain the following

$$
\beta_1 = -c_1 \circ (\vartheta_2 - c_2 \circ \vartheta_1) + (c_1 \circ \vartheta_2 - c_1c_2 \circ \vartheta_1),
$$
$$
\beta_2 = -c_2 \circ \vartheta_1 - c_0 \circ (\vartheta_1 - c_1 \circ (\vartheta_2 - c_2 \circ \vartheta_1)) -
$$
$$
-c_1c_0 \circ \vartheta_2 + (c_2 + c_0(1 + c_1c_2)) \circ \vartheta_1,
$$

and $\beta_k(\vartheta_1, \vartheta_2) = \alpha_k(-\vartheta_1, \vartheta_2)$, for $k = 1, 2$.

3.6.3 Kernel transform

We consider the kernel integer discrete cosine transformation (KIDCT) $\mathcal{K}_2$ of type II, which is defined as the cosine transformation $\mathcal{C}_2$ with respect to the truncation multiplication

$$
\begin{bmatrix} \vartheta_1 \\ \vartheta_2 \end{bmatrix} = \begin{bmatrix} a & a \\ a & -a \end{bmatrix} \circ \begin{bmatrix} x_0 \\ x_1 \end{bmatrix} = \begin{bmatrix} a \circ x_0 + a \circ x_1 \\ a \circ x_0 - a \circ x_1 \end{bmatrix} = \begin{bmatrix} \lfloor ax_0 + 0.5 \rfloor + \lfloor ax_1 + 0.5 \rfloor \\ \lfloor ax_0 + 0.5 \rfloor - \lfloor ax_1 + 0.5 \rfloor \end{bmatrix}.
$$

The calculation of the IDCT can be reduced to the calculation of the KIDCT, when considering that the error data have been tabulated.

Example 3.4
Let the input be $[x_0, x_1] = [10, 20]$. Then, the two-point DCT of this input is defined by

$$\begin{bmatrix} \vartheta'_1 \\ \vartheta'_2 \end{bmatrix} = \begin{bmatrix} a & a \\ a & -a \end{bmatrix} \begin{bmatrix} 10 \\ 20 \end{bmatrix} = 0.7071 \begin{bmatrix} 10 + 20 \\ 10 - 20 \end{bmatrix} = \begin{bmatrix} 21.213 \\ -7.071 \end{bmatrix}$$

and the two-point integer KIDCT is calculated as follows

$$\begin{bmatrix} \vartheta_1 \\ \vartheta_2 \end{bmatrix} = \begin{bmatrix} a & a \\ a & -a \end{bmatrix} \circ \begin{bmatrix} 10 \\ 20 \end{bmatrix} = \begin{bmatrix} \lfloor 10a + 0.5 \rfloor + \lfloor 20a + 0.5 \rfloor \\ \lfloor 10a + 0.5 \rfloor - \lfloor 20a + 0.5 \rfloor \end{bmatrix}$$
$$= \begin{bmatrix} \lfloor 7.571 \rfloor + \lfloor 14.642 \rfloor \\ \lfloor 7.571 \rfloor - \lfloor 14.642 \rfloor \end{bmatrix} = \begin{bmatrix} 7 + 14 \\ 7 - 14 \end{bmatrix} = \begin{bmatrix} 21 \\ -7 \end{bmatrix}.$$

For this given input, the KIDCT coincides with the IDCT because of no error in calculation, i.e., $\alpha_1(10, 20) = 0$ and $\alpha_2(10, 20) = 0$.

The inverse transform is calculated similarly by

$$\begin{bmatrix} y_0 \\ y_1 \end{bmatrix} = \begin{bmatrix} a & a \\ a & -a \end{bmatrix} \circ \begin{bmatrix} 21 \\ -7 \end{bmatrix} = \begin{bmatrix} \lfloor 21a + 0.5 \rfloor - \lfloor 7a + 0.5 \rfloor \\ \lfloor 21a + 0.5 \rfloor + \lfloor 7a + 0.5 \rfloor \end{bmatrix}$$
$$= \begin{bmatrix} \lfloor 15.3491 \rfloor - \lfloor 5.4497 \rfloor \\ \lfloor 15.3491 \rfloor + \lfloor 5.4497 \rfloor \end{bmatrix} = \begin{bmatrix} 15 - 5 \\ 15 + 5 \end{bmatrix} = \begin{bmatrix} 10 \\ 20 \end{bmatrix}.$$

The KIDCT coincides with the IDCT since $\beta_1(21, -7) = 0$ and $\beta_2(21, -7) = 0$. For the considered input, the IDCT can be calculated in the kernel form and the kernel transform is reversible.

For comparison, we consider the direct calculation of the reversible DCT:

$$\vartheta_0 = \begin{bmatrix} 1 & 1 - a^{-1} \end{bmatrix} \circ \begin{bmatrix} x_0 \\ x_1 \end{bmatrix} = x_0 + \lfloor (1 - a^{-1})x_1 + 0.5 \rfloor$$
$$= 10 + \lfloor (1 - \sqrt{2})20 + 0.5 \rfloor \tag{3.52}$$
$$= 10 + \lfloor -8.2843 + 0.5 \rfloor = 10 + \lfloor -7.7843 \rfloor = 10 - 8 = 2$$

$$\vartheta_1 = \begin{bmatrix} 1 & a \end{bmatrix} \circ \begin{bmatrix} x_1 \\ \vartheta_0 \end{bmatrix} = x_1 + \lfloor a \cdot \vartheta_0 + 0.5 \rfloor \tag{3.53}$$
$$= 20 + \lfloor 0.7071 \cdot 2 + 0.5 \rfloor = 20 + \lfloor 1.9142 \rfloor = 20 + 1 = 21,$$

$$\vartheta_2 = \begin{bmatrix} 1 & 1 - a^{-1} \end{bmatrix} \circ \begin{bmatrix} \vartheta_0 \\ \vartheta_1 \end{bmatrix} = \vartheta_0 + \lfloor (1 - a^{-1})\vartheta_1 + 0.5 \rfloor \tag{3.54}$$
$$= 2 + \lfloor (1 - \sqrt{2})21 + 0.5 \rfloor = 2 + \lfloor -8.1985 \rfloor = 2 + (-9) = -7.$$

One can see an advantage of using the kernel IDCT when two operations of multiplication, four additions, and two floor functions are required. One or two additions also may be used in a case of correction of the result, if values of α_1 and α_2 are not zero. For instance, in the range $x_0, x_1 = 0, 1, 2, \ldots, 64$ there are only 124 cases for which two corrections are required. The performance of the IDCT by the direct calculation (3.52)-(3.54) requires three multiplications, six additions, and three floor functions. We use notation $o[3m, 6a, 3f]$ to denote this complexity of the direct calculation of the two-point DCT of type II and $o[2m, 6a, 2f]$ when using the kernel transform. The application of the kernel IDCT requires the look up table method when values of error coefficients are calculated in advance and saved in tables. ◻

Example 3.5
Let input be $[x_0, x_1] = [16, 7]$. Then, the two-point DCT is defined as

$$\begin{bmatrix} \vartheta_1' \\ \vartheta_2' \end{bmatrix} = \begin{bmatrix} a & a \\ a & -a \end{bmatrix} \begin{bmatrix} 16 \\ 7 \end{bmatrix} = \begin{bmatrix} 16.2635 \\ 6.3640 \end{bmatrix}$$

and the two-point integer kernel DCT is calculated as

$$\begin{bmatrix} \vartheta_1 \\ \vartheta_2 \end{bmatrix} = \begin{bmatrix} a & a \\ a & -a \end{bmatrix} \circ \begin{bmatrix} 16 \\ 7 \end{bmatrix} = \begin{bmatrix} 16 \\ 6 \end{bmatrix}.$$

There is no error in calculation, $\alpha_1(16, 17) = 0$ and $\alpha_2(16, 17) = 0$, and the KIDCT coincides with the reversible IDCT. The inverse transform is calculated by the kernel transform as

$$\begin{bmatrix} x_0 \\ x_1 \end{bmatrix} = \begin{bmatrix} a & a \\ a & -a \end{bmatrix} \circ \begin{bmatrix} 16 \\ 6 \end{bmatrix} + \begin{bmatrix} \beta_2 \\ \beta_1 \end{bmatrix} = \begin{bmatrix} 15 \\ 7 \end{bmatrix} + \begin{bmatrix} 1 \\ 0 \end{bmatrix} = \begin{bmatrix} 16 \\ 7 \end{bmatrix}$$

since $\beta_2 = \beta_2(16, 6) = 1$ and $\beta_1 = \beta_1(16, 6) = 0$. In this case, the first output of the inverse KIDCT needs the correction to be equal to the inverse IDCT.
◻

In the general case of inputs, the kernel transform can be used for the reversible DCT. Indeed, the difference of these two transforms is expressed by the simple equation

$$\begin{bmatrix} \vartheta_1 \\ \vartheta_2 \end{bmatrix} = \begin{bmatrix} \vartheta_1' \\ \vartheta_2' \end{bmatrix} + \begin{bmatrix} \alpha_1 \\ \alpha_2 \end{bmatrix}. \tag{3.55}$$

The error functions $\alpha_1(x_0, x_1; a)$ and $\alpha_2(x_0, x_1; a)$ take values of 0 and ± 1, and they can be tabulated and used in calculation by (3.55), for different ranges of x_0 and x_1. The diagram of realization of the IDCT by using the KIDCT is shown in Figure 3.15a, when two tables $T(\alpha_1)$ and $T(\alpha_2)$ with error functions α_1 and α_2 are used.

For the inverse IDCT, we have the expression

$$\begin{bmatrix} x_0 \\ x_1 \end{bmatrix} = \begin{bmatrix} y_0 \\ y_1 \end{bmatrix} + \begin{bmatrix} \beta_2 \\ \beta_1 \end{bmatrix}$$

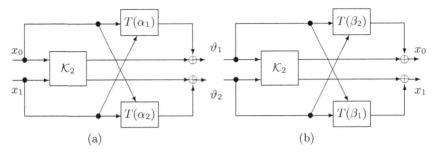

FIGURE 3.15

Block schemes of calculation of the (a) direct and (b) inverse two-point IDCTs of type II by the kernel transform.

and coefficients β_1 and β_2 can be tabulated and used for calculation of the inverse IDCT by KIDCT. The diagram of realization of the inverse IDCT by using the KIDCT is shown in Figure 3.15b, when tabulated data of error functions β_1 and β_2 are used in tables $T(\beta_1)$ and $T(\beta_2)$, respectively. We also note the following simple relations between α- and β-type errors and tables, $\beta_k(\vartheta_1, \vartheta_2) = \alpha_k(-\vartheta_1, \vartheta_2)$, for $k = 1, 2$.

3.6.4   Reversible two-point IDCT of type IV

In this section, we consider the following two-point transformation

$$\mathcal{C}_2 : \begin{bmatrix} x_0 \\ x_1 \end{bmatrix} \rightarrow \begin{bmatrix} \vartheta'_1 \\ \vartheta'_2 \end{bmatrix} = \begin{bmatrix} b & c \\ c & -b \end{bmatrix} \begin{bmatrix} x_0 \\ x_1 \end{bmatrix} \tag{3.56}$$

where b and c are coefficients from the interval $(-1, 1)$. The particular case, when $b = 0.9238$ and $c = 0.3827$, corresponds to the two-point DCT of type IV. By comparing two matrix equations (3.56) and (3.51), we obtain the following equation for the coefficients c_0, c_1, and c_2:

$$\begin{bmatrix} c_1 & 1 + c_1 c_0 \\ 1 + c_1 c_2 & c_0 + c_2(1 + c_1 c_0) \end{bmatrix} = \begin{bmatrix} b & c \\ c & -b \end{bmatrix}$$

and the following system of equations should be solved for $c_1 = b$, $1 + c_1 c_0 = c$, $1 + c_1 c_2 = c$, $c_0 + c_2 c = -b$. From the first three equations, we obtain $c_0 = c_2 = (c-1)/b$, and the last equation $c - 1 + (c - 1)c = -b^2$ leads to condition $c^2 - 1 = -b^2$, which means that point (b, c) lies on the unit circle and $\mathcal{C}_2$ is the Givens rotation. Thus the two-point integer reversible transform in canonical representation (3.46) is defined as

$$\begin{bmatrix} \vartheta_0 \\ \vartheta_1 \\ \vartheta_2 \end{bmatrix} = \begin{bmatrix} 1 & \dfrac{c-1}{b} & 0 & 0 \\ 0 & 1 & b & 0 \\ 0 & 0 & 1 & \dfrac{c-1}{b} \end{bmatrix} \circ \begin{bmatrix} x_0 \\ x_1 \\ \vartheta_0 \\ \vartheta_1 \end{bmatrix} \tag{3.57}$$

for any coefficients c and b such that $c^2 + b^2 = 1$. With the additional error coefficients α_1 and α_2 calculated for given parameters c and b at the point (x_0, x_1), this transform can be written as

$$\begin{bmatrix} \vartheta_0 \\ \vartheta_1 \\ \vartheta_2 \end{bmatrix} = \begin{bmatrix} 1 & \dfrac{c-1}{b} \\ b & c \\ c & -b \end{bmatrix} \circ \begin{bmatrix} x_0 \\ x_1 \end{bmatrix} + \begin{bmatrix} 0 \\ \alpha_1 \\ \alpha_2 \end{bmatrix}$$

and the kernel transform

$$\begin{bmatrix} \vartheta_1 \\ \vartheta_2 \end{bmatrix} = \begin{bmatrix} b & c \\ c & -b \end{bmatrix} \circ \begin{bmatrix} x_0 \\ x_1 \end{bmatrix} + \begin{bmatrix} \alpha_1 \\ \alpha_2 \end{bmatrix}.$$

The inverse two-point integer reversible transform in canonical representation is defined by

$$\begin{bmatrix} \vartheta_0 \\ x_1 \\ x_0 \end{bmatrix} = \begin{bmatrix} 1 & -\dfrac{c-1}{b} & 0 & 0 \\ 0 & 1 & -b & 0 \\ 0 & 0 & 1 & -\dfrac{c-1}{b} \end{bmatrix} \circ \begin{bmatrix} \vartheta_2 \\ \vartheta_1 \\ \vartheta_0 \\ x_1 \end{bmatrix} = \begin{bmatrix} 1 & -\dfrac{c-1}{b} \\ -b & c \\ c & b \end{bmatrix} \circ \begin{bmatrix} \vartheta_2 \\ \vartheta_1 \end{bmatrix} + \begin{bmatrix} 0 \\ \beta_1 \\ \beta_2 \end{bmatrix}.$$

The corresponding inverse kernel transform can be written as

$$\begin{bmatrix} x_1 \\ x_0 \end{bmatrix} = \begin{bmatrix} -b & c \\ c & b \end{bmatrix} \circ \begin{bmatrix} \vartheta_2 \\ \vartheta_1 \end{bmatrix} + \begin{bmatrix} \beta_1 \\ \beta_2 \end{bmatrix}.$$

Error functions β_1 and β_2 are calculated for parameters c and b at point (x_0, x_1). The images of coefficient-functions α_1 and α_2 in the range $x_0, x_1 = 0, 1, 2, \ldots, 64$ are shown in Figure 3.16(a) and (b). The error coefficients take values of 0, -1, and 1 which are shown respectively in images in black, gray, and white colors. These images have regular structures and only periodic parts of images can be used for calculations. Relations between α_1 and β_1, α_2 and β_2 are expressed by operations of transposition and translation of their matrices. Therefore, the same correction tables can be used for both errors α_1 and α_2, and errors β_1 and β_2. Periodicity of error functions for correction of the kernel transform can also be seen in the case of the two-point IDCT of type II, as shown in Figure 3.16(c) and (d).

Example 3.6
Let input be $[x_0, x_1] = [10, 20]$, and let coefficients b and c be equal respectively to 0.9238 and 0.3827. Then $c_1 = 0.9238$ and $c_0 = c_2 = -0.6682$. The two-point DCT equals $[\vartheta_1', \vartheta_2'] = [16.892, -14.649]$, and the kernel IDCT is calculated as

$$\begin{bmatrix} \vartheta_1 \\ \vartheta_2 \end{bmatrix} = \begin{bmatrix} b & c \\ c & -b \end{bmatrix} \circ \begin{bmatrix} 10 \\ 20 \end{bmatrix} = \begin{bmatrix} \lfloor 10b + 0.5 \rfloor + \lfloor 20c + 0.5 \rfloor \\ \lfloor 10c + 0.5 \rfloor - \lfloor 20b + 0.5 \rfloor \end{bmatrix}$$
$$= \begin{bmatrix} \lfloor 9.74 \rfloor + \lfloor 8.15 \rfloor \\ \lfloor 4.33 \rfloor - \lfloor 18.98 \rfloor \end{bmatrix} = \begin{bmatrix} 9 + 8 \\ 4 - 18 \end{bmatrix} = \begin{bmatrix} 17 \\ -14 \end{bmatrix}.$$

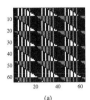

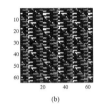

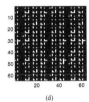

(a)   (b)   (c)   (d)

FIGURE 3.16

Images of error coefficients (a) $\alpha_1(x_0, x_1)$ and (b) $\alpha_2(x_0, x_1)$, for correction of the direct two-point kernel IDCT of type IV. Errors (c) $\alpha_1(x_0, x_1)$ and (d) $\alpha_2(x_0, x_1)$ for correction of the two-point kernel IDCT of type II, when $x_0, x_1 = 0 : 63$.

The kernel IDCT coincides with the IDCT because of no error in calculation, $\alpha_1 = \alpha_1(10, 20) = 0$ and $\alpha_2 = \alpha_2(10, 20) = 0$ for the given input. The kernel IDCT of the input $[17, -14]$ is calculated as follows:

$$\begin{bmatrix} y_0 \\ y_1 \end{bmatrix} = \begin{bmatrix} b & c \\ c & -b \end{bmatrix} \circ \begin{bmatrix} 17 \\ -14 \end{bmatrix} = \begin{bmatrix} \lfloor 16.20 \rfloor - \lfloor 5.86 \rfloor \\ \lfloor 7.01 \rfloor + \lfloor 12.93 + 0.5 \rfloor \end{bmatrix} = \begin{bmatrix} 16 - 5 \\ 7 + 13 \end{bmatrix} = \begin{bmatrix} 11 \\ 20 \end{bmatrix}.$$

This result needs to be corrected as $[x_0, x_1] = [y_0, y_1] + [\beta_2, \beta_1] = [11, 20] + [-1, 0] = [10, 20]$, because $\beta_2 = \beta_2(17, -14) = -1$ and $\beta_1 = \beta_1(17, -14) = 0$. The complexity of the calculation of the two-point IDCT of type IV by using the kernel transform is $o[4m, 8a, 4f]$. However, it is possible to improve the realization of the kernel transform in the following way. The two-point DCT of type IV can be represented as

$$\begin{bmatrix} b & c \\ c & -b \end{bmatrix} \begin{bmatrix} x_0 \\ x_1 \end{bmatrix} = \begin{bmatrix} b(x_0 + x_1) + (c - b)x_1 \\ (c + b)x_0 - b(x_0 + x_1) \end{bmatrix} = \begin{bmatrix} 0 & (c - b) & 1 \\ (c + b) & 0 & -1 \end{bmatrix} \begin{bmatrix} x_0 \\ x_1 \\ b(x_0 + x_1) \end{bmatrix}.$$

Therefore, the 2-D IDCT of type IV can be calculated as

$$\begin{bmatrix} \vartheta_1 \\ \vartheta_2 \end{bmatrix} = \begin{bmatrix} 0 & (c - b) & 1 \\ (c + b) & 0 & -1 \end{bmatrix} \circ \begin{bmatrix} x_0 \\ x_1 \\ b \circ (x_0 + x_1) \end{bmatrix}.$$

Each truncation multiplication uses one operation of multiplication, addition, and floor function. The arithmetical complexity of the two-point IDCT of type IV by the kernel transform equals therefore $o(3\mu, 6a, 3f)$. Two more trivial additions by ± 1 may also be required to correct the kernel transform.

The direct calculation of the RDCT is performed as follows

$$\vartheta_0 = \begin{bmatrix} 1 & \dfrac{c-1}{b} \end{bmatrix} \circ \begin{bmatrix} x_0 \\ x_1 \end{bmatrix} = x_0 + \left\lfloor \dfrac{c-1}{b} x_1 + 0.5 \right\rfloor$$

$$= 10 + \lfloor -13.3636 + 0.5 \rfloor = 10 + \lfloor -12.86836 \rfloor = 10 + \lfloor -13 \rfloor = -3,$$

$$\vartheta_1 = \begin{bmatrix} 1 & b \end{bmatrix} \circ \begin{bmatrix} x_1 \\ \vartheta_0 \end{bmatrix} = x_1 + \lfloor b \cdot \vartheta_0 + 0.5 \rfloor$$

$$= 20 + \lfloor -2.7716 + 0.5 \rfloor = 20 + \lfloor -2.2716 \rfloor = 20 - 3 = 17,$$

$$\vartheta_2 = \begin{bmatrix} 1 & \dfrac{c-1}{b} \end{bmatrix} \circ \begin{bmatrix} \vartheta_0 \\ \vartheta_1 \end{bmatrix} = \vartheta_0 + \left\lfloor \dfrac{c-1}{b}\vartheta_1 + 0.5 \right\rfloor$$

$$= -3 + \lfloor -11.3590 + 0.5 \rfloor = -3 - 11 = -14.$$

The complexity of the direct calculation is $o[3m, 6a, 3f]$, which is almost the same as that for the kernel transform. We now introduce the two-point IDCT that requires fewer operations than the two-point IDCT of type IV does. □

3.6.5 Parameterized two-point IDCT

Since the parameter-point (b, c) lies on the unit circle, we can state that $b = \sin(\varphi)$ and $c = \cos(\varphi)$, where angle $\varphi \in (0, \pi)$. The matrix of transformation C_2 can be parameterized as

$$[C_2(\varphi)] = \begin{bmatrix} b & c \\ c & -b \end{bmatrix} = \begin{bmatrix} \sin(\varphi) & \cos(\varphi) \\ \cos(\varphi) & -\sin(\varphi) \end{bmatrix}.$$

Then the canonical representation in (3.57) can be written as

$$\begin{bmatrix} \vartheta_0 \\ \vartheta_1 \\ \vartheta_2 \end{bmatrix} = \begin{bmatrix} 1 - \tan(\varphi/2) & 0 & 0 \\ 0 & 1 & \sin(\varphi) \\ 0 & 0 & 1 \end{bmatrix} \begin{matrix} \\ -\tan(\varphi/2) \end{matrix} \circ \begin{bmatrix} x_0 \\ x_1 \\ \vartheta_0 \\ \vartheta_1 \end{bmatrix},$$

and the canonical representation of the invertible two-point integer reversible transform as

$$\begin{bmatrix} \vartheta_0 \\ x_1 \\ x_0 \end{bmatrix} = \begin{bmatrix} 1 & \tan(\varphi/2) & 0 & 0 \\ 0 & 1 & -\sin(\varphi) & 0 \\ 0 & 0 & 1 & \tan(\varphi/2) \end{bmatrix} \circ \begin{bmatrix} \vartheta_2 \\ \vartheta_1 \\ \vartheta_0 \\ x_1 \end{bmatrix}.$$

The $\varphi = 3\pi/8$ case corresponds to the known two-point IDCT of type IV. We now consider the case when angle $\varphi = \varphi_0 = \tan^{-1}(2)$. Let us assume that $b > c$ and consider the following matrix equation

$$[C_2(\varphi)] = \begin{bmatrix} b & c \\ c & -b \end{bmatrix} = \begin{bmatrix} c & c \\ c & -c \end{bmatrix} + (b - c) \begin{bmatrix} 1 & 0 \\ 0 & -1 \end{bmatrix}.$$

One can note the simple case, when the above matrix is defined by only one parameter c, i.e., when $b - c = c$, or $b = 2c$. Because point (b, c) lies on the unit circle, we obtain $c = 1/\sqrt{5} = 0.4472$. In this case, $b = 0.8944$, $\varphi_0 = 1.1071$ and the transform C_2 is calculated as follows

$$[C_2(\varphi_0)] \circ \begin{bmatrix} x_0 \\ x_1 \end{bmatrix} = c \begin{bmatrix} 2 & 1 \\ 1 & -2 \end{bmatrix} \circ \begin{bmatrix} x_0 \\ x_1 \end{bmatrix}$$

$$= c \circ \begin{bmatrix} 2x_0 + x_1 \\ x_0 - 2x_1 \end{bmatrix} = \begin{bmatrix} \lfloor c(2x_0 + x_1) + 0.5 \rfloor \\ \lfloor c(x_0 - 2x_1) + 0.5 \rfloor \end{bmatrix}.$$

The implementation of this transform requires only two operations of multiplication by c, four additions, and two floor functions, instead of three multiplications, six additions, and three floor functions in the direct method.

We also consider another case, when coefficients b and c are defined by the angle equal to the golden number, $\varphi_* = \Phi = 1.61803399\ldots$ (or $92.70652°$). In this case $b = 0.9989$ and $c = -0.0472$, and we call the transformation $C_2(\varphi_*)$ the golden two-point DCT. It is interesting to note that the golden angle provides the minimum error of approximation of the 2-D DCT $C_2(\varphi)$ by the reversible IDCT. Figure 3.17 shows the error, $E(\varphi)$, of approximation of $C_2(\varphi)$ by its reversible IDCT, where φ varies in the interval $(0, \pi)$. Errors were calculated over more than 10000 inputs $[x_0, x_1]$, where $x_0, x_1 = 0 : 100$. The angle-point $3\pi/8$ (or $67.50°$) corresponding to the two-point DCT of type IV is shown by the bullet and the angle-point $\varphi_0 = 1.1071$ (or $63.43°$) by the open circle.

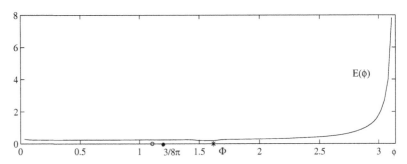

FIGURE 3.17

Square-root-mean error of approximation of the parameterized two-point DCT, $C_2(\phi)$, by the reversible IDCT.

3.6.6 Codes for the integer 2-point DCT

Below are MATLAB-based codes for computing the two-point integer discrete cosine transforms discussed above.

```
% ----------------------------------------------------------------------
% demo_dct2.m file for MATLAB 7 (library of codes of Grigoryans)
% List of codes for integer approximation of Givens rotations (and DCTs)
%    integer approximation of the Givens rotation  -  'cell_dct2phi.m'
%    inverse transforms of this approximation      -  'cell_idct2phi.m'
%    special case of the rotation by angle atan(2) -  'art_dct2c.m'
%    truncation multiplication (xoc)               -  'fl_byc.m'
% ----------------------------------------------------------------------
%    The main program for calculating the 2-point DCTs:
```

```
        a05=0.5;
        x0=17; x1=20;     X=[17 20];      % input vector
%       ----------------------------------------------
%       I. 2-point DCT of type IV
        phi=3*pi/8;        % 1.1781
%       direct calculation:
        b=sin(phi); c=cos(phi);  % 0.9239, 0.3827
        A=[ b   c
            c  -b];
        X_dct4=X*A;        % 23.3596, -11.9720
%       integer 2-point DCT of type IV
        Y_dct4=cell_dct2phi(x0,x1,phi,a05); % 24, -12
%       inverse transforms
        X1=cell_idct2phi(Y_dct4(1),Y_dct4(2),phi,a05);
%       ----------------------------------------------
%       II. Integer 2-point DCT of type II
        phi2=pi/4;         % 0.7854
        Y_dct=cell_dct2phi(x0,x1,phi2,a05); % 26, -2
%       inverse transform
        X2=cell_idct2phi(Y_dct(1),Y_dct(2),phi2,a05);
%       ----------------------------------------------
%       III. Integer 2-point DCT when angle=atan(2)
        phi0=atan(2);      % 1.1071
        Y_dct0=cell_dct2phi(x0,x1,phi0,a05); % or use
        Y_dct0=art_dct2c(x0,x1,a05);  % 24 -10
%       inverse transform (use one of these commands)
        X0=cell_idct2phi(Y_dct0(1),Y_dct0(2),phi0,a05);
        X0=art_dct2c(Y_dct0(1),Y_dct0(2),a05);
%       ----------------------------------------------
%       IV. Golden 2-point DCT
        phi_g=(sqrt(5)+1)/2;   %  1.6180 (in radian)
        phi_gr=phi_g/pi*180;   %  92.7065 (in angle)
%       1. direct transform
        b=sin(phi_g); c=cos(phi_g);   % 0.9989, -0.0472
        A=[ b   c
            c  -b];
        C=X*A;                 % 16.0366, -20.7804
%       2. integer approximation
        Yg_dct=cell_dct2phi(x0,x1,phi_g,a05); % 16 -21
%       3. inverse transform
        Xg=cell_idct2phi(Yg_dct(1),Yg_dct(2),phi_g,a05);
%       end of main program
%       ----------------------------------------------
        function Y=cell_dct2phi(x0,x1,phi,a05)
        b=sin(phi);  c=cos(phi);  c0=(c-1)/b;
        t0 = x0 + fl_byc(x1,c0,a05);
        y1 = x1 + fl_byc(t0,b, a05);
        y2 = t0 + fl_byc(y1,c0,a05);
        Y=[y1 y2];
```

```
%    ---------------------------------------------------
     function X=cell_idct2phi(y1,y2,phi,a05)
     b=sin(phi);  c=cos(phi);  c0=(c-1)/b;
     t0 = y2 + fl_byc(y1,-c0,a05);
     x1 = y1 + fl_byc(t0, -b,a05);
     x0 = t0 + fl_byc(x1,-c0,a05);
     X=[x0 x1];
%    ---------------------------------------------------
     function Y=art_dct2c(x0,x1,a05)
     c=1/sqrt(5);
     t1 = bitshift(x0,1) + x1;
     t2 = x0 - bitshift(x1,1);
     y1 = fl_byc(t1,c,a05);
     y2 = fl_byc(t2,c,a05);
     Y=[y1 y2];
%    ---------------------------------------------------
     function y=fl_byc(x,c,a05)
        if x==0 y=0;
        else
           y=floor(c*x + a05);
        end;
%    ----------------------------------------------------------
```

3.6.7   Four- and eight-point IDCTs

We consider the application of the two-point kernel IDCT for calculating the
N-point DCT of type II for the $N = 8$ case. For that, we can use the known
method [41] of decomposition of the eight-point DCT by two-point DCTs and
one four-point DCT as shown in Figure 3.11. Eight two-point DCTs are used
in this decomposition, namely, seven two-point DCTs of type II each and one
two-point DCT of type IV. This diagram can be used for realization of the
eight-point IDCT, if we consider all short DCTs to be integer transforms.

The four-point DCT of type IV is used to define the transform outputs at
points 1, 3, 5, and 7. The coefficients of the four-point IDCT can be found
from the canonical representation similar to the two-point IDCT case. In
canonical representation, the four-point IDCT of type IV is written as

$$
\begin{bmatrix} \vartheta_0 \\ \vartheta_1 \\ \vartheta_2 \\ \vartheta_3 \\ \vartheta_4 \end{bmatrix} = \begin{bmatrix} 1 & c_0 & c_1 & c_2 & 0 & 0 & 0 & 0 \\ 0 & 1 & c_4 & c_5 & c_3 & 0 & 0 & 0 \\ 0 & 0 & 1 & c_7 & c_6 & c_7 & 0 & 0 \\ 0 & 0 & 0 & 1 & -c_3 & -c_5 & c_4 & 0 \\ 0 & 0 & 0 & 0 & 1 & -c_2 & c_1 & -c_0 \end{bmatrix} \circ [x_0, x_1, x_2, x_3, \vartheta_0, \vartheta_1, \vartheta_2, \vartheta_3]' \quad (3.58)
$$

where the coefficients $c_0 = -0.594161$, $c_1 = -4.025361$, $c_2 = -1.435140$,
$c_3 = 0.693520$, $c_4 = 3.184516$, $c_5 = 1.133248$, $c_6 = 0.441342$, and $c_7 = 0.211380$. The operation $\circ$ is a vector operation of truncation multiplication
$c \circ x = \lfloor cx + 0.5 \rfloor$, where cx is considered to be the multiplication of vector-

row coefficient $c = [c_1, c_2, c_3, c_4]$ and vector $x = [x_1, x_2, x_3, x_4]'$, i.e.,

$$cx = [c_1, c_2, c_3, c_4] \cdot [x_1, x_2, x_3, x_4]' = c_1 x_1 + c_2 x_2 + c_3 x_3 + c_4 x_4$$

for coefficients c_k and integer inputs x_k, $k = 1, 2, 3, 4$.

The inverse four-point DCT of type IV in canonical representation can be written as

$$
\begin{bmatrix} \vartheta_0 \\ x_3 \\ x_2 \\ x_1 \\ x_0 \end{bmatrix}
=
\begin{bmatrix}
1 & c_0 & -c_1 & c_2 & 0 & 0 & 0 & 0 \\
0 & 1 & -c_4 & c_5 & c_3 & 0 & 0 & 0 \\
0 & 0 & 1 & -c_7 & -c_6 & -c_7 & 0 & 0 \\
0 & 0 & 0 & 1 & -c_3 & -c_5 & -c_4 & 0 \\
0 & 0 & 0 & 0 & 1 & -c_2 & -c_1 & -c_0
\end{bmatrix}
\circ
\begin{bmatrix} \vartheta_4 \\ \vartheta_3 \\ \vartheta_2 \\ \vartheta_1 \\ \vartheta_0 \\ x_3 \\ x_2 \\ x_1 \end{bmatrix}.
\tag{3.59}
$$

The complexity of the four-point IDCT by the direct calculation in (3.58) is estimated as $o[15m, 20a, 5f]$. Therefore, the complexity of the eight-point IDCT with decomposition shown in Figure 2.10 can be estimated as follows: $8 \times o[3m, 6a, 3f] + o[15m, 20a, 5f] = o[39m, 68a, 29f]$. The complexity of the eight-point IDCT, when using kernel transforms for calculation of two-point IDCTs, is estimated by $7 \times o[2m, 6a, 2f] + o[3m, 8a, 3f] + o[15m, 20a, 5f] = o[32m, 70a, 22f]$, and what's more 16 additions by 0 and ± 1 are considered to be trivial. Thus seven multiplications and seven floor functions can be saved when using the two-point kernel IDCTs. One can use the parameterized two-point DCT $\mathcal{C}_2(\varphi_0)$ instead of the two-point IDCT of type IV. Such substitution will save one multiplication, two additions, and one floor function.

The kernel transform can be used for calculating the eight-point reversible Walsh-Hadamard transform (RWHT). The calculation cost for the 8-point RWHT can be estimated as $12 \times o[2m, 6a, 2f] = o[24m, 72a, 24f]$, instead of $12 \times o[3m, 6a, 3f] = o[36m, 72a, 36f]$ in the known algorithm [41]. Thus, twelve multiplications and twelve operations of the floor function can be saved when applying the kernel transforms in the proposed look-up table method.

Problems

We first formulate two problems related to the decomposition of the discrete cosine transforms by projection operators, and then list a few problems as exercises on the above discussed topics of the paired transform and integer DCT.

Problem 3.1 For an arbitrary transformation, the decomposition of its matrix assumes the expression by defined components (or eigenfunctions of the matrix). The projection operators characterize the properties of the transform and we consider such operators on the $N = 3$ example.

Let us consider the following matrix of the three-point discrete cosine transform, DCT of type II:

$$C = \begin{pmatrix} 1 & 1 & 1 \\ \cos(w) & 0 - \cos(w) \\ \cos(2w) & -1 & \cos(2w) \end{pmatrix} = \begin{pmatrix} 1 & 1 & 1 \\ \frac{\sqrt{3}}{2} & 0 & -\frac{\sqrt{3}}{2} \\ \frac{1}{2} & -1 & \frac{1}{2} \end{pmatrix},$$

where $w = \pi/6$. The determinant of this matrix $\det(C) = -2.5981$ and three eigenvalues equal $\lambda_{1,2} = \pm\sqrt{2\cos(w)} = \pm 3^{\frac{1}{4}} = \pm 1.3161$, $\lambda_3 = 1.5$.

A. Show that the projection operators P_1, P_2, and P_3 of this matrix are defined by the following matrices:

$$P_1 = U_1 + V_1 = \frac{1}{11}\begin{pmatrix} 10 & 0 & 10 \\ 4 & 0 & 4 \\ 1 & 0 & 1 \end{pmatrix} + \frac{3}{11}\begin{pmatrix} 2 & 0 & 2 \\ 3 & 0 & 3 \\ -2 & 0 & -2 \end{pmatrix},$$

$$P_2 = U_2 + V_2 = \frac{1}{6+4b}\begin{pmatrix} 1+b & 0 & -2-b \\ -a-b & 0 & a+2b \\ -1-b & 0 & 2+b \end{pmatrix} + \begin{pmatrix} 0 & -(2b)^{-1} & 0 \\ 0 & 2^{-1} & 0 \\ 0 & (2b)^{-1} & 0 \end{pmatrix},$$

$$P_3 = U_3 + V_3 = \frac{1}{6-4b}\begin{pmatrix} 1-b & 0 & -2+b \\ b(1-b) & 0 & b(-2+b) \\ -1+b & 0 & 2-b \end{pmatrix} + \frac{1}{2}\begin{pmatrix} 0 & b^{-1} & 0 \\ 0 & 1 & 0 \\ 0 & -b^{-1} & 0 \end{pmatrix},$$

where $a = \sqrt{3}$ and $b = \lambda_1 = 3^{\frac{1}{4}}$.

B. Show that the following properties hold for the projection operators:

p1. $U_1^2 = U_1$, $V_1^2 = 0$, $U_1 V_1 = 0$, $V_1 U_1 = V_1$.
p2. $U_n^2 = \frac{1}{2}U_n$, $V_n^2 = \frac{1}{2}V_n$, $n = 2, 3$.
p3. $U_n V_n = \frac{1}{2}V_n$, $V_n U_n = \frac{1}{2}U_n$, $n = 2, 3$.

C. Obtain the following properties:

p4. $V_2 V_3 + V_3 V_2 = \frac{1}{2}(V_2 + V_3)$,
p5. $U_2 V_3 + U_3 V_2 = -\frac{1}{2}(V_2 + V_3)$,
p6. $U_2 U_3 = -V_2 V_3$, $V_2 V_3 = -U_2 V_3$.

Problem 3.2 Since $P_1 = I - U_2 - U_3 - V_2 - V_3$, the four matrices U_2, U_3, V_2, and V_3 define the discrete cosine transform. In general, we denote by $\mathcal{C}(3)$ the set of four matrices which possess the above listed properties. It is possible to substitute the given four matrices by four others from this set, in order to obtain an integer matrix (3×3) which would be similar to the DCT matrix. Construct three-point integer transforms by using matrices from $\mathcal{C}(3)$.

Problem 3.3 Compute the discrete cosine transform of type II of the signal $\mathbf{f} = \{2, 1, 4, 3, 8, 2, 3, 5, 2, 4, 1, 6, 2, 4, 3, 1\}$ from the DFT.

Problem 3.4 Compute the four-point DCT of type II of the signal $\mathbf{f} = \{2, 1, 4, 3\}$, by using the decomposition by the paired transform as shown in Figure 3.2.

Problem 3.5 Derive the formula of decomposition of the inverse four-point DCT of type II by the inverse paired transform.

Problem 3.6 Draw the signal-flow graph for calculating the inverse four-point DCT of type II by the inverse paired transform.

Problem 3.7 Compute the four-point DCT of type IV of the signal $\mathbf{f} = \{2, 1, 4, 3\}$.

Problem 3.8 Compute the eight-point DCT of type II of the signal $\mathbf{f} = \{2, 1, 4, 3, 5, 1, 7, 3\}$, by using the decomposition by the paired transform as shown in Figure 3.1.

Problem 3.9 Compute the first splitting-signal $f_{T_1'}$ of the signal $\mathbf{f} = \{2, 1, 4, 3, -5, -1, -7, -3\}$.

Problem 3.10 Compute the eight-point DCTs of type II of the signals

$$\mathbf{f}^r = \{3, 2, 1, 4, 3, 5, 1, 7\}, \quad \text{and} \quad \mathbf{f}^l = \{1, 4, 3, 5, 1, 7, 3, 2\},$$

and compare these transforms with the DCT of the signal $\mathbf{f} = \{2, 1, 4, 3, 5, 1, 7, 3\}$.

Problem 3.11 Compute the matrix of the 16-point paired transformation χ_{16}' with respect to the cosine transform.

Problem 3.12 Draw the basis functions of the 16-point paired transformation χ_{16}' with respect to the cosine transform.

Problem 3.13 Compute the first two splitting-signals $f_{T_1'}$ and $f_{T_2'}$ of the signal $\mathbf{f} = \{2, 1, 4, 3, 5, 1, 7, 3, -3, 7, -1, 5, -3, 4, -1, 2\}$.

Problem 3.14 Calculate the integer four-point DCT of type II of the signal $\mathbf{f} = \{2, 1, 4, 3\}$, by using the method of control bits in the paired algorithm.

Problem 3.15 Calculate the integer four-point DCT of type IV of the signal $\mathbf{f} = \{2, 1, 4, 3\}$, by using the method of control bits in the paired algorithm. For that use one of the signal-flow graphs shown in Figures 3.9 and 3.10.

Problem 3.16* Compute the eight-point integer DCT of type II of the signal $\mathbf{f} = \{2, 1, 4, 3, 5, 1, 7, 3\}$, by using the method of control bits and lifting schemes in the paired algorithm, as shown in Figure 3.11.

Problem 3.17 Compute the four-point reversible integer DCT of type II of the signal $\mathbf{f} = \{10, 20, 30, 40\}$, by using the method of nonlinear equations, when parameter $\alpha = 0.5$.

Problem 3.18 Compute the eight-point reversible integer DCT of type II of the signal $\mathbf{f} = \{1, 2, 3, 4, 5, 6, 7, 8\}$, by using the method of nonlinear equations (see §3.5), when parameter $\alpha = 0.5$ and 0.45.

Problem 3.19 Compute the eight-point reversible integer DCT of type II of the signal $\mathbf{f} = \{31, 52, 13, 41, 25, 16, 37, 28\}$, by using the method of nonlinear equations, when parameter $\alpha = 0.5$ and 0.45.

4

Hadamard Transform

It is well known that the discrete Hadamard transform (DHdT) is computationally advantageous over the fast Fourier transform. Being orthonormal and taking value 1 or -1 in each point, the Hadamard functions can be used for a series expansion of the signal. The DHdT has found useful applications in signal and image processing, communication systems, image coding, image enhancement, pattern recognition, and general two-dimensional filtering [4],[65]-[68]. We describe properties of the Hadamard transformation and a class of the so-called bit-and transformations, as well as the class of mixed transformations.

4.1 The Walsh and Hadamard transform

In this section, we consider a discrete unitary transform, namely, the Hadamard transform, which can be constructed recursively from the matrix (2×2)

$$[\mathcal{A}_2] = \begin{bmatrix} 1 & -1 \\ 1 & 1 \end{bmatrix} \qquad \left([\mathcal{F}_2] = \begin{bmatrix} 1 & 1 \\ 1 & -1 \end{bmatrix}\right) \tag{4.1}$$

which is different from the matrix $[\mathcal{F}_2]$ of the two-point DFT.

The basic functions of the Hadamard transform are derived from the following continuous in time functions which are called Radamacher functions and based on the quantization of the sinusoidal waves $\sin(2^n \pi t)$, where n is a nonnegative integer, $rad(n, t) = \text{sign}(\sin(2^n \pi t))$, $t \in [0, 1)$, and the sign function $\text{sign}(x) = \pm 1$ depending on $x \geq 0$ or < 0. We here remind readers that the paired transform can also be derived by quantization of cosine waves (see Figure 1.7). As an example, Figure 4.1 shows the first six Radamacher functions $rad(n, t)$, $n = 0 : 5$, in the interval $[0, 1)$. All these functions are periodic with the common period $T = 1$.

The periodic Radamacher functions can be calculated recursively in the interval $[0, 1)$. For $n > 1$, the following holds:

$$rad(n, t) = rad(1, 2^{n-1}t) = rad(1, 2^{n-2}[t/2]) = rad(n-1, t/2),$$

when $rad(1, t) = \text{sign}(0.5 - t)$, i.e., $rad(1, t) = 1$ for $t \in [0, 1/2)$ and $rad(1, t) = -1$ for $t \in [0, 1/2)$.

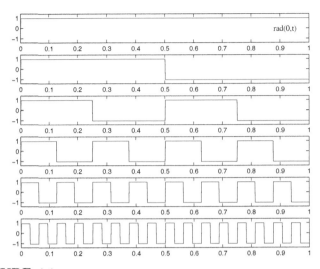

FIGURE 4.1
The first six Radamacher functions.

The discrete N-point Radamacher functions $Rad(n,k)$, $k = 0 : (N-1)$, are calculated by sampling the corresponding functions $rad(n, k/N)$ for N discrete values of t in the interval $[0, 1]$. For instance, for $N = 8$, the first four discrete functions represent rows of the following matrix (4×8)

$$[Rad] = \begin{bmatrix} 1 & 1 & 1 & 1 & 1 & 1 & 1 & 1 \\ 1 & 1 & 1 & 1 & -1 & -1 & -1 & -1 \\ 1 & 1 & -1 & -1 & 1 & 1 & -1 & -1 \\ 1 & -1 & 1 & -1 & 1 & -1 & 1 & -1 \end{bmatrix}. \tag{4.2}$$

These functions compose an incomplete system of basic functions, since they are odd (when $n \geq 1$). The discrete paired functions compose a complete set. However, Radamacher functions are used to compose a complete system of the Hadamard transformation, which are called Walsh functions.

We first consider the Walsh functions [64] which are defined by the operation called the Gray code which is illustrated in Table 4.1 for integers $n = 0 : 7$.

TABLE 4.1
Binary and Gray codes

n	0	1	2	3	4	5	6	7
Binary code	000	001	010	011	100	101	110	111
Gray code	000	001	011	010	110	111	101	100

The Gray code is calculated as follows. Let $a_{r-1}a_{r-2}...a_2a_1$ be the binary code of integer n. The binary digits b_k of n in the Gray code are calculated as $b_k = a_k \oplus a_{k+1}$, when $k = 0 : (r - 2)$, and $b_{r-1} = a_{r-1}$. For instance, let $n = 3$, or $n = {}'a_2 a_1 a_0' = {}'011'$ in the binary representation. We have $a_2 = 0$, $a_1 = 1$, and $a_0 = 1$. Then, $b_2 = a_2 = 0$, $b_1 = a_1 \oplus a_2 = 1 \oplus 0 = 1$, and $b_0 = a_0 \oplus a_1 = 1 \oplus 1 = 0$. Therefore, the Gray code of n is $'b_2 b_1 b_0' = {}'010'$.

The complete system of Walsh functions for $N = 8 = 2^3$ is defined from three functions $rad(1, t), rad(2, t)$, and $rad(3, t)$ as follows:

$$wal(0, t) = 1, \quad wal(1, t) = rad(1, t), \quad wal(2, t) = rad(1, t) \cdot rad(2, t),$$
$$wal(3, t) = rad(2, t), \quad wal(4, t) = rad(2, t) \cdot rad(3, t),$$
$$wal(5, t) = rad(1, t) \cdot rad(2, t) \cdot rad(3, t),$$
$$wal(6, t) = rad(1, t) \cdot rad(3, t), \quad wal(7, t) = rad(3, t).$$

The rule is the following: the basic function $wal(n, t)$ is the product of $rad(b_k = 1, t)$ functions, where b_k are units in the Gray code of n. For instance, $n = 5$ in the Gray code is $'111'$, and the basic function $wal(5, t)$ is thus equal to $rad(1, t) rad(2, t) rad(3, t)$. The following property holds for the functions: $wal(n, t) wal(m, t) = wal(n \oplus m, t)$.

Figure 4.2 shows the first eight basic Walsh functions in the interval $[0, 1)$. The discrete Walsh transformation is defined by the sampled Walsh continuous in time functions. The basic functions of the Hadamard transformation are defined from the Walsh functions $wal(n, t)$ whose indices n are ordered in accordance with the Gray code. In the above $N = 8$ case, the order of Walsh

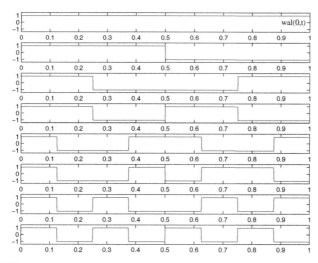

FIGURE 4.2
The first eight Walsh functions.

functions is $[0, 1, 3, 2, 7, 6, 4, 5]$.

Let $N = 2^r$, $r > 1$, and let $\mathcal{A}_N$ be the N-point discrete Hadamard transformation (DHdT), the image of which over a sequence $f = \{f_n\}$ of length N is written as

$$A_p = (\mathcal{A}_N \circ f)_p = \sum_{n=0}^{N-1} f_n a(p, n), \quad p = 0 : (N - 1),$$

where the kernel of the transformation is defined by

$$a(p, n) = a_N(p, n) = (-1)^{n_0 p_0 + n_1 p_1 + \cdots + n_{r-1} p_{r-1}}, \quad n = 0 : (N - 1).$$

n_i, p_i are coefficients of expansions of numbers n and p in the binary representation

$$k = k_0 + k_1 2^1 + \ldots + k_{r-1} 2^{r-1} \rightarrow (k_0, k_1, \ldots, k_{r-1}),$$

where $k_n = 0$ or 1, for $n = 1 : (r - 1)$.

As an example, Figure 4.3 shows a signal of length 512 and its DHdT.

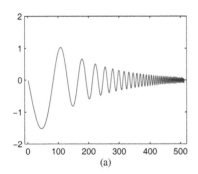
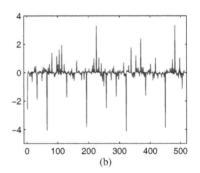

FIGURE 4.3

(a) Discrete-time signal and (b) the Hadamard transform of the signal.

Example 4.1

Let $\mathcal{A}_8$ be the eight-point transformation and

$$A_p = (\mathcal{A}_8 \circ f)_p = \sum_{n=0}^{7} f_n (-1)^{n_0 p_0 + n_1 p_1 + n_2 p_2}, \quad p = 0 : 7.$$

Here $n = n_0 + 2^1 n_1 + 2^2 n_2$, $p = p_0 + 2^1 p_1 + 2^2 p_2$, $n_i, p_i = 0$ or 1, for all $i = 0, 1, 2$. The construction of the matrix $[\mathcal{A}_8]$ is illustrated in Table 4.2.

TABLE 4.2

Construction of the Hadamard matrix (8×8)

p, n	$p_0\ p_1\ p_2$	$n_0\ n_1\ n_2$	$[\mathcal{A}_8] = \|\ (-1)^{n_0p_0+n_1p_1+n_2p_2}\ \|$							
7	1 1 1	1 1 1	1	-1	-1	1	-1	1	1	-1
6	0 1 1	0 1 1	1	1	-1	-1	-1	-1	1	1
5	1 0 1	1 0 1	1	-1	1	-1	-1	1	-1	1
4	0 0 1	0 0 1	1	1	1	1	-1	-1	-1	-1
3	1 1 0	1 1 0	1	-1	-1	1	1	-1	-1	1
2	0 1 0	0 1 0	1	1	-1	-1	1	1	-1	-1
1	1 0 0	1 0 0	1	-1	1	-1	1	-1	1	-1
0	0 0 0	0 0 0	1	1	1	1	1	1	1	1

One can see that the matrix $[\mathcal{A}_8]$ is the block matrix consisting of the Hadamard matrices (4×4), namely:

$$[\mathcal{A}_8] = \begin{bmatrix} [\mathcal{A}_4] & -[\mathcal{A}_4] \\ [\mathcal{A}_4] & [\mathcal{A}_4] \end{bmatrix}, \qquad [\mathcal{A}_4] = \begin{bmatrix} 1 & -1 & -1 & 1 \\ 1 & 1 & -1 & -1 \\ 1 & -1 & 1 & -1 \\ 1 & 1 & 1 & 1 \end{bmatrix}. \qquad (4.3)$$

Hence, the conditions of the orthogonality of rows and columns of the matrix $[\mathcal{A}_8]$ directly follow from the orthogonality of $[\mathcal{A}_4]$. ☐

The representation of the Hadamard matrix $[\mathcal{A}_N]$ for the other arbitrary power of two $N = 2^r$, $r \geq 1$, is similar to (4.3), namely

$$[\mathcal{A}_N] = \begin{bmatrix} [\mathcal{A}_{N/2}] & -[\mathcal{A}_{N/2}] \\ [\mathcal{A}_{N/2}] & [\mathcal{A}_{N/2}] \end{bmatrix}.$$

In the paired representation, the Hadamard matrix can be composed similarly to the matrix of the discrete Fourier transformation in the form of

$$[\mathcal{A}_N] = \left(\bigoplus_{n=1}^{r} [\mathcal{A}_{N/2^n}] \oplus 1 \right) [\chi'_N]$$

where $[\chi'_N]$ is the matrix of the N-point discrete paired transformation. Therefore, the full decomposition of the Hadamard matrix by matrices of the paired transformation is described as $[\mathcal{A}_N] = X_N^r \cdot X_N^{r-1} \cdots X_N^2 \cdot [\chi'_N]$, where the matrices X_N^{k+1}, $k = 1 : (r - 1)$, are defined by

$$X_N^{k+1} = [\chi'_{N/2^{k+1}}] \oplus [\chi'_{N/2^{k+2}}] \oplus \cdots \oplus [\chi'_2] \oplus I_{N(1-2^{1-k})+2}.$$

Thus, with the aid of $N/2^k$-point paired transformations, the matrix of the N-point DHdT is decomposed as the product of the binary sparse matrices.

For instance, In the $N = 4$ case,

$$[\mathcal{A}_4] = X_N^2 \cdot [\chi_4'] = \begin{bmatrix} 1 & -1 & & \\ 1 & 1 & & \\ & & 1 & \\ & & & 1 \end{bmatrix} \begin{bmatrix} 1 & 0 & -1 & 0 \\ 0 & 1 & 0 & -1 \\ 1 & -1 & 1 & -1 \\ 1 & 1 & 1 & 1 \end{bmatrix} = \begin{bmatrix} 1 & -1 & -1 & 1 \\ 1 & 1 & -1 & -1 \\ 1 & -1 & 1 & -1 \\ 1 & 1 & 1 & 1 \end{bmatrix}.$$

The number of operations of addition and subtraction required for calculating the N-point DHDT equals $6N - r^2 - 3r - 6$. The N-point paired transform uses $2N - 2$ additions (subtractions).

Figure 4.4 shows the signal-flow graph of calculation of the 8-point DHdT by the paired transforms. One can see that the graph is similar to the one used for the 8-point DFT (see Figure 1.8), only without the coefficients W^t.

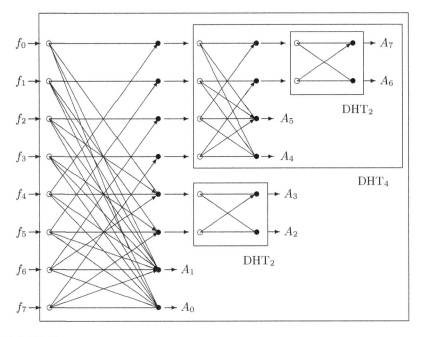

FIGURE 4.4
Signal-flow graph of the fast eight-point Hadamard transform.

Example 4.2
Let $N = 8$ and let f_n be the signal $\{1, 2, 2, 4, 5, 3, 1, 3\}$. The paired transform of this signal results in four splitting-signals as follows:

$$\chi_8' : f_n \rightarrow \{\{-4, -1, 1, 1\}, \{3, -2\}, \{-3\}, \{21\}\}$$

The calculation of the Hadamard transform of the signal f_n is performed as

$$
\begin{bmatrix} A_7 \\ A_{3\to 6} \\ A_5 \\ A_{1\to 4} \\ A_{6\to 3} \\ A_2 \\ A_{4\to 1} \\ A_0 \end{bmatrix} = \left[\begin{array}{cc} \left[\begin{array}{cc} [\mathcal{A}_2] & \\ & 1 \\ & 1 \end{array} \right] \left[\chi_4' \right] & \\ & \begin{array}{c} [\mathcal{A}_2] \\ 1 \\ 1 \end{array} \end{array} \right] \left[\chi_8' \right] \begin{bmatrix} 1 \\ 2 \\ 2 \\ 4 \\ 5 \\ 3 \\ 1 \\ 3 \end{bmatrix} = \begin{bmatrix} -3 \\ -7 \\ -3 \\ -3 \\ 5 \\ 1 \\ -3 \\ 21 \end{bmatrix},
$$

where the permutation of the components A_p is performed as $p = (p_0, p_1, p_2) \to (p_2, p_1, p_0)$, and

$$
[\mathcal{A}_2] = [\mathcal{F}_2] = \begin{bmatrix} 1 & -1 \\ 1 & 1 \end{bmatrix}.
$$

Thus, the following decomposition is valid:

$$
[\mathcal{A}_8] = \left[\begin{array}{cc} [\mathcal{A}_4] & \\ & \begin{array}{c} [\mathcal{A}_2] \\ 1 \\ 1 \end{array} \end{array} \right] \left[\chi_8' \right] = \left[\begin{array}{cc} \left[\begin{array}{cc} [\mathcal{A}_2] & \\ & 1 \\ & 1 \end{array} \right] \left[\chi_4' \right] & \\ & \begin{array}{c} [\mathcal{A}_2] \\ 1 \\ 1 \end{array} \end{array} \right] \left[\chi_8' \right].
$$

☐

A similar result is valid for the 1-D DHdT (up to a permutation),

$$
A_{\overline{(2m+1)2^k}} = \sum_{t=0}^{2^{r-k-1}-1} f'_{p,t} a(m, t), \quad m = 0 : (2^{r-k-1} - 1),
$$

and the splitting of the 2^r-point DHdT equals

$$
\mathcal{R}(\mathcal{A}_{2^r}; \sigma') = \{ \mathcal{A}_{2^{r-1}}, \mathcal{A}_{2^{r-2}}, \mathcal{A}_{2^{r-3}}, ..., \mathcal{A}_2, 1, 1 \}.
$$

For the discrete Hadamard transform of order 2^r, on the average no more than six additions per sample are used, when using the paired transforms.

4.1.1   Codes for the paired DHdT

Below are simple examples of MATLAB-based codes for computing the discrete Hadamard transform of order $N = 2^r$, $r > 1$, by the paired transform. The code "paired_1dfft.m" from §1.4 can also be used for DHdT, by making all twiddle coefficients one.

```
% ------------------------------------------------------------------------
%  run_dhdt.m file of programs, MATLAB (library of codes of Grigoryans)
%  Demo code for computing the discrete Hadamard transform, when N=2^r.
%  List of codes used in this program:
%  fht_1d.m   - computing the DHdT by the paired transform
%  had_Matp.m - computing the matrix of the DHdT by paired transform
%  fst_1d.m   - fast paired transform (or fastpaired_1d.m from pfft.m)
%
     %1. Integer 8-point test-signal
     N=8;
     x=[18,32,3,15,4,31,1,25];
     % Apply the paired algorithm for the Hadamard transform:
     h1=fht_1d(x,N);
     % h1=[1,23,25,7,-5,41,-77,129]
     %2. Calculate the matrix (NxN) of the DHdT
     T1=had_Matp(N);
     % the matrix T1 equals H(N:-1:1,:)
     % when comparing with the matrix H of the DHdT from MATLAB
     H=hadamard(N);
     y=H*x';
     % y'=[129,-77,41,-5,7,25,23,1], or y(N:-1:1)=h1.
% --------------------------------------------- end of the code ---
     function B1=fht_1d(A1,ND)
     MD=log2(ND);
     NK1=ND;
     B1=A1;
     for I=1:MD
        LK=NK1/2;            % LK=bitshift(NK1,-1);
        for J1=1:2*NK1:ND  % 1:bitshift(NK1,1):ND
           N_1=NK1+J1-1;
           B2=zeros(1,NK1);
           B2=B1(J1:N_1);
           B1(J1:N_1)=fst_1d(B2);
        end;
        NK1=LK;
     end;
% -----------------------------
     function A=fst_1d(A)
     ND=length(A);
     MD=log2(ND);
     LK=0; NK=ND;
     I=1;  II=1;
     while I <= MD
        NK=NK/2;            % NK=bitshift(NK,-1);
        LK=LK+NK; J=II;
        while J <= LK
           J1=J+NK;
           T=A(J1); T1=A(J);
           A(J1)=T1+T;
```

```
      A(J) =T1-T; J=J+1;
    end;
    II=LK+1; I=I+1;
  end;
% --------------------------
  function T=had_Matp(N)
  T=zeros(N);
  for i1=1:N
      y=zeros(1,N);
      y(i1)=1;
      T(:,i1)=fht_1d(y,N);
  end;
```

4.2 Mixed Hadamard transformation

The discrete Fourier and Hadamard transforms have interesting properties related to their roots or powers. To describe these properties, we start with the Hadamard transform, but in the definition of the transform we consider the normalized coefficient $1/\sqrt{N}$. For instance, the matrices of the two- and four-point transforms are considered to be equal

$$[\mathcal{A}_2] = \frac{1}{\sqrt{2}} \begin{bmatrix} 1 & -1 \\ 1 & 1 \end{bmatrix}, \qquad [\mathcal{A}_4] = \frac{1}{2} \begin{bmatrix} 1 & -1 & -1 & 1 \\ 1 & 1 & -1 & -1 \\ 1 & -1 & 1 & -1 \\ 1 & 1 & 1 & 1 \end{bmatrix}.$$

We consider a concept of combined, or mixed transformations which are the transformations in the time and frequency domains. For simplicity of our calculation, we denote by $A = [\mathcal{A}_N]$ the matrix of the DHdT, the order of which (N is a power of two) will be omitted. A mixed transformation is defined as a transformation whose matrix is a linear combination of the identity matrix I and A, i.e., $S = aI + bA$, where coefficients a and b are real or complex numbers. We call a and b *mixed parameters* for S. The $a = 0$ case corresponds to the Hadamard transform. If $a \neq 0$ and $b \neq 0$, then the mixed transformation S is referred to as a transformation in the time-and-frequency "domain."

The Hadamard matrix is a square root of the identity matrix, $A^2 = A \cdot A = I$. The inverse matrix S^{-1} is thus defined as $S^{-1} = \frac{1}{\Delta}(aI - bA)$, where $\Delta = a^2 - b^2$. Indeed, the following calculations hold:

$$S^{-1} \cdot S = \frac{1}{\Delta}(aI - bA) \cdot (aI + bA) = \frac{1}{\Delta}\left(a^2 I - b^2 A^2\right) = \frac{1}{\Delta}\left(a^2 - b^2\right) I = I.$$

The coefficients a and b by now are arbitrary, but satisfy the condition $\Delta \neq 0$. The choice of these parameters depends on what we want to do with mixed transforms, and here many questions arise. Can we find parameters that will result in a more effective application of the mixed transform than the original Hadamard transform, when solving specific problems in signal processing? Can we find the roots of the Hadamard transform in the space $\mathcal{L}(I, A)$ spanned on the transforms I and A?

Example 4.3

Let $x = x_n$ be an input signal and let $y = y_n$ be the transformed signal $y = Sx = ax + bAx$. The mean value of this signal

$$m_y = \frac{1}{N} \sum_{n=0}^{N-1} y_n = am_x + bm_{Ax}$$

is defined as a linear combination of the means of the input and its transform. It should be noted that the sum of elements along each column (except the first one) in the matrix of the Hadamard transform equals zero. Therefore $m_{Ax} = \frac{1}{\sqrt{N}} x_0$, in other words, the mean value of the Hadamard transform is defined by only the first element of the input.

If we want to make the mean of the output be equal to a given number m_0, i.e., $am_x + bm_{Ax} = m_0$, then we obtain

$$b = \frac{m_0 - am_x}{m_{Ax}} = \sqrt{N} \frac{m_0 - am_x}{x_0} \qquad (4.4)$$

if $x_0 \neq 0$. Assuming the value of m_0 to be zero, we obtain $b = -\sqrt{N}(am_x)/x_0$. Another interesting case is when the value of m_0 equals m_x. In this case, $b = \sqrt{N}(1 - a)m_x/x_0$. As an example, Figure 4.5 shows the original signal of length 512 and the mixed Hadamard transform defined for $a = 0.5$. The value of x_0 equals 1, $b = 162.0547$, and $m_y = m_x = 14.3237$. The value of the parameter a can be found from another condition, for instance, when the

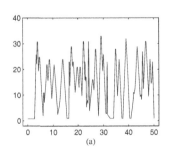
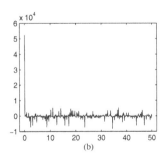

FIGURE 4.5
(a) Signal of length 512 and (b) the mixed transform defined for $a = 0.5$.

variance of y equals a given number.   ▯

Example 4.4
We now analyze the mean square deviation of the output signal $y = Sx = ax + bAx$. The following calculations hold:

$$\overline{y^2} = y'y = \frac{1}{N}(ax' + bx'A')(ax + bAx) = \frac{1}{N}(a^2 + b^2)x'x + \frac{2}{N}abx'A'x$$

since $A' = A^{-1}$. The mean value of y^2 can thus be calculated by

$$m_{y^2} = (a^2 + b^2)m_{x^2} + \frac{2}{N}abm_{x'A'x}$$

and the mean square deviation of the signal y is defined by

$$D(y) = m_{y^2} - (m_y)^2 = (a^2+b^2)m_{x^2} + \frac{2}{N}abm_{x'A'x} - \left(am_x + \frac{b}{\sqrt{N}}x_0\right)^2. \quad (4.5)$$

It is interesting to find such parameters a and b, for which the mixed Hadamard transform leads to a smaller deviation $D(y)$ than the Hadamard transform does. For such parameters, the following inequality should hold:

$$\frac{(a^2 + b^2)m_{x^2} + \frac{2}{N}abm_{x'A'x} - \left(am_x + \frac{b}{\sqrt{N}}x_0\right)^2}{m_{x^2} - \frac{1}{N}x_0^2} < 1, \quad (4.6)$$

where we use (4.5) with $a = 0$ and $b = 1$ for the deviation of the Hadamard transform. Since the scaling $x \rightarrow \lambda x$ does not change this ratio, we reduce the domain of parameters a and b to the unit circle $\{(a, b); |a|^2 + |b|^2 = 1\}$. Coefficients a and b can be parameterized by angle ϕ as $a = \sin(\phi)$ and $b = \cos(\phi)$, as well as the mixed transformation $S = S(\phi) = \sin(\phi)I + \cos(\phi)A$, where $\phi \in [0, 2\pi)$. In the $\phi = 0$ case, $S = A$, and when $\phi = \pi/2$, $S = I$. The condition in (4.6) takes the form of

$$ctg(\phi) < L = \frac{1}{2}\frac{Nm_{x^2} - x_0^2}{x'Ax - \sqrt{N}m_x x_0}. \quad (4.7)$$

As an example, Figure 4.6 shows the graph of $ctg(\phi)$ function in the interval $[-\pi/4, \pi/4]$ and the horizontal line $L = -13.87$ calculated for the signal of Figure 4.5a. The condition of (4.7) holds for small negative angles. Another horizontal line $L = 9.99$ has been calculated for a cosine-type wave, and the mentioned condition holds in this case for large positive angles.   ▯

We now consider the application of the mixed Hadamard transform in the following problem. Let $y = S(\phi)x$ be the mixed transform of the input x. We consider a random noise added to the transformed signal,

$$z = y + n = S(\phi)x + n \quad (4.8)$$

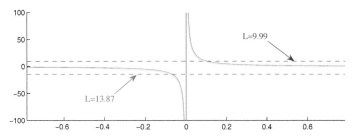

FIGURE 4.6
Cotangent function.

where, for instance, the noise n is uniform with values in $[0, r]$. We can analyze the inverse transform $\hat{y} = S(\phi)^{-1}z$ and find values of the angle ϕ, for which the error of reconstruction is small (in metric L_0)

$$\varepsilon_0(\phi) = \varepsilon_0(\hat{y}, y) = \max_n |\hat{y}_n - y_n|, \tag{4.9}$$

or (in the metric L_2)

$$\varepsilon_2(\phi) = \varepsilon_2(\hat{y}, y) = \frac{1}{N} \sqrt{\sum_{n=0}^{N-1} |\hat{y}_n - y_n|^2}. \tag{4.10}$$

As an example, Figure 4.7 shows the graph of the function $\varepsilon_0(\phi)$ calculated for the signal of Figure 4.5a in part a, along with the curve of $\varepsilon_2(\phi)$ in b. The angle ϕ runs the interval $[0, \pi/2]$. The random noise is in the interval $[0, 0.1]$. In both cases, the mixed transform $S(\phi)$, when $\phi \geq 0.8$, results in errors which are smaller than errors for the Hadamard transform (when $\phi = 0$).

4.2.1   Square roots of mixed transformations

Different order roots of the mixed transform can be derived in a simple form. For instance, we can define coefficients p and s for the square root of S in the space $\mathcal{L}(I, A)$, i.e., $S^{[1/2]} = pI + sA$. From the matrix multiplication

$$S = aI + bA = S^{[1/2]} \cdot S^{[1/2]} = (p^2 + s^2)I + 2spA,$$

we obtain the following system of equations for coefficients p and s: $p^2 + s^2 = a$, $2ps = b$. Since $a + b = (p+s)^2$ and $a - b = (p-s)^2$, the system of equations can be written as

$$\begin{cases} p + s = \pm\sqrt{a+b} \\ p - s = \pm\sqrt{a-b} \end{cases}.$$

Therefore

$$p = \frac{1}{2}\left(\pm\sqrt{a+b} \pm \sqrt{a-b}\right), \quad s = \frac{1}{2}\left(\pm\sqrt{a+b} \mp \sqrt{a-b}\right)$$

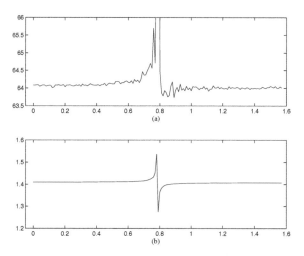

FIGURE 4.7
(a) The errors (a) $\varepsilon_0(\phi)$ and (b) $\varepsilon_2(\phi)$ of signal reconstruction.

and there are two pairs of matrices for the square root $S^{[1/2]}$, which differ by only the sign. These two matrices are

$$S^{[1/2]} = \frac{1}{2}\left(\sqrt{a+b}+\sqrt{a-b}\right)I + \frac{1}{2}\left(\sqrt{a+b}-\sqrt{a-b}\right)A, \qquad (4.11)$$

$$S^{[1/2]} = \frac{1}{2}\left(\sqrt{a+b}-\sqrt{a-b}\right)I + \frac{1}{2}\left(\sqrt{a+b}+\sqrt{a-b}\right)A.$$

For instance, if $(a, b) = (2, 1)$ and $(a, b) = (5, 1)$ we obtain respectively

$$S = 2I + A, \quad S^{[1/2]} = \frac{1}{2}\left[\left(\sqrt{3}+1\right)I + \left(\sqrt{3}-1\right)A\right],$$

$$S = 5I + A, \quad S^{[1/2]} = \frac{1}{2}\left[\left(\sqrt{6}+2\right)I + \left(\sqrt{6}-2\right)A\right].$$

As an example, Figure 4.8 shows the signal x sampled with frequency 512Hz in the interval $[0, 50]$ in part a, along with the 512-point Hadamard transform in b, the mixed transform S with parameters $(a, b) = (2, 1)$ in c, and the root of the transform in d.

If parameters a and b are such that $a - b < 0$, then the matrix $S^{[1/2]}$ defines complex unitary transforms which are roots of the mixed transformation. Similarly, other roots of mixed transforms can be defined. We now consider roots of the Hadamard transformation.

Example 4.5
Let us define such parameters a and b so that the matrix S of the mixed

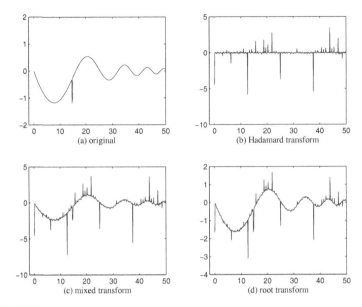

FIGURE 4.8

(a) Original signal of length 512, (b) 512-point DHdT, (c) mixed transform with parameters $(a, b) = (2, 1)$, and (d) root of the transform.

transformation satisfies the condition $S^2 = A$. If we open this equation

$$S^2 = (aI + bA)(aI + bA) = (a^2 + b^2)I + 2abA = A$$

we obtain the following system of equations for parameters a and b: $a^2 + b^2 = 0$ and $2ab = 1$, or $(a + b)^2 = 1$ and $(a - b)^2 = -1$. Therefore $a = (\pm 1 \pm j)/2$ and $b = (\pm 1 \mp j)/2$. Taking the cases $a = (1 + j)/2$ and $b = (1 - j)/2$, and $a = (1 - j)/2$ and $b = (1 + j)/2$, we obtain the following matrices

$$S = S_1 = \frac{1+j}{2}I + \frac{1-j}{2}A, \quad S = S_1^* = \frac{1-j}{2}I + \frac{1+j}{2}A,$$

such that $S^2 = A$, $S \cdot S^* = I$. Each of these matrices defines a complex unitary mixed transformation, which is a root of the Hadamard transformation. These transformations can also be defined in complex form as follows:

$$S = S_1 = \frac{1}{2}(I + A) + j\frac{1}{2}(I - A), \quad S = S_1^* = \frac{1}{2}(I + A) - j\frac{1}{2}(I - A).$$

As an example, Figure 4.9 shows the real and imaginary parts of the square root S_1 of the 512-point DHdT calculated over the signal of Figure 4.8a.

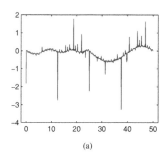
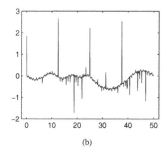

(a)   (b)

FIGURE 4.9

(a) Real and (b) imaginary parts of the square root Hadamard transform.
(The transform has been shifted cyclicly to the center.)

The above obtained square roots of the Hadamard transform are not unique
and other roots exist. We leave to the reader to verify this fact, by solving
Problem 4.2. Other roots of the DHdT can also be considered.

Example 4.6

Let $S = aI + bA$ be such a matrix of the mixed transformation that $S^3 = A$.
Then, from equation $(aI + bA)^3 = A$, the following system of equations can
be derived for parameters a and b

$$\begin{cases} a^2 + 3b^2 = 0 \\ 3a^2b + b^3 = 1 \end{cases} \Rightarrow \quad b^3 = -\frac{1}{8} \quad \text{and} \quad a = \pm j\sqrt{3}b.$$

There are three solutions for b and for simplicity, we consider the $b = -1/2$
case, when $S = S_1 = j[\sqrt{3}I - A]/2$, or $S = S_2 = -j[\sqrt{3}I - A]/2$. Then
$a = \pm j\sqrt{3}/2$, $a^2 - b^2 = -1$, and the following interesting property holds

$$S \cdot S^* = (aa^* + b^2)I + b(a + a^*)H = \left(\frac{3}{4} + \frac{1}{4}\right)I = I. \qquad (4.12)$$

Other cases also can be considered, when $b = b_2 = -\frac{1}{2}e^{j\frac{2\pi}{3}}$, and $b = b_3 = -\frac{1}{2}e^{-j\frac{2\pi}{3}}$. For instance, for parameters $b = b_2$ and $a = j\sqrt{3}b$, we obtain the
matrix $S = S_3 = b\left(j\sqrt{3}I + A\right) = -\frac{1}{2}e^{j\frac{2\pi}{3}}\left(j\sqrt{3}I + A\right)$. This matrix does not
satisfy condition (4.12), since $S \cdot S^* = (aa^* + b^2)I = 4b^2I = e^{-j\frac{2\pi}{3}}I \neq I$.

As an example, Figure 4.10 shows the real and imaginary parts of the
3rd degree root S_3 of the Hadamard transform calculated over the signal of
Figure 4.8a. ⧠

4.2.2 High degree roots of the DHdT

In order to find the high degree roots of the Hadamard matrix in the general
case, we define two matrices $B = (I - A)/2$, and $C = (I + A)/2$, which have the

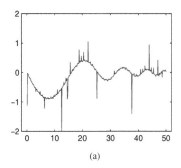

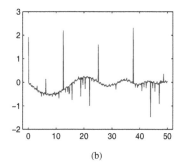

(a)   (b)

FIGURE 4.10
(a) Real and (b) imaginary parts of the 3rd degree root Hadamard transform.
(The transform has been shifted cyclicly to the center.)

following properties: $B^2 = B$, $C^2 = C$, $B+C = I$, $BC = \frac{1}{4}(I-A^2) = 0$.
The last two equalities mean that the mixed transformations defined by these
two matrices transfer a linear space of vectors into orthogonal subspaces which
divide the space. It follows by recursion that $B^n = B$ and $C^n = C$, for any
integer $n \geq 1$.

LEMMA 4.1
Let $w = w_n$ be the nth degree root of the minus unit, i.e., $w^n = -1$. Then,
the matrix $S = C+wB$ is the nth degree root of the Hadamard matrix, which
we denote by $A^{[1/n]}$.
   Proof. Since $CB = 0$, we obtain the following

$$S^n = (C + wB)^n = C^n + w^n B^n = C + w^n B = C - B = \frac{I + A}{2} - \frac{I - A}{2} = A.$$

Thus, we found n roots of the nth degree of the Hadamard matrix, as many
as -1 has. One can also see that the matrix $S = C - wB$ satisfies condition
$S^n = I$, when n is an odd integer, and $S^n = A$, when n is even. Figure 4.11
shows the real and imaginary parts of the 5rd root of the Hadamard transform
in parts a and b, along with the 12th root in (c) and (d), and 24th root in (e)
and (f). The signal of Figure 4.8a has been used for those transforms. The
root $w = \exp(j\pi/n)$ has been used for each case $n = 5, 12$, and 24.

LEMMA 4.2
Let $w = w_n$ be the n-th root of the minus unit. Then, the matrix $S^{-1} =
S^* = C + \bar{w}B$ is the inverse matrix to the nth root of the Hadamard matrix
$S = C + wB$.
   Proof: $SS^* = (C + wB)(C + \bar{w}B) = C^2 + |w|^2 B^2 = C + B = I.$

We now can generalize the concept of the power of the Hadamard matrix.

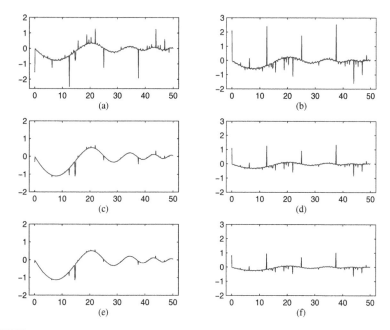

FIGURE 4.11

Real and imaginary parts of roots of the Hadamard transform: (a,b) 5th root, (c,d) 12th root, and (e,f) 24th root.

LEMMA 4.3

The matrix $S = C + (-1)^k B$, $(k \geq 0)$, is the kth power of the Hadamard matrix, i.e., $S = A^k$.

Proof. Indeed, the following calculations hold: $C + B = I$, when $k = 0$, $C - B = A$, when $k = 1$, and by recursion we obtain $C + (-1)^k B = (C - B)(C + (-1)^{k-1}B) = AA^{k-1} = A^k$, when $k > 1$.

As a conclusion of these lemmas, one can see that the linear space $\mathcal{L}(I, A)$ spanned on two operators I and A contains roots and powers of the Hadamard transformation.

4.2.3   S-x transformation

For a given sequence or discrete-time signal $x = \{x_0, x_1, ..., x_{N-1}\}$, where $N > 0$, we consider the following transformation of this signal into the space of matrices $(N \times N)$. Let S be the above considered Nth root of the Hadamard transformation, $S = A^{[1/N]}$. We define the transformation by

$$x \to S_x = \sum_{n=0}^{N-1} x_n S^n. \tag{4.13}$$

This transformation which we call the *S-x transformation* has the following properties:

P1: $S_{x+y} = S_y + S_x$.

P2: $S_x S_y = S_y S_x$.

P3:

$$S_x S_y = \left[\sum_{k=0}^{N-1} x_k S^k\right]\left[\sum_{m=0}^{N-1} y_m S^m\right] = \sum_{n=0}^{N-1} z_n S^n = S_z$$

where z_n is the cyclic convolution of x_n and y_n.

We now analyze the existence of the transformation (4.13). This transformation S_x, which looks complicated at a glance, can be written as

$$S_x = \sum_{n=0}^{N-1} x_n S^n = \sum_{n=0}^{N-1} x_n(C + w^n B)$$

$$= \left(\sum_{n=0}^{N-1} x_n\right) C + \left(\sum_{n=0}^{N-1} x_n w^n\right) B = c(x)C + b(x)B$$

where

$$c(x) = \sum_{n=0}^{N-1} x_n, \qquad b(x) = \sum_{n=0}^{N-1} x_n w^n. \tag{4.14}$$

As an example, Figure 4.12 shows the real and imaginary parts of the transform S_x performed over the signal x_n of length $N = 512$ shown in Figure 4.8. The following transform has been performed: $x \rightarrow S_x(x)$. In other words, the signal generates the transform which is applied over the signal. In this case $c(x) = -0.1969$ and $b(x) = -77.9208 + j75.8981$, and the root $w = \exp(-j\pi/512)$.

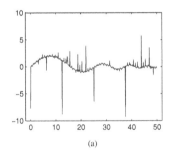
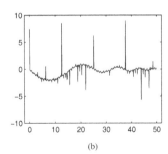

(a)   (b)

FIGURE 4.12

(a) Real and (b) imaginary parts of the 512-point transform $S_x(x)$ performed over the signal of Figure 4.8.

The construction of the matrix S_x includes the concept of the Fourier transform of the signal x_n. Indeed, let $\hat{X}_p$, $p = 0 : (2N-1)$ be the $2N$-point DFT of the sequence $\hat{x}_n$ obtained from x_n by the up-sampling operation by factor 2. Thus, the new signal $\hat{x}_n$ of length $2N$ equals $\{x_0, 0, x_1, 0, x_2, 0, \ldots, x_{N-1}, 0\}$. Coefficients $c(x)$ and $b(x)$ defined in (4.14) can be written as

$$c(x) = \hat{X}_0, \qquad b(x) = \sum_{n=0}^{N-1} x_n e^{-j\frac{\pi}{N}n} = \sum_{n=0}^{2N-1} \hat{x}_n e^{-j\frac{2\pi}{2N}n} = \hat{X}_1,$$

when the Nth root of -1 equals $w = \exp(-j\pi/N)$. If the Nth root of -1 equals $w = \exp(-j3\pi/N)$, then $b(x) = \hat{X}_3$. Taking other roots of -1, we receive for $b(x)$ different values of the DFT of the up-sampled signal.

Given a signal y, the S-x transform of y can be written as

$$S_x(y) = c(x)Cy + b(x)By = \frac{c(x) + b(x)}{2}y + \frac{c(x) - b(x)}{2}Ay$$
$$= \frac{\hat{X}_0 + \hat{X}_1}{2}y + \frac{\hat{X}_0 - \hat{X}_1}{2}Ay = ay + bAy.$$

The mixed parameters a and b for the S-x transform are defined by the first two values of the Fourier transform of the signal-generator x (namely, $\hat{x}$).

Since $C + B = I$, we will try to find the inverse matrix in the form $S_x^{-1} = pC + sB$, for which the following equations should be solved: $c(x)p = 1$ and $b(x)s = 1$. Indeed,

$$S_x S_x^{-1} = (c(x)C + b(x)B)(pC + sB) = c(x)pC + b(x)sB = I = C + B$$

and we consider that $p = 1/c(x)$, $s = 1/b(x)$, which requires conditions $c(x) \neq 0$ and $b(x) \neq 0$. The first condition means that the mean of x is not zero, and the second one shows that the value of $\hat{X}_1$ is not equal to zero.

4.3 Generalized bit-and transformations

In all examples described above with the mixed Hadamard transformation, we have used the input signals of length N being a power of two. The Hadamard transformation exists for many other orders, for instance for orders multiple to 4 (for more information see *"A Library of Hadamard Matrices"* by N. J. A. Sloane at http://www.research.att.com/~njas/hadamard/).[*] In this section, we define and analyze the concept of the Hadamard-like transformations for any order N.

Given an integer $N \geq 1$, let $\mathcal{A}(N) = \{A\}$ be a class of matrices $(N \times N)$ which satisfy the following four properties:

---

[*]Our library of Hadamard matrices is available at *http://www.fasttransforms.com*.

P1. All elements of the matrix A consist only of ± 1.

P2. The matrix A is not singular.

P3. The inverse matrix coincides with the transposed matrix (up to the normalized coefficient), i.e., $A^{-1} = A'$.

P4. The matrix is symmetric, $A = A'$.

The first property allows for performing all the multiplication of any matrix of this class by an input signal, by using only operations of addition and subtraction. There is no zero in coefficients of the matrix, and each component of the input contributes to the output. The next property requires the existence of the inverse matrix (or the transform). Property P3 simplifies greatly the calculation of the inverse matrix (or transformation). The inverse matrix contains the same elements as the original matrix. Property P3 is thus very important and desirable in many practical applications. It is clear that the last property relates to the form of the matrix, and a permutation of a row (or a column) in the matrix may destroy this property. Two transformations whose matrices are equal up to a permutation of rows are considered to be equivalent; from the computation point of view both transforms are equal. Therefore, we may not require this property for all matrices of $\mathcal{A}(N)$, for instance, for the $N = 12$ case. In the general N case, P3 is a strong and maybe un-practicable requirement to matrices $(N \times N)$ with coefficients ± 1. We will thus try to substitute this property with a small "deviation," but one which is simple and does not require many operations.

We here focus on such a class of Hadamard-like transformations, namely, bit-and transformations, whose matrices are composed by elements calculated by this simple and elegant formula:

$$a_{n,m} = (-1)^{n \oplus m}, \quad n, m = 0 : (N-1),$$

where $n \oplus m = n_0 m_0 + n_1 m_1 + \ldots + n_r m_r$, and n_k and m_k are coefficients of the binary representation of numbers n and m, respectively. These coefficients compose the infinite table (see Table 4.3) of multiplication similar to Table 4.2

TABLE 4.3

Multiplication table for bit-and matrices

$\|(-1)^{n_0 p_0 + n_1 p_1 + n_2 p_2 + n_3 p_3 + \ldots}\|$								
1	1	1	1	1	1	1	1	.
1	-1	1	-1	1	-1	1	-1	.
1	1	-1	-1	1	1	-1	-1	.
1	-1	-1	1	1	-1	-1	1	.
1	1	1	1	-1	-1	-1	-1	.
1	-1	1	-1	-1	1	-1	1	.
1	1	-1	-1	-1	-1	1	1	.
1	-1	-1	1	-1	1	1	-1	.
.	.	.	.	.	.	.	.	.

but with the starting point $(0,0)$ at the left top corner of the table.

For a given size $N \times M$, we define the bit-and matrix as the matrix $(N \times M)$ composed by the first N rows and M columns of the above table. We denote this matrix as $A_{[N,M]}$, or $A_{[N]}$ when $M = N$. For instance, for $N = 1$ and $M = 2$, we obtain

$$A_{[1,2]} = \begin{bmatrix} 1 & 1 \end{bmatrix}, \quad A_{[2,1]} = \begin{bmatrix} 1 \\ 1 \end{bmatrix}, \quad A_{[2,1]} A_{[1,2]} = \begin{bmatrix} 1 & 1 \\ 1 & 1 \end{bmatrix}.$$

Example 4.7

For $N = 2$, we have the following matrices:

$$A_{[2]} = A_{[2,2]} = \begin{bmatrix} 1 & 1 \\ 1 & -1 \end{bmatrix}, \quad A_{[2]}^{-1} = \frac{1}{2} \begin{bmatrix} 1 & 1 \\ 1 & -1 \end{bmatrix}.$$

The matrix $A_{[2]}$ coincides with the matrix (2×2) of the Fourier transform considered above in (4.1). □

Example 4.8

The bit-and matrix (3×3) and its inverse are defined as follows:

$$A_{[3]} = \begin{bmatrix} 1 & 1 & 1 \\ 1 & -1 & 1 \\ 1 & 1 & -1 \end{bmatrix}, \quad A_{[3]}^{-1} = \frac{1}{2} \begin{bmatrix} 0 & 1 & 1 \\ 1 & -1 & 0 \\ 1 & 0 & -1 \end{bmatrix}.$$

The calculation of the direct matrix requires five operations of addition and subtraction, and three such operations for the inverse matrix.

The bit-and matrix can be written as the following block-matrix

$$A_{[3]} = \begin{bmatrix} A_{[2]} & A_{[2,1]} \\ A_{[1,2]} & -A_{[1]} \end{bmatrix}.$$

This matrix can also be considered as the submatrix of bit-and matrix (4×4)

$$A_{[4]} = \begin{bmatrix} 1 & 1 & 1 & 1 \\ 1 & -1 & 1 & -1 \\ 1 & 1 & -1 & -1 \\ 1 & -1 & -1 & 1 \end{bmatrix} = \begin{bmatrix} & & & 1 \\ & A_{[3]} & & -1 \\ & & & -1 \\ 1 & -1 & -1 & 1 \end{bmatrix}$$

which is orthogonal. The orthogonality is expressed by equations

$$\sum_{k=0}^{3} \left(A_{[4]}\right)_{m,k} \left(A_{[4]}\right)_{n,k} = 4\delta_{m,n}, \quad m, n = 0 : 3.$$

For the bit-and matrix (3×3), similar equations can be written as

$$\sum_{k=0}^{2} \left(A_{[3]}\right)_{m,k} \left(A_{[3]}\right)_{n,k} = \sum_{k=0}^{2} \left(A_{[4]}\right)_{m,k} \left(A_{[4]}\right)_{n,k} = 4\delta_{m,n} - \left(A_{[4]}\right)_{m,3} \left(A_{[4]}\right)_{n,3},$$

which in matrix form equals

$$A_{[3]}^2 = 4I_{[3]} - B_{[3]}. \tag{4.15}$$

Coefficients of the matrix $B_{[3]}$ are calculated by

$$\left(B_{[3]}\right)_{n,m} = \left(A_{[4]}\right)_{m,4} \left(A_{[4]}\right)_{n,4} = (-1)^{m \oplus 3 + n \oplus 3} = (\phi\phi')_{n,m}$$

where the vector $\phi = (1, -1, -1)'$. Thus

$$B_{[3]} = \phi\phi' = \begin{bmatrix} 1 & -1 & -1 \\ -1 & 1 & 1 \\ -1 & 1 & 1 \end{bmatrix}.$$

It should be noted that $A_{[3]}\phi = -\phi$, i.e., ϕ is the eigenvector of the bit-and matrix with eigenvalue -1. Therefore, the following holds: $A_{[3]}B_{[3]} = -B_{[3]}$, $A_{[3]}^{-1}B_{[3]} = -B_{[3]}$ and (4.15) can be written as $A_{[3]} = 4A_{[3]}^{-1} - A_{[3]}^{-1}B_{[3]} = 4A_{[3]}^{-1} + B_{[3]}$. The inverse Hadamard matrix can thus be calculated by

$$A_{[3]}^{-1} = \frac{1}{4}\left(A_{[3]} - B_{[3]}\right). \tag{4.16}$$

This equation can be considered as a measure of violation of the orthogonality in the bit-and matrix, which relates to the eigenvector $\phi = (1, -1, -1)'$. We here note that there are also another two eigenvectors $\phi = (2, 1, 1)'$ and $\phi = (0, 1, -1)'$, which can also be used to define the matrix. The characteristic equation for the matrix $A_{[3]}$ is the following: $A^3 + A^2 - 4A - 4I = (A^2 - 4I)(A + I) = 0$.

The considered bit-and matrix (3×3) has the following properties:

P1. All elements of the matrix A consist only of ± 1.

P2. The matrix A is not singular, and $\det(A) = -4$.

P3'. $A^{-1} \neq A'$ (but elements of A^{-1} consist only of 0 and ± 1).

P4. $A = A'$.

Property P3 in its original assumption has been broken; however, we can express the inverse matrix through the original matrix with a deviation. Indeed, by substituting zeros in the inverse matrix by the units, we obtain the following:

$$\left(2A_{[3]}^{-1} = \begin{bmatrix} 0 & 1 & 1 \\ 1 & -1 & 0 \\ 1 & 0 & -1 \end{bmatrix}\right) + \left(T = \begin{bmatrix} 1 & 0 & 0 \\ 0 & 0 & 1 \\ 0 & 1 & 0 \end{bmatrix}\right) = \begin{bmatrix} 1 & 1 & 1 \\ 1 & -1 & 1 \\ 1 & 1 & -1 \end{bmatrix} = A_{[3]}.$$

Therefore property P3' can be expressed as $A_{[3]}^{-1} = \frac{1}{2}\left(A_{[3]} - T\right)$, and now the permutation matrix T describes a measure of violation of the orthogonality in the bit-and matrix $A_{[3]}$. This deviation requires only three additional operations of addition, instead of five operations when using the matrix $B_{[3]}$ in (4.16). ☐

Example 4.9

The bit-and matrix (5×5) and its inverse are defined as follows:

$$A_{[5]} = \begin{bmatrix} 1 & 1 & 1 & 1 & 1 \\ 1 & -1 & 1 & -1 & 1 \\ 1 & 1 & -1 & -1 & 1 \\ 1 & -1 & -1 & 1 & 1 \\ 1 & 1 & 1 & 1 & -1 \end{bmatrix}, \quad A_{[5]}^{-1} = \frac{1}{4}\begin{bmatrix} -1 & 1 & 1 & 1 & 2 \\ 1 & -1 & 1 & -1 & 0 \\ 1 & 1 & -1 & -1 & 0 \\ 1 & -1 & -1 & 1 & 0 \\ 2 & 0 & 0 & 0 & -2 \end{bmatrix}.$$

The determinant of the direct matrix equals -32, and the matrix can be written as

$$A_{[5]} = \begin{bmatrix} A_{[3]} & A_{[3,2]} \\ A_{[2,3]} & A_{[2]} \end{bmatrix}, \quad \text{where } A_{[2,3]} = \begin{bmatrix} 1 & -1 & -1 \\ 1 & 1 & 1 \end{bmatrix}, A_{[3,2]} = A'_{[2,3]} = \begin{bmatrix} 1 & 1 \\ -1 & 1 \\ -1 & 1 \end{bmatrix}.$$

We also can express the matrix $A_{[5]}$ through the orthogonal matrix $A_{[4]}$ as

$$A_{[5]} = \begin{bmatrix} A_{[4]} & A_{[1,4]} \\ A_{[4,1]} & -1 \end{bmatrix} = \begin{bmatrix} & & & & 1 \\ & & & & 1 \\ & A_{[4]} & & & 1 \\ & & & & 1 \\ 1 & 1 & 1 & 1 & -1 \end{bmatrix}$$

and the inverse matrix by

$$A_{[5]}^{-1} = \begin{bmatrix} & & 0 & & \\ & & 0 & & \\ & A_{[4]} & 0 & & \\ & & 0 & & \\ 0 & 0 & 0 & 0 & 0 \end{bmatrix} + 2\begin{bmatrix} -1 & & 1 \\ & 0 & \\ 1 & & -1 \end{bmatrix}.$$

The following characteristic equation holds for the matrix $A_{[5]}$:

$$A^5 + A^4 - 12A^3 - 12A^2 + 32A + 32I = (A+I)(A-2I)(A+2I)(A^2 - 8I) = 0.$$

▯

Example 4.10

The bit-and matrix (6×6) and its inverse are defined as follows:

$$A_{[6]} = \begin{bmatrix} 1 & 1 & 1 & 1 & 1 & 1 \\ 1 & -1 & 1 & -1 & 1 & -1 \\ 1 & 1 & -1 & -1 & 1 & 1 \\ 1 & -1 & -1 & 1 & 1 & -1 \\ 1 & 1 & 1 & 1 & -1 & -1 \\ 1 & -1 & 1 & -1 & -1 & 1 \end{bmatrix}, \quad A_{[6]}^{-1} = \frac{1}{4}\begin{bmatrix} 0 & 0 & 1 & 1 & 1 & 1 \\ 0 & 0 & 1 & -1 & 1 & -1 \\ 1 & 1 & -1 & -1 & 0 & 0 \\ 1 & -1 & -1 & 1 & 0 & 0 \\ 1 & 1 & 0 & 0 & -1 & -1 \\ 1 & -1 & 0 & 0 & -1 & 1 \end{bmatrix}.$$

These matrices can be written as

$$A_{[6]} = \begin{bmatrix} A_{[2]} & A_{[2]} & A_{[2]} \\ A_{[2]} & -A_{[2]} & A_{[2]} \\ A_{[2]} & A_{[2]} & -A_{[2]} \end{bmatrix} = A_{[3]} \otimes A_{[2]}$$

$$A_{[6]}^{-1} = \begin{bmatrix} 0 & A_{[2]} & A_{[2]} \\ A_{[2]} & -A_{[2]} & 0 \\ A_{[2]} & 0 & -A_{[2]} \end{bmatrix} = A_{[3]}^{-1} \otimes A_{[2]}^{-1}.$$

It should be noted that $A_{[6]} \neq A_{[2]} \otimes A_{[3]}$. We now consider the bit-and matrix (6×6) as a part of the orthogonal matrix (8×8),

$$A_{[8]} = \begin{bmatrix} & & & & 1 & 1 \\ & & & & 1 & -1 \\ & & & & -1 & -1 \\ & A_{[6]} & & & -1 & 1 \\ & & & & -1 & -1 \\ & & & & -1 & 1 \\ \begin{bmatrix} a'_6 = 1 & 1 & -1 & -1 & -1 & -1 \\ a'_7 = 1 & -1 & -1 & 1 & -1 & 1 \end{bmatrix} & 1 & 1 \\ & & & & 1 & -1 \end{bmatrix},$$

where we denote the last row-vectors of length six by a_6 and a_7,

$$a_6 = (1, 1, -1, -1, -1, -1)' = \{(-1)^{6 \oplus n}; n = 0 : 5\}$$
$$a_7 = (1, -1, -1, 1, 1, -1)' = \{(-1)^{7 \oplus n}; n = 0 : 5\}.$$

It is not difficult to see that $A_{[6]}a_6 = -(a_7 + a_6)$, $A_{[6]}a_7 = (a_7 - a_6)$, and if we denote by $B_{[6]}$ the matrix

$$B_{[6]} = \frac{1}{2}(a_6 a'_6 + a_7 a'_7) = \begin{bmatrix} 1 & 0 & -1 & 0 & 0 & -1 \\ 0 & 1 & 0 & -1 & -1 & 0 \\ -1 & 0 & 1 & 0 & 0 & 1 \\ 0 & -1 & 0 & 1 & 1 & 0 \\ 0 & -1 & 0 & 1 & 1 & 0 \\ -1 & 0 & 1 & 0 & 0 & 1 \end{bmatrix},$$

then we obtain the following:

$$A_{[6]}^2 = 2(I - B_{[6]}) \quad \text{and} \quad A_{[6]}^2 B_{[6]} = 2B_{[6]}.$$

Therefore the following formula holds for the inverse bit-and matrix $A_{[6]}$:

$$A_{[6]}^{-1} = \frac{1}{2} A_{[6]} \left(I + B_{[6]} \right), \qquad (4.17)$$

where the matrix $B_{[6]}$ defines the deviation of the bit-and matrix from being orthogonal. ▯

Example 4.11

The bit-and matrix (7×7) is defined as follows:

$$
A_{[7]} = \begin{bmatrix}
1 & 1 & 1 & 1 & 1 & 1 & 1 \\
1 & -1 & 1 & -1 & 1 & -1 & 1 \\
1 & 1 & -1 & -1 & 1 & 1 & -1 \\
1 & -1 & -1 & 1 & 1 & -1 & -1 \\
1 & 1 & 1 & 1 & -1 & -1 & -1 \\
1 & -1 & 1 & -1 & -1 & 1 & -1 \\
1 & 1 & -1 & -1 & -1 & -1 & 1
\end{bmatrix} = \begin{bmatrix} A_{[4]} & A_{[4,3]} \\ A_{[3,4]} & -A_{[3]} \end{bmatrix}.
$$

$\det(A_{[7]}) = -512$ and the inverse matrix equals

$$
A_{[7]}^{-1} = \frac{1}{4} \begin{bmatrix}
1 & 0 & 0 & 1 & 0 & 1 & 1 \\
0 & 0 & 1 & -1 & 1 & -1 & 0 \\
0 & 1 & 0 & -1 & 1 & 0 & -1 \\
1 & -1 & -1 & 1 & 0 & 0 & 0 \\
0 & 1 & 1 & 0 & 0 & -1 & -1 \\
1 & -1 & 0 & 0 & -1 & 1 & 0 \\
1 & 0 & -1 & 0 & -1 & 0 & 1
\end{bmatrix}.
$$

The matrix $A_{[7]}$ can be considered as the submatrix of the Hadamard matrix (8×8), and the calculations similar to the above mentioned for the case $N = 3$ can be used to derive the analytical formula for the inverse bit-and matrix $A_{[7]}^{-1}$ by its eigenvectors. Indeed the orthogonality of the matrix (8×8)

$$
\sum_{k=0}^{7} \left(A_{[8]} \right)_{m,k} \left(A_{[8]} \right)_{n,k} = 8\delta_{m,n}, \quad n, m = 0 : 7,
$$

leads to the equality $A_{[7]}^2 = 8I_{[7]} - B_{[7]}$. Coefficients of the matrix $B_{[7]}$ are calculated by $\left(B_{[7]} \right)_{n,m} = \left(A_{[8]} \right)_{m,7} \left(A_{[8]} \right)_{n,7} = (-1)^{m\oplus 7 + n\oplus 7} = (\phi\phi')_{n,m}$, i.e.,

$$
B_{[7]} = \begin{bmatrix}
1 & -1 & -1 & 1 & -1 & 1 & 1 \\
-1 & 1 & 1 & -1 & 1 & -1 & -1 \\
-1 & 1 & 1 & -1 & 1 & -1 & -1 \\
1 & -1 & -1 & 1 & -1 & 1 & 1 \\
-1 & 1 & 1 & -1 & 1 & -1 & -1 \\
1 & -1 & -1 & 1 & -1 & 1 & 1 \\
1 & -1 & -1 & 1 & -1 & 1 & 1
\end{bmatrix}.
$$

The vector $\phi = (1, -1, -1, 1, -1, 1, 1)'$ is the eigenvector of $A_{[7]}$ with eigenvalue 1, and $A_{[7]} B_{[7]} = B_{[7]}$. Therefore the inverse bit-and transform can be calculated by $A_{[7]}^{-1} = \frac{1}{8} \left[A_{[7]} + B_{[7]} \right]$, where $B_{[7]}$ defines a measure of violation of the orthogonality in the bit-and matrix (7×7). Matrix $B_{[7]}$ consists only of ± 1 and its determinant $\det(B_{[7]}) = 0$, as all violation matrices $B_{[N]}$ considered in the above examples for $N = 3, 5$, and 6. $\quad \Box$

One can see from the given examples that although the bit-and matrix in general is not orthogonal, the inverse matrix can be calculated directly from the original matrix and a matrix of "violation," which has a simple form and does not require many operations.

In conclusion, Table 4.4 shows the characteristic equations for the first sixteen bit-and matrices $A_{[N]}$.

TABLE 4.4
Characteristic equations of matrices $A_{[N]}$

$N = 1$	$(x - 1) = 0$
$N = 2$	$(x^2 - 2) = 0$
$N = 3$	$(x^2 - 4)(x + 1) = 0$
$N = 4$	$(x^2 - 4)^2 = 0$
$N = 5$	$(x^2 - 8)^1(x^2 - 4)(x + 1) = 0$
$N = 6$	$(x^2 - 8)^2(x^2 - 2) = 0$
$N = 7$	$(x^2 - 8)^3(x - 1) = 0$
$N = 8$	$(x^2 - 8)^4 = 0$
$N = 9$	$(x^2 - 16)^1(x^2 - 8)^3(x + 1) = 0$
$N = 10$	$(x^2 - 16)^2(x^2 - 8)^2(x^2 - 2) = 0$
$N = 11$	$(x^2 - 16)^3(x^2 - 8)(x^2 - 4)(x - 1) = 0$
$N = 12$	$(x^2 - 16)^4(x^2 - 4)^2 = 0$
$N = 13$	$(x^2 - 16)^5(x^2 - 4)(x - 1) = 0$
$N = 14$	$(x^2 - 16)^6(x^2 - 2) = 0$
$N = 15$	$(x^2 - 16)^7(x + 1) = 0$
$N = 16$	$(x^2 - 16)^8 = 0$

One can notice the rule between the dimension N and the sum of the eigenvalues of the bit-and matrix. This rule for $N = 5, 11, 12$, and 15 is shown in Table 4.5. The roots λ of the characteristic equations of the bit-and

TABLE 4.5
Rules for eigenvalues

$N = 5$	$(x^2 - \underline{8})(x^2 - \underline{4})(x + \underline{1}) = 0$	$5 = 8 - 4 + 1$
$N = 11$	$(x^2 - \underline{16})^3(x^2 - \underline{8})(x^2 - \underline{4})(x - \underline{1}) = 0$	$11 = 16 - 8 + 4 - 1$
$N = 12$	$(x^2 - \underline{16})^4(x^2 - \underline{4})^2 = 0$	$12 = 16 - 4$
$N = 15$	$(x^2 - \underline{16})^7(x + \underline{1}) = 0$	$15 = 16 - 1$

matrices $(N \times N)$ are equal to the plus or minus square root of powers of two, i.e., $\lambda_n = \pm\sqrt{2^m}$, where m are integers.

4.3.1 Projection operators

In this section, the characteristic equations of the bit-and matrices $(N \times N)$ are described. These equations can be used to construct roots and powers of the matrices in the general case $N \neq 2^r$.

Let A be the bit-and matrix $(N \times N)$, and let λ_i, $i = 1 : k$, be eigenvalues of the matrix, which we consider for simplicity to be different $(k = N)$. We denote by P the characteristic polynomial of the matrix

$$P(\lambda) = \prod_{i=1}^{k}(\lambda - \lambda_i), \qquad \left(\prod_{i=1}^{k}(A - \lambda_i I) = 0\right).$$

Denoting by P_n and G_n the polynomials

$$P_n(\lambda) = \prod_{i \neq n}(\lambda - \lambda_i), \quad G_n = \prod_{i \neq n}(A - \lambda_i I), \quad n = 1 : k, \qquad (4.18)$$

we obtain the following: $G_n A = \lambda_n G_n$, $G_n G_m = 0$, $n \neq m$, and therefore

$$G_n^2 = G_n \prod_{i \neq n}(A - \lambda_i I) = \prod_{i \neq n}(\lambda_n - \lambda_i)G_n = P_n(\lambda_n)G_n.$$

Matrices G_n are proportional to the projection operators

$$P_n = \frac{1}{P_n(\lambda_n)}G_n, \quad (P_n^2 = I), \quad n = 1 : k, \qquad (4.19)$$

which compose a decomposition of the identity operator $\sum_{n=1}^{N} P_n = I$, $P_n P_m = 0$, $n \neq m$. As a result we obtain a decomposition of the bit-and matrix and its powers by the projection operators:

$$A = \sum_{n=1}^{N} \lambda_n P_n, \quad A^l = \sum_{n=1}^{N} \lambda_n^l P_n, \quad l \neq 0.$$

Example 4.12
Let A be the following bit-and matrix (3×3) :

$$A = A_{[3]} = \begin{bmatrix} 1 & 1 & 1 \\ 1 & -1 & 1 \\ 1 & 1 & -1 \end{bmatrix}.$$

The eigenvalues of this matrix are $-1, -2$, and 2, and the characteristic polynomial is $(A + I)(A + 2I)(A - 2I)$. Therefore, $AP_1 = -P_1$, $AP_2 = -2P_2$, $AP_3 = 2P_3$, and $P_1(-1) = -3$, $P_2(-2) = 4$, and $P_3(2) = 12$.

From (4.18) and (4.19), we obtain the following projection operators:

$$P_1 = \frac{G_1}{-3} = \frac{A^2 - 4I}{-3} = \frac{1}{3}\begin{bmatrix} 1 & -1 & -1 \\ -1 & 1 & 1 \\ -1 & 1 & 1 \end{bmatrix},$$

$$P_2 = \frac{G_2}{4} = \frac{A^2 - A - 2I}{4} = \frac{1}{2}\begin{bmatrix} 0 & 0 & 0 \\ 0 & 1 & -1 \\ 0 & -1 & 1 \end{bmatrix},$$

$$P_3 = \frac{G_3}{12} = \frac{A^2 + 3A + 2I}{12} = \frac{1}{6}\begin{bmatrix} 4 & 2 & 2 \\ 2 & 1 & 1 \\ 2 & 1 & 1 \end{bmatrix}.$$

The decomposition of the bit-and matrix and its powers by the projection operators can be written as $A = -P_1 - 2P_2 + 2P_3$, $A^l = (-1)^l P_1 + (-2)^l P_2 + 2^l P_3$. In the $l = -1$ and 0.5 cases, the matrices A^l are calculated as follows: $A^{-1} = -P_1 - \frac{1}{2}P_2 + \frac{1}{2}P_3$, $A^{1/2} = \sqrt{-1}P_1 + \sqrt{-2}P_2 + \sqrt{2}P_3$. One can notice that in the above equations, it is enough to use only two projection operators, for instance, P_1 and P_2, since $P_3 = I - P_1 - P_2$. ▯

4.4   T-decomposition of Hadamard matrices

In this section, we demonstrate the construction of square roots of the Hadamard matrices $(2^r \times 2^r)$, $r > 1$, by using a nonlinear method based on the Kronecker sum of matrices, which we denote by $\oplus$. Our goal is to analyze the case when the following statement is valid: If $R_1, R_2, \ldots, R_n$ are roots (or powers) of the Hadamard matrices, then their sum $R = R_1 \oplus R_2 \oplus \cdots \oplus R_n$ is also a root (or power) of a Hadamard matrix. If that is true, then the matrix R has at least n^2 square roots. Moreover, for many matrices there exist an infinite number of roots of different degrees.

We start with the case (2×2) and consider for that the following four matrices

$$T_{0,0} = \begin{bmatrix} 1 & 0 \\ 0 & 1 \end{bmatrix}, \ T_{0,1} = \begin{bmatrix} 1 & 0 \\ 0 & -1 \end{bmatrix}, \ T_{1,0} = \begin{bmatrix} 0 & 1 \\ 1 & 0 \end{bmatrix}, \ T_{1,1} = \begin{bmatrix} 0 & 1 \\ -1 & 0 \end{bmatrix}.$$

It is not difficult to see that the set of matrices obtained by multiplication from these four matrices compose a multiplicative group, since

$$T_{m_1,n_1}T_{m_2,n_2} = (-1)^{m_1 n_2}T_{m_1+n_1,m_2+n_2}, \quad m_1, m_2, n_1, n_2 = 0, 1,$$

where the sum of indices is considered by modulo 2.

The product of more than two T matrices can be written as

$$\prod_{i=1}^{N} T_{m_i,n_i} = (-1)^{a(\bar{m},\bar{n})}T_{\sum m_i, \sum n_i}$$

where $\bar{m} = (m_1, m_2, ..., m_N)$, $\bar{n} = (n_1, n_2, ..., n_N)$, $N \geq 2$, and

$$a(\bar{m}, \bar{n}) = \sum_{i=2}^{N} n_i \sum_{j=1}^{i-1} m_j .$$

The matrices T_{m_i, n_i} compose an orthogonal basis in the linear space of matrices (2×2), wherein the inner product is defined as

$$(B, C) = \sum_{i=0}^{1} \sum_{j=0}^{1} b_{i,j} c_{i,j}, \quad B = \begin{bmatrix} b_{0,0} & b_{0,1} \\ b_{1,0} & b_{1,1} \end{bmatrix}, \quad C = \begin{bmatrix} c_{0,0} & c_{0,1} \\ c_{1,0} & c_{1,1} \end{bmatrix}.$$

The coefficients $a(\bar{m}, \bar{n})$ compose the matrix $A = BC$. Let $d_{i,j}$ be coefficients of decomposition of B by matrices T,

$$B = d_{0,0} T_{0,0} + d_{0,1} T_{0,1} + d_{1,0} T_{1,0} + d_{1,1} T_{1,1}.$$

We call this decomposition T-decomposition of the matrix B. Because of the orthogonality of matrices T, $(T_{m_1,n_1}, T_{m_2,n_2}) = 2\delta_{m_1,m_2}\delta_{n_1,n_2}$, the coefficients of decomposition of the matrix B can be defined from its coefficients by the following two systems of equations:

$$\begin{cases} d_{0,0} = b_{0,0} + b_{1,1} \\ d_{0,1} = b_{0,0} - b_{1,1} \end{cases}, \quad \begin{cases} d_{1,0} = b_{0,1} + b_{1,0} \\ d_{1,1} = b_{0,1} - b_{1,0} \end{cases}.$$

These systems are defined by the Hadamard matrix as

$$\begin{bmatrix} d_{0,0} \\ d_{0,1} \end{bmatrix} = \begin{bmatrix} 1 & 1 \\ 1 & -1 \end{bmatrix} \begin{bmatrix} b_{0,0} \\ b_{1,1} \end{bmatrix}, \quad \begin{bmatrix} d_{1,0} \\ d_{1,1} \end{bmatrix} = \begin{bmatrix} 1 & 1 \\ 1 & -1 \end{bmatrix} \begin{bmatrix} b_{0,1} \\ b_{1,0} \end{bmatrix}.$$

It is not difficult to see that when taking the square of B, the coefficients of the matrix decomposition by matrices T are transformed as follows:

$$\mathcal{D}: \begin{bmatrix} d_{0,0} \\ d_{0,1} \\ d_{1,0} \\ d_{1,1} \end{bmatrix} \rightarrow \begin{bmatrix} d_{0,0}^2 + d_{0,1}^2 + d_{1,0}^2 - d_{1,1}^2 \\ 2d_{0,0}d_{0,1} \\ 2d_{0,0}d_{1,0} \\ 2d_{0,0}d_{1,1} \end{bmatrix}. \qquad (4.20)$$

This nonlinear transformation requires seven multiplications and three additions. The first component of the square matrix is defined by the norm in the Minkovsky metric. The transformation $\mathcal{D}: (d_{0,0}, d_{0,1}, d_{1,0}, d_{1,1}) \rightarrow (a_{0,0}, a_{0,1}, a_{1,0}, a_{1,1})$ is invertible. Indeed, this transformation can be rewritten in the following form of biquadratic equation:

$$d_{0,0}^4 - d_{0,0}^2 a_{0,0} + \frac{a_{0,1}^2 + a_{1,0}^2 + a_{1,1}^2}{4} = 0,$$

which leads to the solutions

$$
\begin{cases}
d_{0,0}^2 = \dfrac{a_{0,0} \pm \sqrt{a_{0,0}^2 - a_{0,1}^2 - a_{1,0}^2 - a_{1,1}^2}}{2} \\
d_{0,1} = a_{0,1}/(2d_{0,0}) \\
d_{1,0} = a_{1,0}/(2d_{0,0}) \\
d_{1,1} = a_{1,1}/(2d_{0,0}).
\end{cases}
$$

4.4.1 Square roots of the Hadamard transformation

We now consider the transformation, $\mathcal{D}$, to define square roots of the Hadamard matrix $A = A_{[2]}$. Let B be a matrix which satisfies the equation $B^2 = A$, i.e., $B = A^{[1/2]}$. It is not difficult to see that in the decomposition $A = a_{0,0}T_{0,0} + a_{0,1}T_{0,1} + a_{1,0}T_{1,0} + a_{1,1}T_{1,1}$ the coefficients equal $a_{0,0} = 0$, $a_{0,1} = \frac{1}{\sqrt{2}}$, $a_{1,0} = \frac{1}{\sqrt{2}}$, $a_{1,1} = 0$. Therefore, the transformation $\mathcal{D}$ leads to the solution of the following system of equations:

$$
\begin{cases}
d_{0,0}^2 + d_{0,1}^2 + d_{1,0}^2 - d_{1,1}^2 = 0 \\
2d_{0,0}d_{0,1} = \dfrac{1}{\sqrt{2}} \\
2d_{0,0}d_{1,0} = \dfrac{1}{\sqrt{2}} \\
2d_{0,0}d_{1,1} = 0
\end{cases}
\qquad \approx \qquad
\begin{cases}
d_{0,0}^2 + d_{0,1}^2 + d_{1,0}^2 = 0 \\
2d_{0,0}d_{0,1} = \dfrac{1}{\sqrt{2}} \\
2d_{0,0}d_{1,0} = \dfrac{1}{\sqrt{2}} \\
d_{1,1} = 0.
\end{cases}
$$

This system has four solutions

$$
\begin{cases}
d_{0,0} = d_k = \dfrac{1}{\sqrt{2}} e^{j\frac{(2k+1)\pi}{4}} \quad (k = 0:3) \\
d_{0,1} = d_{1,0} = \dfrac{1}{2\sqrt{2}d_k}, \quad d_{1,1} = 0.
\end{cases}
$$

Square roots of the Hadamard matrix can therefore be calculated as follows:

$$
A^{[1/2]} = d_{0,0}T_{0,0} + d_{0,1}T_{0,1} + d_{1,0}T_{1,0} = d_k T_{0,0} + \frac{1}{2\sqrt{2}d_k}(T_{0,1} + T_{1,0})
$$

$$
= d_k \begin{bmatrix} 1 & 0 \\ 0 & 1 \end{bmatrix} + \frac{1}{2d_k} \cdot \frac{1}{\sqrt{2}} \left(\begin{bmatrix} 1 & 0 \\ 0 & -1 \end{bmatrix} + \begin{bmatrix} 0 & 1 \\ 1 & 0 \end{bmatrix} \right) = d_k I + \frac{1}{2d_k} A,
$$

where $k = 0, 1, 2, 3$. Thus, we obtain four square roots of the Hadamard matrix, by using the decomposition of the matrix by the basis composed of matrices $T_{m,n}$. This result coincides with the results of §4.2.1, where the concept of mixed transformations is used. It is interesting to note that the basic matrices T that are used in the decomposition of the matrices considered above are very similar to the Dirac matrices σ_k used in quantum mechanics,

$$
\sigma_0 = \begin{bmatrix} 0 & 1 \\ 1 & 0 \end{bmatrix} = T_{1,0}, \quad \sigma_1 = \begin{bmatrix} 0 & -j \\ j & 0 \end{bmatrix} = jT_{1,1}, \quad \sigma_2 = \begin{bmatrix} 1 & 0 \\ 0 & -1 \end{bmatrix} = T_{0,1}.
$$

4.4.2 Square roots of the identity transformation

By means of the basis $\{T_{n,m}\}$, the square roots and roots of high orders of other transformations can be found.

Example 4.13

Let B be a matrix which is a square root of the identity matrix, i.e., $B^2 = I$, or $B = I^{[1/2]}$. In the decomposition $I = a_{0,0}T_{0,0} + a_{0,1}T_{0,1} + a_{1,0}T_{1,0} + a_{1,1}T_{1,1}$ the coefficients equal $a_{0,0} = 1$, $a_{0,1} = a_{1,0} = a_{1,1} = 0$. Therefore, the transformation $\mathcal{D}$ leads to the solution of the following system of equations:

$$
\begin{cases} d_{0,0}^2 + d_{0,1}^2 + d_{1,0}^2 - d_{1,1}^2 = 1 \\ d_{0,0}d_{0,1} = 0 \\ d_{0,0}d_{1,0} = 0 \\ d_{0,0}d_{1,1} = 0 \end{cases} \approx \quad \begin{cases} d_{0,1}^2 + d_{1,0}^2 - d_{1,1}^2 = 1 \\ d_{0,0} = 0. \end{cases} \tag{4.21}
$$

The first equation describes a rotation in the 3-D space with the Minkovsky metric. Rotations in this space are well known as the Lorenz group. The $d_{1,1} = 0$ case relates to the 2-D rotation. The trivial solution of this system is $d_{0,1} = d_{1,0} = d_{1,1} = 0$, $d_{0,0}^2 = 1$, and we will not consider it.

The solution of (4.21) can be parameterized and written as $d_{0,0} = 0$, $d_{1,1} = \sinh(\vartheta)$, $d_{0,1} = \cosh(\vartheta)\cos(\varphi)$, $d_{1,0} = \cosh(\vartheta)\sin(\varphi)$, where parameters $\varphi \in [0, 2\pi)$ and $\vartheta \in (-\infty, \infty)$ (we consider only the real case). Then, the decomposition of the square root matrix $B = d_{0,0}T_{0,0} + d_{0,1}T_{0,1} + d_{1,0}T_{1,0} + d_{1,1}T_{1,1}$ can be written as

$$
B = B(\vartheta, \varphi) = \begin{bmatrix} \cosh(\vartheta)\cos(\varphi) & \sinh(\vartheta) + \cosh(\vartheta)\sin(\varphi) \\ -\sinh(\vartheta) + \cosh(\vartheta)\sin(\varphi) & -\cosh(\vartheta)\cos(\varphi) \end{bmatrix}
$$

$$
= \cosh(\vartheta) \begin{bmatrix} \cos(\varphi) & \sin(\varphi) \\ \sin(\varphi) & -\cos(\varphi) \end{bmatrix} + \sinh(\vartheta) \begin{bmatrix} 0 & 1 \\ -1 & 0 \end{bmatrix}.
$$

For instance, for values of parameters $(\vartheta, \varphi) = (-1, \pi/4)$ and $(0, \pi/3)$, the square root matrices equal

$$
B(-1, \pi/4) = \begin{bmatrix} 1.0911 & -0.0841 \\ 2.2663 & -1.0911 \end{bmatrix}, \quad B(0, \pi/3) = \begin{bmatrix} 0.5 & 0.8660 \\ 0.8660 & -0.5 \end{bmatrix}.
$$

Thus, we receive the two-parameter representation of the square roots of the identity matrix (2×2). The number of such square roots is infinite. The Hadamard transform is a particular case of such roots, when $\vartheta = 0$ and $\varphi = \pi/4$. In other words, $H = B(0, \pi/4)$.

Square roots of the identity matrix $(N \times N)$, where $N = 2^r$, $r > 1$, can be defined by

$$
B_N = B_N(\vartheta, \varphi) = B(\vartheta, \varphi)^{\oplus r} = \bigotimes_{k=1}^{r} B(\vartheta, \varphi).
$$

As an example, Figure 4.13 shows the 512-point discrete square root transforms of the signal f_n of Figure 4.3, when using the Hadamard matrix in
part a, and matrices $B_{512}(\vartheta, \varphi)$ with parameters $(\vartheta, \varphi) = (0, \pi/3)$ in b,
$(\vartheta, \varphi) = (0.5, \pi/4)$ in c, and $(\vartheta, \varphi) = (-1, \pi/4)$ in d.

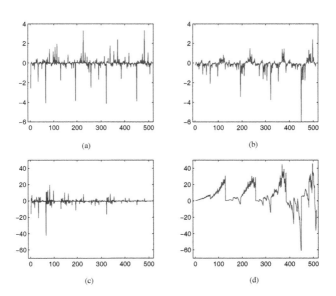

(a)   (b)

(c)   (d)

FIGURE 4.13
Square root transform of the signal, when using (a) the Hadamard matrix,
and (b)-(d) matrices $B_{512}(\vartheta, \varphi)$, for $(\vartheta, \varphi) = (0, \pi/3)$, $(\vartheta, \varphi) = (0.5, \pi/4)$, and
$(\vartheta, \varphi) = (-1, \pi/4)$, respectively.

Below are examples of MATLAB-based codes for computing the N-point
discrete square root transforms of the signal, as shown in Figure 4.13 in parts
a and d.

```
% --------------------------------------------------------------------
% demo_gdsrt.m file of programs (library of codes of Grigoryans)
% List of codes for processing signals of the length N>1:
%    N-point discrete square root transform (DSRT)   -  'm_hadN.m'
%    2-point discrete square root transform          -  'm_had2.m'
%
%    The main program for computing the 512-point square root transforms
%    1. Composition of the signal of length N=512 in the interval [0,50]
     N=512;
     step_1=50/(N-1);
     x=0:step_1:50;
     y=2*exp(-x/16).*(sin((pi)*pow2(x/8)));
%    2. Square root of the identity matrix when using the Hadamard matrix:
```

```
        a1=0;  a2=pi/4;
        H=m_sqrtN(N,a1,a2);
        x1=H*y';
%    3. Square root of the identity matrix when using the matrix B(-1,pi/4):
        a1=-1; a2=pi/4;
        H=m_sqrtN(N,a1,a2);
        x4=H*y';
%    4. Print the transforms as in Figure 4.13 in parts a and d:
        figure;
        subplot(2,2,1);  plot(x1);
        axis([-10,520,-6,4]);
        h_x=xlabel('(a)');
        subplot(2,2,4);  plot(x4);
        axis([-10,520,-70,50]);
        h_x=xlabel('(d)');
%    end of the main program.
%    ---------------------------------------------
        function H=m_sqrtN(N,a1,a2)
            H2=m_had2(a1,a2);
            r=log2(N)-1;
            H=H2;;
            for k=1:r
                H=kron(H2,H);
            end;
%    ---------------------------------------------
        function H2=m_sqrt2(a1,a2)
            c=cos(a2);      s=sin(a2);
            ch=cosh(a1);  sh=sinh(a1);
            b=ch*s;
            h1=ch*c;
            H2=[ h1      sh+b
                -sh+b, -h1];
%    ---------------------------------------------
```

Square roots of the identity matrix $(N \times N)$ can also be defined by square roots of order 2 with different parameters

$$B_N = B_N(\bar{\vartheta}, \bar{\varphi}) = \bigotimes_{k=1}^{r} B(\vartheta_k, \varphi_k),$$

where $\varphi_k \in [0, 2\pi)$ and $\vartheta_k \in (-\infty, +\infty)$. We here denote vector-parameters by $\bar{\varphi} = (\varphi_1, \varphi_2, \ldots, \varphi_r)$ and $\bar{\vartheta} = (\vartheta_1, \vartheta_2, \ldots, \vartheta_r)$. In this case, the code "m_sqrtN.m" can be substituted by the code given below.

```
%    General N-point (DSRT) with vector parameters a1 and a2 - 'm_hadNV.m'
        function H=m_sqrtNV(N,a1,a2)
            r=log2(N)-1;
            H=m_had2(a1(1),a2(1));
            for k=1:r
```

```
            k1=k+1;
            H2=m_had2(a1(k1),a2(k1));
            H=kron(H2,H);
        end;
%   --------------------------------------------------------------
```

As an example, Figure 4.14 shows the 512-point transforms $B_{512}f$ of the signal f of Figure 4.3 in parts (a)-(d), respectively, for the following sets of vector-parameters:

$$
\begin{cases}
\vartheta_k = 0, & \varphi_k = \dfrac{\pi}{4} + (k-1)0.1, \\[2mm]
\vartheta_k = 0, & \varphi_k = \dfrac{\pi}{3} + (k-1)0.1, \\[2mm]
\vartheta_k = 0.1k, & \varphi_k = \dfrac{\pi}{4} + (k-1)0.1, \\[2mm]
\vartheta_k = 0.1k, & \varphi_k = \dfrac{\pi}{3} + (k-1)0.1.
\end{cases}
\tag{4.22}
$$

where $k = 1 : 9$. One can observe different pictures of these transforms. The number of square roots of the identity matrix is infinite. The selection of parameters of those transforms should be made depending on an application.
□

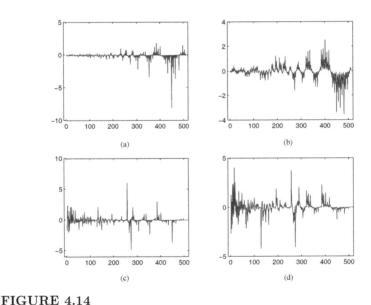

(a) (b) (c) (d)

FIGURE 4.14

512-point square roots transforms for different vector parameters $\bar{\varphi} = (\varphi_1, \ldots, \varphi_9)$ and $\bar{\vartheta} = (\vartheta_1, \ldots, \vartheta_9)$ given in (4.22).

For comparison of two types of the square roots, Figure 4.15 shows the row-functions of the matrix $B_8(0, \pi/3)$ in part a, along with the row-functions of the square root $B_8(0, \bar{\vartheta})$, where three different values of ϑ are used, namely, $\vartheta_1 = 1.0472, \vartheta_2 = 1.2472$, and $\vartheta_3 = 1.4472$.

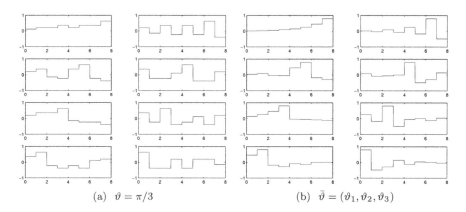

(a) $\vartheta = \pi/3$ (b) $\bar{\vartheta} = (\vartheta_1, \vartheta_2, \vartheta_3)$

FIGURE 4.15
Eight basic functions of the square roots of the identity transform of order 8, when (a) $B_8(0, \pi/3)$ matrix is used, and (b) $B_8(0, \bar{\vartheta})$ is used with $\bar{\vartheta} = (1.0472, 1.2472, 1.4472)$.

Example 4.14
We consider the problem described in (4.8), when the transformed signal of length N is degraded by the random noise,

$$z = y + n = B_N(\vartheta, \varphi)x + n. \qquad (4.23)$$

We consider the values of parameters (ϑ, φ) for which the inverse transform $\hat{y} = B_N(\vartheta, \varphi)^{-1}z$ is maximally close to the original signal. The closeness is with respect to the metric L_0 or L_2 defined as in (4.9) and (4.10), respectively,

$$\varepsilon_{0;\vartheta,\varphi} = \varepsilon_0(\hat{y}, y) = \max_n |\hat{y}_n - y_n| \qquad (4.24)$$

and

$$\varepsilon_{2;\vartheta,\varphi} = \varepsilon_2(\hat{y}, y) = \frac{1}{N}\sqrt{\sum_{n=0}^{N-1} |\hat{y}_n - y_n|^2}. \qquad (4.25)$$

It is also interesting to know how close or far the Hadamard transform is to the transforms providing a minimum error of reconstruction.

As an example, we consider the $\vartheta = 0$ case which includes the Hadamard transform (when $\varphi = \varphi_0 = \pi/4$). The signal x is considered to be the signal of length 512, which is shown in Figure 4.5a. Figure 4.16 shows a realization of the random degraded transformed signal $z = B_{256}(0, \varphi_0)x + n$ with a random noise n of amplitude distributed uniformly in the interval $[0, 0.1]$ in part a, along with the inverse transform $\hat{y} = B_{256;0,\varphi_0}^{-1} z$ of this realization in b. The graph of error $\varepsilon_{0;0,\varphi}(\hat{y}, y)$, where φ runs the interval $[0, \pi/2]$ with sampling interval 0.0025 is shown in c. The horizontal line corresponds to the error of reconstruction when the Hadamard transform is used. This transform maximizes the error ε_0 of signal reconstruction. In part d, the graph of error $\varepsilon_{2;0,\varphi}(\hat{y}, y)$ is shown. The horizontal line along the middle of the graph

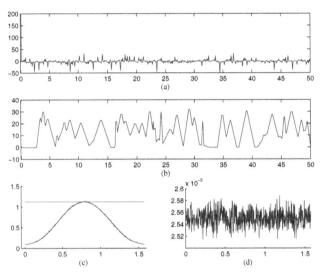

FIGURE 4.16 (See color insert following page 242.)
(a) Random realization of the degraded transform, (b) the inverse transform $\hat{y}$, (c) error $\varepsilon_{0;0,\varphi}(\hat{y}, y)$, and (d) error $\varepsilon_{2;0,\varphi}(\hat{y}, y)$, when $\varphi \in [0, \pi/2]$.

corresponds to the error when the Hadamard transformation is applied; the error is in the middle of the error-interval $[2.52, 2.58] \times 10^{-5}$. In both cases, the Hadamard transformation is not the best transformation to be used for signal reconstruction. Such a result of reconstruction by the Hadamard transform is given in Figure 4.17, when errors $\varepsilon_0 = 1.136917$ and $\varepsilon_2 = 0.002566$. The random noise n in this case is distributed uniformly in the interval $[0, 0.2]$. It is difficult to recognize a difference between this result and the one shown in Figure 4.16b. ▯

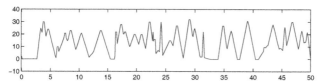

FIGURE 4.17
Inverse Hadamard transform of the random degraded transform.

4.4.3 The 4th degree roots of the identity transformation

Decomposition of matrices by basic matrices $T_{m,n}$ allows for constructing roots of other matrices, too. As an example, we consider roots of the 4th degree of the identity matrix.

Example 4.15
Let C be a matrix satisfying the equation $C^4 = I$, i.e., $C = I^{[1/4]}$. Let $d_{i,j}$ be coefficients of the T-decomposition of this matrix

$$C = d_{0,0}T_{0,0} + d_{0,1}T_{0,1} + d_{1,0}T_{1,0} + d_{1,1}T_{1,1}. \tag{4.26}$$

According to the transform $\mathcal{D}$ given in (4.20), which we need to use twice, coefficients of this decomposition are transformed as follows:

$$\begin{bmatrix} d_{0,0} \\ d_{0,1} \\ d_{1,0} \\ d_{1,1} \end{bmatrix} \xrightarrow{\mathcal{D}} \begin{bmatrix} ||d||^2 \\ 2d_{0,0}d_{0,1} \\ 2d_{0,0}d_{1,0} \\ 2d_{0,0}d_{1,1} \end{bmatrix} \xrightarrow{\mathcal{D}} \begin{bmatrix} ||d||^4 + 4d_{0,0}^2(d_{0,1}^2 + d_{1,0}^2 - d_{1,1}^2) \\ 4||d||^2 d_{0,0}d_{0,1} \\ 4||d||^2 d_{0,0}d_{1,0} \\ 4||d||^2 d_{0,0}d_{1,1} \end{bmatrix} \tag{4.27}$$

where we denote the norm $||d||^2 = d_{0,0}^2 + d_{0,1}^2 + d_{1,0}^2 - d_{1,1}^2$. For the $C^4 = I$ case, the transformation $\mathcal{D}^2 : (d_{0,0}, d_{0,1}, d_{1,0}, d_{1,1}) \to (1, 0, 0, 0)$ leads to the solution of the following system of equations:

$$\begin{cases} ||d||^4 + 4d_{0,0}^2(d_{0,1}^2 + d_{1,0}^2 - d_{1,1}^2) = 1 \\ ||d||^2 d_{0,0}d_{0,1} = ||d||^2 d_{0,0}d_{1,0} = ||d||^2 d_{0,0}d_{1,1} = 0. \end{cases}$$

We consider the most interesting and not trivial case of this system, when $4d_{0,0}^2(d_{0,1}^2 + d_{1,0}^2 - d_{1,1}^2) = 1$ and $||d||^2 = 0$. The system is reduced to the following one: $4d_{0,0}^4 = -1$, $d_{0,0}^2 + d_{0,1}^2 + d_{1,0}^2 - d_{1,1}^2 = 0$, and therefore

$$d_{0,0} = \pm(1 \pm j)/2, \quad d_{0,1}^2 + d_{1,0}^2 - d_{1,1}^2 = \mp j/2.$$

This solution can also be parameterized and written as

$$\begin{cases} d_{0,0} = \pm(1 \pm j)/2 \\ d_{0,1} = d_{0,0}\sinh(\vartheta)\cos(\varphi) \\ d_{1,0} = d_{0,0}\sinh(\vartheta)\sin(\varphi) \\ d_{1,1} = d_{0,0}\cosh(\vartheta) \end{cases}$$

where parameters $\varphi \in [0, 2\pi)$ and $\vartheta \in (-\infty, +\infty)$. T-decomposition of the root matrix in (4.26) can therefore be written as

$$C = C(\vartheta, \varphi) = d_{0,0} \begin{bmatrix} 1 + \sinh(\vartheta)\cos(\varphi) & \sinh(\vartheta)\sin(\varphi) + \cosh(\vartheta) \\ \sinh(\vartheta)\sin(\varphi) - \cosh(\vartheta) & 1 - \sinh(\vartheta)\cos(\varphi) \end{bmatrix}$$

$$= d_{0,0} \left(\sinh(\vartheta) \begin{bmatrix} \cos(\varphi) & \sin(\varphi) \\ \sin(\varphi) & -\cos(\varphi) \end{bmatrix} + \begin{bmatrix} 1 & \cosh(\vartheta) \\ -\cosh(\vartheta) & 1 \end{bmatrix} \right).$$

For instance, when $d_{0,0} = (1+j)/2$, $\vartheta = 0.1$, and $\varphi = \pi/4$, the matrix is

$$C = C(0, \pi/4) = \begin{bmatrix} 0.5354 - j0.5354 & 0.5379 - j0.5379 \\ -0.4671 + j0.4671 & 0.4646 - j0.4646 \end{bmatrix}. \tag{4.28}$$

We also consider the following three matrices calculated for $d_{0,0} = (1-j)/2$:

$$C = C(1, \pi/3) = \begin{bmatrix} 0.7938 - j0.7938 & 1.2804 - j1.2804 \\ -0.2627 + j0.2627 & 0.2062 - j0.2062 \end{bmatrix},$$

$$C = C(-0.5, \pi/5) = \begin{bmatrix} 0.2892 - j0.2892 & 0.4107 - j0.4107 \\ -0.7170 + j0.7170 & 0.7108 - j0.7108 \end{bmatrix}, \tag{4.29}$$

$$C = C(-0.25, \pi/5) = \begin{bmatrix} 0.3978 - j0.3978 & 0.4415 - j0.4415 \\ -0.5899 + j0.5899 & 0.6022 - j0.6022 \end{bmatrix}.$$

We receive the two-parameter representation of an infinite number of roots of the 4th degree of the identity matrix (2×2). Similar to the square root matrices, the 4th degree roots of the identity matrix $(N \times N)$, when $N = 2^r$, $r > 1$, can be defined by $C_N = C_N(\vartheta, \varphi) = C(\vartheta, \varphi)^{\oplus r} = \bigoplus_{k=1}^{r} C(\vartheta, \varphi)$, or, when using matrices $C(\vartheta, \varphi)$ with different parameters as $C_N = C_N(\bar{\vartheta}, \bar{\varphi}) = \bigoplus_{k=1}^{r} C(\vartheta_k, \varphi_k)$, where $\varphi_k \in [0, 2\pi)$ and $\vartheta_k \in (-\infty, +\infty)$.

As an example, Figure 4.18 shows the eight row-basic functions of the matrix $C_8(0.1, \pi/4)$ in part a. In this illustration, the functions are not normalized and only real parts of the functions are shown. The Fourier matrix is also the 4th degree root of the identity matrix. For comparison, the row-functions of the Fourier transform are shown in b.

Figure 4.19 shows the 512-point discrete root transforms of the signal f_n of Figure 4.3, when matrices $C_{512}(\vartheta, \varphi)$ of the transforms are defined by the matrices $C(\vartheta, \varphi)$ shown in (4.28) and (4.29). It is interesting to see that by changing the parameters ϑ and φ, we may achieve different forms of the transformed signal.

Figure 4.20 shows the sinusoidal-type signal in part a. The transform with the matrix $C_{512}(0.1, \pi/4)$ has been applied on the signal. The amplitude of this transform is shown in part b, and the Fourier transform in c, for comparison.   ☐

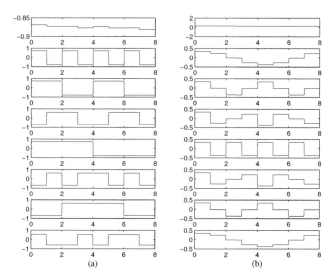

FIGURE 4.18

Real parts of basic functions of the 4th degree of the identity transformation of order 8, when (a) $C_8(0.1, \pi/4)$ matrix is used, and (b) the matrix of the 8-point DFT.

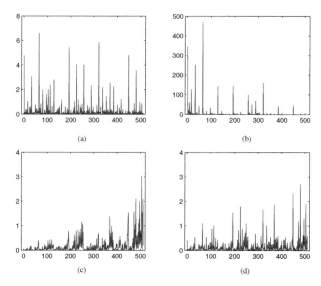

FIGURE 4.19

The 4th degree root transforms of the signal, when using matrices $C_{512}(\vartheta, \varphi)$, with parameters (a) $(\vartheta, \varphi) = (0, \pi/4)$, (b) $(\vartheta, \varphi) = (1, \pi/3)$, (c) $(\vartheta, \varphi) = (-0.5, \pi/5)$, and (d) $(\vartheta, \varphi) = (-0.25, \pi/6)$.

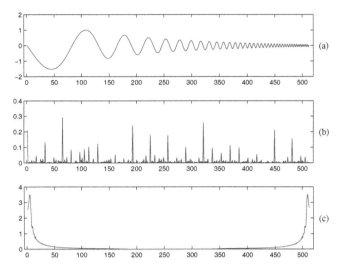

FIGURE 4.20
(a) The signal of length 512, (b) the 4th degree root transform of the signal, and (c) the 512-point DFT. The transforms are shown in the amplitude scale.

4.5   Mixed Fourier transformations

In this section, the concept of the combined time-and-frequency transformations with respect to the Fourier transformations is considered, which is based on an approach similar to the mixed transformations described in §4.2 for the Hadamard transformation. This concept allows for defining square and high degree roots of the Fourier transformation, and other transformations as well. We first try to define such a mixed transformation, whose matrix is represented as a linear combination of the following three matrices ($N \times N$) $S = aI + bE + cF$ where a, b, and c are coefficients, F is the matrix of the discrete Fourier transformation, and $E = F^2 = F \cdot F$. We call S *the mixed transformation* with respect to the Fourier transformation. Note that $EE = FFFF = I$, $\quad EF = FFF = FE$.

We now try to find an inverse matrix S^{-1} in the similar form $S^{-1} = pI + sE + tF$. The direct calculation of the product

$$I = SS^{-1} = (aI + bE + cF)(pI + sE + tF)$$
$$= (ap + bs)I + (as + bp + ct)E + (at + pc)F + (cs + bt)EF$$

leads to the system of four equations with six unknown variables

$$\begin{cases} ap + bs = 1, & as + bp + ct = 0, \\ at + pc = 0, & cs + bt = 0. \end{cases} \tag{4.30}$$

Considering the first and the two last equations, we obtain the following:

$$\begin{cases} ap + bs = 1 \\ at + pc = 0 \\ cs + bt = 0 \end{cases} \rightarrow \begin{bmatrix} a & b & 0 \\ c & 0 & a \\ 0 & c & b \end{bmatrix} \begin{bmatrix} p \\ s \\ t \end{bmatrix} = \begin{bmatrix} 1 \\ 0 \\ 0 \end{bmatrix}$$

where the determinant of the matrix (3×3) equals $-c(a^2 + b^2) \neq 0$. The solution of this system is

$$p = \frac{a}{\Delta}, \quad s = \frac{b}{\Delta}, \quad t = -\frac{c}{\Delta}, \quad (\Delta = a^2 + b^2),$$

and from the second equation of (4.30), it is not difficult to obtain that $c^2 - 2ab = 0 \rightarrow c = \pm\sqrt{2ab}$. Thus the inverse matrix of $S = aI + bE + cF$ equals

$$S^{-1} = \frac{1}{\Delta}(aI + bE - cF)$$

with the constraint $c = \pm\sqrt{2ab}$. In this case, the matrices S and S^{-1} are squares of matrices of the mixed transformations,

$$S = \left(\sqrt{a}I \pm \sqrt{b}F\right)^2, \quad S^{-1} = \frac{1}{\Delta}\left(\sqrt{a}I \mp \sqrt{b}F\right)^2.$$

4.5.1 Square roots of the Fourier transformation

To find a general solution of the equation

$$S^2 = F, \tag{4.31}$$

we consider the linear space $\mathcal{L}(F)$ spanned on the transforms $I, E = FF, F$, and F^*. Let S be a matrix in the form of

$$S = aI + bE + cF + dF^*,$$

where F^* is the matrix complex conjugate to F. It follows from (4.31) that the coefficients a, b, c, and d satisfy the following system of equations:

$$\begin{cases} a^2 + b^2 + 2cd = 0 \\ c^2 + d^2 + 2ab = 0 \\ 2(ac + bd) = 1 \\ 2(ad + bc) = 0. \end{cases} \tag{4.32}$$

By summing and subtracting equations from each other, this system can be reduced to the following system:

$$\begin{cases} a + b + c + d = s \\ a + b - c - d = w \\ ja - jb + c - d = p \\ -ja + jb + c - d = t \end{cases} \text{where} \quad \begin{aligned} s^2 &= 1 \\ w^2 &= -1 \\ p^2 &= j \\ t^2 &= -j. \end{aligned}$$

In matrix form, this system can be written as

$$
\begin{bmatrix}
1 & 1 & 1 & 1 \\
1 & 1 & -1 & -1 \\
j & -j & 1 & -1 \\
-j & j & 1 & -1
\end{bmatrix}
\begin{bmatrix}
a \\
b \\
c \\
d
\end{bmatrix}
=
\begin{bmatrix}
s \\
w \\
p \\
t
\end{bmatrix}.
$$

The above matrix (4×4), which we denote by D_4, satisfies the equation $D_4 D_4' = 4I$. Therefore, the coefficients $a, b, c,$ and d are calculated by

$$
\begin{bmatrix}
a \\
b \\
c \\
d
\end{bmatrix}
=
\frac{1}{4}
\begin{bmatrix}
1 & 1 & -j & j \\
1 & 1 & j & -j \\
1 & -1 & 1 & 1 \\
1 & -1 & -1 & -1
\end{bmatrix}
\begin{bmatrix}
s \\
w \\
p \\
t
\end{bmatrix},
\quad
a = \frac{s + w - jp + jt}{4}, \quad b = \frac{s + w + jp - jt}{4},
$$
$$
c = \frac{s - w + p + t}{4}, \quad d = \frac{s - w - p - t}{4}.
$$

The solution is not unique, because of ambiguity of square roots $s, w, p,$ and t.

Case 1: $[+, +, +, +]$ We consider the following values of the square roots:

$$
s = 1, \ w = j, \ p = \sqrt{j} = \frac{1 + j}{\sqrt{2}}, \ t = \sqrt{-j} = \frac{1 - j}{\sqrt{2}}.
$$

Then, we obtain the coefficients

$$
a = \frac{1 + \sqrt{2} + j}{4}, \ b = \frac{1 - \sqrt{2} + j}{4}, \ c = \frac{1 + \sqrt{2} - j}{4}, \ d = \frac{1 - \sqrt{2} - j}{4}. \quad (4.33)
$$

It is not difficult to check that the mixed matrix $S = aI + bE + cF + dF^* = aI + bE + a^*F + b^*F^*$ is the square root of the Fourier matrix.

Case 2: $[-, +, +, +]$ We consider the following values of the square roots:

$$
s = -1, \ w = j, \ p = \sqrt{j} = \frac{1 + j}{\sqrt{2}}, \ t = \sqrt{-j} = \frac{1 - j}{\sqrt{2}}.
$$

The coefficients are calculated by

$$
a = \frac{-1 + \sqrt{2} + j}{4}, \ b = \frac{-1 - \sqrt{2} + j}{4}, \ c = a^*, \ d = b^*. \quad (4.34)
$$

The matrix $S = aI + bT + cF + dF^*$ with these coefficients is another square root matrix of F. If we consider other cases: $[+, -, +, +], [-, -, +, +]$, we will receive matrices equal to the above considered two square roots.

It is interesting to note that by replacing the coefficients c and d in the square root $F^{[1/2]}$, we obtain a square root of the inverse Fourier matrix, i.e., $(F^*)^{[1/2]} = aI + bT + dF + cF^*$. The Fourier matrix is the 4th order root of the identity matrix, and the matrix $F^{[1/2]}$ is thus the 8th order root of the identity matrix, i.e., $F^{[1/2]} = I^{[1/8]}$. As an example, Figure 4.21 shows the original signal of length 512 in part a, along with amplitude of the Fourier transform

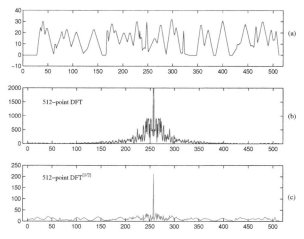

FIGURE 4.21

(a) Original signal of length 512, (b) the Fourier transform, and (c) the square root Fourier transform. (The transforms are plotted in the absolute scale and shifted to the center.)

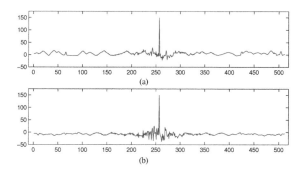

FIGURE 4.22

The square root Fourier transform: (a) real and (b) imaginary parts.

of this signal in b, and the square root Fourier transform in c. The square root DFT is not symmetric, but has small amplitude, when compared with the Fourier transform. Figure 4.22 shows separately the real and imaginary parts of the 512-point square root Fourier transform. We call the transformation whose matrix is calculated by (4.33) *the 1st square root discrete Fourier transformation (1-SQ DFT),* and the transformation whose matrix is calculated by (4.34) is *the 2nd square root discrete Fourier transformation (2-SQ DFT).*

Example 4.16

To illustrate all parts of the two above defined square roots of the Fourier transformation, we consider the following sinusoidal wave in the exponential envelope: $x(t) = 2e^{-t/16}\sin(\pi t^2/64)$, $t \in [0, 50]$. The discrete-time signal x_n of length 512 is composed from $x(t)$ by sampling uniformly in the interval $[0, 50]$. Figure 4.23 shows the sampled signal x_n in part a, along with the real parts of the DFT of this signal in b, 1-SQ DFT in c, and 2-SQ DFT in d.

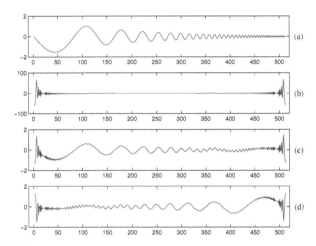

FIGURE 4.23

(a) Discrete-time signal of length 512, and the real parts of the (b) DFT, (c) 1-SQ DFT, and (d) 2-SQ DFT.

Figure 4.24 shows the imaginary part of the DFT of the signal x_n in part a, along with the imaginary part of the 1-SQ DFT in b. The 2-SQ DFT and 1-SQ DFT have the same imaginary parts. Figure 4.25 shows the phase in

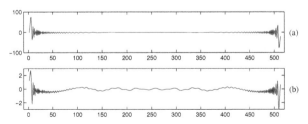

FIGURE 4.24

The imaginary parts of the (a) DFT and (b) 1-SQ DFT.

radians of the DFT of the signal $x(t)$ in part a, along with the phase of the 1-SQ DFT in b, and 2-SQ DFT in c.　☐

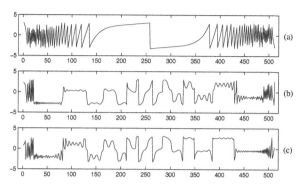

FIGURE 4.25
The phase of the (a) DFT, (b) 1-SQ DFT, and (c) 2-SQ DFT.

4.5.2   Series of Fourier transforms

In this section, we consider a simple method of calculation of the square roots of the Fourier transformation, which uses the cyclic convolution. The linear space $\mathcal{L}(F)$ considered above has the following four basic matrices:

$$I = F^0,\ F = F^1,\ E = F^2,\ F^* = F^3,$$

which compose the multiplicative group of order 4 (since $F^4 = I$). Therefore, any matrix S of $\mathcal{L}(F)$ is represented by

$$S = \sum_{n=0}^{3} a_k F^k \tag{4.35}$$

with real or complex coefficients a_k, $k = 0 : 3$. The square of this matrix can be written as

$$S^2 = \left(\sum_{n=0}^{3} a_k F^k \right)^2 = \sum_{n=0}^{3} b_k F^k, \tag{4.36}$$

where the coefficients b_n are calculated by the cyclic convolution

$$b_n = \sum_{k=0}^{3} a_k a_{n-k \bmod 4}, \quad n = 0, 1, 2, 3. \tag{4.37}$$

Let S be a square root of F. Then, the above cyclic convolution is written as

$$\sum_{k=0}^{3} a_k a_{n-k \bmod 4} = \delta_{n;1}, \quad n = 0, 1, 2, 3, \tag{4.38}$$

where $\delta_{1;1} = b_1 = 1$ and $\delta_{n;1} = b_n = 0$ if $n = 0, 2, 3$. In the frequency domain, this convolution has a form $|A_p|^2 = e^{\frac{2\pi j}{4} p}$, $p = 0, 1, 2, 3$, where A_p is the four-point discrete Fourier transform of the vector-coefficient $\bar{a} = \{a_0, a_1, a_2, a_3\}$. Thus, the inverse four-point DFT of the vector with elements calculated by $A_p = \pm \exp(j\pi p/4)$, i.e.,

$$A_0 = \pm 1, \quad A_1 = \pm \frac{1}{\sqrt{2}}(1 + j), \quad A_2 = \pm j, \quad A_3 = \pm \frac{1}{\sqrt{2}}(-1 + j),$$

results in the square root matrices of F.

We can extend this method for calculating roots of other degrees of the Fourier matrix. The equation $S^n = F$, for $n \geq 2$, can be reduced to the solution of the equation $|A_p|^n = e^{\frac{2\pi j}{4} p}$, $p = 0, 1, 2, 3$, in the frequency domain. In general, instead of an integer n, we can use any number $r > 0$.

Example 4.17

Let the mixed matrix S defined in (4.35) be a square root of the identity matrix, i.e.,

$$S^2 = \left(\sum_{n=0}^{3} a_k F^k \right)^2 = \sum_{n=0}^{3} b_k F^k = I.$$

It is clear that $S = \pm I$ and $S = \pm E$ are such square roots, but these are trivial cases. We now try to find other solutions, when there are at least two coefficients a_k not equal to zero.

The cyclic convolution of the vector with coefficients a_k is written as

$$\sum_{k=0}^{3} a_k a_{n-k \bmod 4} = \delta_n, \quad n = 0, 1, 2, 3,$$

where $\delta_n = 1$ if $n = 0$, and $\delta_n = 0$ if $n = 1, 2, 3$. In the frequency domain, this convolution has a form $|A_p|^2 = 1$, $p = 0, 1.2.3$, Thus, the coefficients c_k can be defined from the inverse four-point DFT of the sequence(s) $A_p = \pm 1$, $p = 0, 1, 2, 3$. The set of coefficients $A = (A_0, A_1, A_2, A_3)$ leads to a square root of the identity matrix, whose coefficients are calculated as:

$$A = (-1, 1, 1, 1) \xrightarrow{\mathcal{F}^{-1}} c_0 = 0.5, \ c_1 = -0.5, \ c_2 = -0.5, \ c_3 = -0.5. \tag{4.39}$$

As a result, we obtain the following square root of the identity matrix: $S = I^{[1/2]} = \frac{1}{2}(I - E - F - F^*)$. In the $N = 4$ and 8 cases, this matrix equals respectively

$$\frac{1}{\sqrt{4}} \begin{bmatrix} -1 & -1 & -1 & -1 \\ -1 & 1 & 1 & -1 \\ -1 & 1 & -1 & 1 \\ -1 & -1 & 1 & 1 \end{bmatrix},$$

$$\frac{1}{\sqrt{8}} \begin{bmatrix} -1 & -1 & -1 & -1 & -1 & -1 & -1 & -1 \\ -1 & 0.7071 & 0 & 0.7071 & 1 & 0.7071 & 0 & -2.1213 \\ -1 & 0 & 2.4142 & 0 & -1 & 0 & -0.4142 & 0 \\ -1 & 0.7071 & 0 & 0.7071 & 1 & -2.1213 & 0 & 0.7071 \\ -1 & 1 & -1 & 1 & -1 & 1 & -1 & 1 \\ -1 & 0.7071 & 0 & -2.1213 & 1 & 0.7071 & 0 & 0.7071 \\ -1 & 0 & -0.4142 & 0 & -1 & 0 & 2.4142 & 0 \\ -1 & -2.1213 & 0 & 0.7071 & 1 & 0.7071 & 0 & 0.7071 \end{bmatrix}.$$

One can notice the regular structure of this matrix; if we remove the 1st and 5th lines and columns, then the submatrices (3×3) lying on diagonals are equal. Basic functions of the transformation defined by this square root orthogonal matrix are shown in Figure 4.26. Basic functions of the matrix (16×16) are illustrated in Figure 4.27.

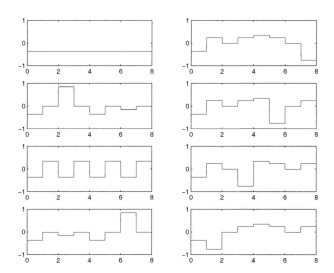

FIGURE 4.26
Basic functions of the square root of the 8-point identity transformation. (The functions are ordered from left to right, and top to bottom.)

Thus, in the space $\mathcal{L}(F)$ of the mixed transformations with respect to the Fourier transformation, we found six square roots of the identity ma-

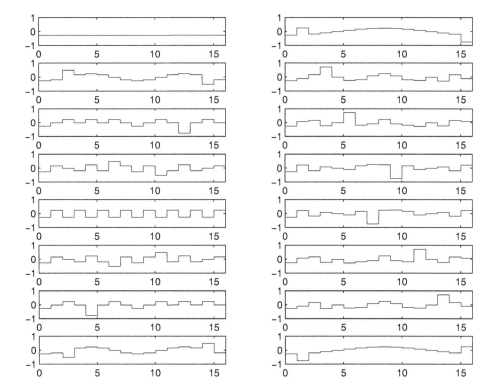

FIGURE 4.27
Basic functions of the square root of the 16-point identity transformation.

trix $S_0 = \pm I$, $S_1 = \pm E$, $S_2 = \pm\frac{1}{2}(I - E - F - F^*)$. We call the discrete transformation defined by the mixed matrix S_2 *the square root of discrete identity transformation,* or SR-DIT. The transform $S_1 = E$ is a square root of I, because the Fourier transformation is the 4th degree root of the identity transformation. This root E is a permutation of the input. For instance, for $N = 4$, the matrix of this root equals

$$E = \begin{bmatrix} 1 & 0 & 0 & 0 \\ 0 & 0 & 0 & 1 \\ 0 & 0 & 1 & 0 \\ 0 & 1 & 0 & 0 \end{bmatrix}.$$

In the $N = 8$ case, the rows of the matrix E are shown in Figure 4.28.

The square root S_2 differs much from S_1 and contains the cosine transformation $C = (F + F^*)/2$,

$$S_2 = \frac{1}{2}(I - E - F - F^*) = \frac{1}{2}(I - E) - \frac{F + F^*}{2} = \frac{1}{2}(I - E) - C,$$

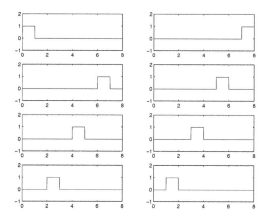

FIGURE 4.28

The basic functions of the square of the Fourier transformation.

and the difference of transformations equals $S_2 - S_1 = \frac{1}{2}(I - 3E) + C$. As an example, Figure 4.29 shows the original signal of length 512 in part a, along with the square root of discrete identity transform in b. For comparison, the real part of the square of the Fourier transform of this signal in given in c.

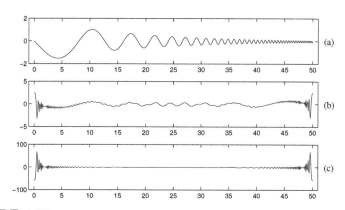

FIGURE 4.29

(a) Signal of length 512, (b) the 512-point SR-DIT and (c) real part of the square of the 512-point DFT of this signal.

In conclusion, we describe the direct method of calculation of the square root of the identity matrix, which is similar to that described in the beginning of §4.5.

Example 4.18

Let S be a mixed matrix which is a square root of the identity matrix, i.e., $S = aI + bE + cF + dF^*$, $S^2 = I$. Then, coefficients a, b, c, and d satisfy the following system of equations:

$$\begin{cases} a^2 + b^2 + 2cd = 1, & c^2 + d^2 + 2ab = 0 \\ 2(ac + bd) = 0, & 2(ad + bc) = 0. \end{cases} \tag{4.40}$$

The first two equations result in $(a+b)^2 + (c+d)^2 = 1$, and thus $a+b = \sin\alpha$, $c + d = \cos\alpha$, $\alpha \in [0, 2\pi)$. From the last two equations of (4.40), we receive $2\alpha = \pi n$, where n is an integer. We consider the $n = 1$ case, i.e., when $\alpha = \pi/2$. From the above equations, we obtain the following values of the coefficients: $a = 0.5$, $b = 0.5$, and $d = -c$, where the coefficient c is to be determined. The matrix S can thus be written as

$$S = \frac{1}{2}\big(I + E + c(F - F^*)\big).$$

The direct calculation of the square S^2 results in the following: $S^2 = \left(\frac{1}{2} - 2c^2\right)I + \left(\frac{1}{2} + 2c^2\right)E$, and thus $c = \pm j/2$, and

$$S = \frac{1}{2}(I + E) + \pm\frac{1}{2}j(F - F^*) = \frac{1}{2}\big(I + E \pm j(F - F^*)\big). \tag{4.41}$$

As an example, Figure 4.30 shows the raw-functions of the matrix S defined for $c = j/2$, in the $N = 8$ case.   ▯

4.6   Mixed transformations: Continuous case

We here discuss briefly a concept of mixed transformations which are defined as transformations in the time and frequency domains simultaneously. In sections 4.2-4.5, we have considered linear combinations of the identity transformation and powers of the Fourier or Hadamard transformations in the discrete case. However, the discrete case is easy to explain by using matrix forms of transformations. Many difficulties arise when considering the continuous case of functions in the time or frequency domain.

We first start with the simple example, when a real function $f(t)$ with the Fourier transform $F(\omega)$ is transformed as

$$m_f(t) = f(t) + F^*(\lambda t), \quad t \in (-\infty, +\infty), \tag{4.42}$$

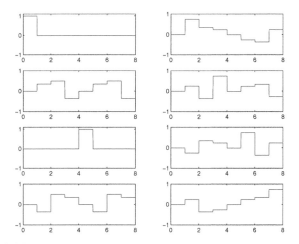

FIGURE 4.30
The basic functions of the square of the Fourier transformation.

where λ is a scalar parameter to be chosen. The function $m_f(t)$ in the frequency domain is described as

$$M_f(\omega) = F(\omega) + \frac{2\pi}{\lambda} f\left(\frac{\omega}{\lambda}\right), \quad \omega \in (-\infty, +\infty).$$

By substituting ω/λ by t, we can write the last equation as follows:

$$M_f(\lambda t) = F(\lambda t) + \frac{2\pi}{\lambda} f(t), \quad t \in (-\infty, +\infty). \tag{4.43}$$

If $\lambda = 2\pi$, equations (4.42) and (4.43) can be written as

$$\begin{cases} m_f(t) = f(t) + F^*(2\pi t) \\ M_f(2\pi t) = F(2\pi t) + f(t). \end{cases} \tag{4.44}$$

The transformation $f(t) \rightarrow m_f(t) = f(t) + F^*(2\pi t)$ is called *the time-and-frequency mixed transformation*, or simply *the mixed transformation*. The spectrum of the function $m_f(t)$ is expressed thus by the function

$$M_f(2\pi t) = m_f^*(t), \quad t \in (-\infty, +\infty). \tag{4.45}$$

In other words, the transform of this function coincides with its complex conjugate function scaled in time by factor of $1/(2\pi)$. We define the space of such functions by $M_{1,1}$. The variable t in the left part of the equality in (4.45) can be considered as the frequency in Hz, and $2\pi t$ then as the frequency in rad/sec. We thus consider t as the time and frequency at the same time, and, then, state that the Fourier transform of the function $m_f(t)$ equals the complex conjugate function itself.

Example 4.19
Consider the rectangle pulse signal, $f(t) = \text{rect}(t)$, which is 1 if $t \in [-1/2, 1/2]$, and 0 otherwise. The Fourier transform of $f(t)$ equals $\text{sinc}(\omega/2)$. The mixed transform of this function and its Fourier transform are calculated as

$$m_f(t) = \text{rect}(t) + \text{sinc}(\pi t) \xrightarrow{\mathcal{F}} M_f(\omega) = \text{sinc}\left(\frac{\omega}{2}\right) + \text{rect}\left(\frac{\omega}{2\pi}\right).$$

Taking $\omega = 2\pi t$, we obtain $M_f(2\pi t) = \text{sinc}(\pi t) + \text{rect}(t) = m_f(t)$. ⬜

Example 4.20
Let $f(t)$ be the delta function $\delta(t)$. Then, the mixed transform of $f(t)$ and its Fourier transform are calculated as follows:

$$m_f(t) = \delta(t) + 1 \xrightarrow{\mathcal{F}} M_f(\omega) = 1 + 2\pi\delta(\omega).$$

Therefore, substituting $\omega = 2\pi t$, we obtain $M_f(2\pi t) = 1 + \delta(t) = m_f(t)$. ⬜

Example 4.21
For the cosine wave $f(t) = \cos(\pi t)$ with frequency π, the mixed transform and its Fourier transform are calculated respectively as

$$m_f(t) = \cos(\pi t) + \pi \left[\delta\left(2\pi t - \pi\right) + \delta\left(2\pi t + \pi\right)\right]$$

$$M_f(\omega) = \pi[\delta(\omega - \pi) + \delta(\omega - \pi)] + \frac{1}{2}\left[e^{-j\omega/2} + e^{j\omega/2}\right].$$

Therefore, substituting $\omega = 2\pi t$, we obtain

$$M_f(2\pi t) = \pi\left[\delta(2\pi t - \pi) + \delta(2\pi t + \pi)\right] + \cos(\pi t) = m_f(t).$$

If $f(t)$ is periodic with the fundamental period $\omega_0 = 2\pi f_0$, then in the space $M_{1,1}$, we obtain the following representation of the function:

$$f(t) \rightarrow m_f(t) = \sum_{n=0,\pm 1,\dots} c_n \left[e^{jn\omega_0 t} + \delta(t - nf_0)\right], \qquad (4.46)$$

where c_n are coefficients of the Fourier series of $f(t)$. ⬜

It is interesting to analyze if the function $f(t)$ can be reconstructed from its mixed transform $m_f(t)$, or $M_f(2\pi t)$. It follows directly from (4.44) that $M_f(2\pi t) - m_f(t) = F(2\pi t) - F^*(2\pi t) = 2j\text{Im}\, F(2\pi t)$ and

$$\text{Im}\, F(2\pi t) = \frac{M_f(2\pi t) - m_f(t)}{2j} = -\text{Im}\, m_f(t).$$

Thus, the imaginary part of the Fourier transform of the original function $f(t)$ can be defined from the mixed transform. For the real part of $F(2\pi t)$, we can derive the following:

$$M_f(2\pi t) + m_f(t) = F(2\pi t) + F^*(2\pi t) + 2f(t) = 2\text{Re}\, F(2\pi t) + 2f(t)$$

and

$$\mathrm{Re}\, F(2\pi t) + f(t) = \frac{M_f(2\pi t) + m_f(t)}{2} = \mathrm{Re}\, m_f(t). \qquad (4.47)$$

Since $f(t)$ is the real function, the real part of the Fourier transform is defined by the even part of the function

$$\mathrm{Re}\, F(\omega) = \int_0^\infty [f(t) + f(-t)]\cos(\omega t)dt.$$

If the function is odd, this integral equals zero, and from (4.47), we obtain $f(t) = \mathrm{Re}\, m_f(t)$.

There are many other functions for which the Fourier transform is expressed simply by the function as $m_f(t)$. For instance, for a real function $f(t)$ with the Fourier transform $F(\omega)$ we can define the transform as $m_f(t) = f(t) - F^*(2\pi t)$, $t \in (-\infty, +\infty)$. Indeed, the spectrum of the function $m_f(t)$ is expressed by the function as $M_f(2\pi t) = -m_f^*(t)$. And it is not possible to reconstruct the original function $f(t)$ from the mixed transform. To make the mixed transform invertible, we consider this concept in a more general sense. Given two parameters $|a| \ne |b|$, let the mixed transform of a real function $f(t)$ be defined as

$$m_f(t) = af(t) + bF^*(2\pi t), \quad t \in (-\infty, +\infty). \qquad (4.48)$$

We denote by $M_{a,b}$ the space of such functions $m_f(t)$. For instance, the mixed transform of the periodic function $f(t)$ with frequency ω_0 can be written as

$$m_f(t) = \sum_{n=0,\pm 1,\dots} c_n \left[ae^{jn\omega_0 t} + b\delta(t - nf_0) \right],$$

a particular case of which has been given above in (4.46).

For the Fourier transform of this mixed transform, the following holds: $M_f(2\pi t) = aF(2\pi t) + bf(t)$, and since $f(t)$ is periodic

$$M_f^*(2\pi t) = a^* F^*(2\pi t) + b^* f(t). \qquad (4.49)$$

Values of parameters a and b can be taken arbitrarily under condition $|a|^2 - |b|^2 \ne 0$. The Fourier transform of the function $m_f(t)$ is an element of the space M_{a^*,b^*}. The following matrix equation holds from (4.48) and (4.49):

$$\begin{pmatrix} m_f(t) \\ M_f^*(2\pi t) \end{pmatrix} = \begin{pmatrix} a & b \\ b^* & a^* \end{pmatrix} \begin{pmatrix} f(t) \\ F^*(2\pi t) \end{pmatrix}. \qquad (4.50)$$

Therefore, the original function $f(t)$ and its Fourier transform can be defined by the mixed transform as follows:

$$\begin{pmatrix} f(t) \\ F^*(2\pi t) \end{pmatrix} = \frac{1}{D} \begin{pmatrix} a^* & -b \\ -b^* & a \end{pmatrix} \begin{pmatrix} m_f(t) \\ M_f^*(2\pi t) \end{pmatrix}$$

where the determinant of the matrix (2×2) in (4.50) equals $D = |a|^2 - |b|^2$.

4.6.1   Linear convolution

We consider the operation of linear convolution in the space $M_{1,1}$. Let $x(t)$ be a function from this space, which is convoluted with a function $h(t)$. The linear convolution $y(t) = x(t) * h(t)$ in the frequency-or-time domain equals $Y(t) = X(t)H(t) = x^*(t/2\pi) H(t)$. If the function $h(t)$ is also from the space $M_{1,1}$, then the linear convolution of these functions equals $Y(t) = x^*(t/2\pi) h^*(t/2\pi)$, or $Y(2\pi t) = x^*(t)h^*(t)$. In other words, the linear convolution of function in this space is reduced to the multiplication of the functions, and there is not much sense in using the linear convolution.

We now consider the space $M_{a,b}$. Let $f(t)$ be a function of this space, which is convoluted with a function $h(t)$. The mixed transforms of $f(t)$ and the linear convolution $y(t) = f(t) * h(t)$ are defined respectively as

$$\begin{cases} m_f(t) = af(t) + bF^*(2\pi t) \\ m_y(t) = ay(t) + bF^*(2\pi t)H^*(2\pi t), \end{cases}$$

and the Fourier transform of the mixed transform $m_y(t)$ equals

$$M_y^*(2\pi t) = a^* F^*(2\pi t)H^*(2\pi t) + b^* y(t).$$

In matrix form, the last two equations can be written as

$$\begin{pmatrix} m_y(t) \\ M_y^*(2\pi t) \end{pmatrix} = \begin{pmatrix} aL_h & bH^*(2\pi t) \\ b^* L_h & a^* H^*(2\pi t) \end{pmatrix} \begin{pmatrix} f(t) \\ F^*(2\pi t) \end{pmatrix} \qquad (4.51)$$

where L_h denotes the operator of linear convolution with the function $h(t)$. The matrix (2×2) of this equation can be written as

$$\begin{pmatrix} a & b \\ b^* & a^* \end{pmatrix} \begin{pmatrix} L_h & 0 \\ 0 & H^*(2\pi t) \end{pmatrix}$$

Then, the inverse transform can be derived from equation (4.51) as follows:

$$\begin{pmatrix} f(t) \\ F^*(2\pi t) \end{pmatrix} = \frac{1}{D} \begin{pmatrix} L_h^{-1} & 0 \\ 0 & \dfrac{1}{H^*(2\pi t)} \end{pmatrix} \begin{pmatrix} a^* & -b \\ -b^* & a \end{pmatrix} \begin{pmatrix} m_y(t) \\ M_y^*(2\pi t) \end{pmatrix} \qquad (4.52)$$

where L_h^{-1} denotes the operator inverse to the linear convolution, i.e., it is the operator which is reduced in the frequency domain to the multiplication of the spectrum by $1/H(t)$.

Problems

Problem 4.1 Consider the discrete-time signal $\mathbf{x}$ sampled from the continuous-time signal

$$x(t) = 2e^{-t/16} \sin(\pi(t/16)^2), \quad t \in [0:50],$$

with frequency 512Hz. Plot the signal, and calculate and plot the Hadamard transform of $\mathbf{x}$, the mixed transform S with parameters $(a, b) = (1, -0.5)$, and the square root of the transform, which is defined by Eq. 4.11.

Problem 4.2 In Example 4.5 square roots $S = S_1$ and $S = S_1^*$ are not only the square roots of the Hadamard matrix (2×2). Show that there are other roots which are not mixed transforms. For instance, define such numbers x and y that the matrix

$$S = xA + yA' = x \begin{bmatrix} 1 & -1 \\ 1 & 1 \end{bmatrix} + y \begin{bmatrix} 1 & 1 \\ -1 & 1 \end{bmatrix}$$

is the square root of the Hadamard matrix A.

Problem 4.3 In Example 4.6, the 3rd degree roots $S = S_1$ and $S = S_2$ of the Hadamard matrix (2×2) have been defined. Show that there are other 3rd degree roots which are not mixed transforms. As an example, consider the matrix

$$S = xA + yA' = x \begin{bmatrix} 1 & -1 \\ 1 & 1 \end{bmatrix} + y \begin{bmatrix} 1 & 1 \\ -1 & 1 \end{bmatrix}$$

and define the required numbers x and y, for S to be the 3rd degree root of the Hadamard matrix A.

Problem 4.4 Let $N = 2^r$, where $r > 1$, and let S_k, $k = 0 : r$, be square root matrices of the Hadamard matrix A. Show that the matrix

$$S_N = S_1 \otimes S_2 \otimes \cdots \otimes S_r$$

is the square root of the Hadamard matrix $(N \times N)$.

Problem 4.5 Compute the matrix of the bit-and transforms as shown in Table 4.2. Verify if the matrix (12×12), $A_{[12]}$, coincides with the known Hadamard matrix (12×12). Calculate the determinant and inverse matrix of the matrix $A_{[12]}$.

Problem 4.6 Consider the bit-and matrix (20×20) and find which of the following equations is valid for this matrix:

$$A_{[20]} = A_{[5]} \otimes A_{[4]}, \quad \text{and} \quad A_{[20]}^{-1} = A_{[5]}^{-1} \otimes A_{[4]}^{-1},$$

or

$$A_{[20]} = A_{[4]} \otimes A_{[5]}, \quad \text{and} \quad A_{[20]}^{-1} = A_{[4]}^{-1} \otimes A_{[5]}^{-1}.$$

Problem 4.7 Compute the characteristic polynomials of Table 4.3. Extend this table by calculating the corresponding polynomials for $N = 17, 18, ..., 21$.

Problem 4.8* Let $p(x)$ be the characteristic polynomial of the bit-and matrix $(N \times N)$,

$$p(x) = p_N(x) = \prod_{n=1}^{l} (x - \lambda_n)^{t_n},$$

where $l = l(N)$ is the number of different roots and $t_n = t_n(N)$ is the multiplicity of the root λ_n. As we know, all roots λ of the polynomial are equal to the plus or minus square root of powers of two, i.e., $\lambda_n = p_{\pm m} = \pm\sqrt{2^m}$, where m are integers. Show that the polynomial $p_{2N+1}(x)$ has always one root being 1 or -1.

Problem 4.9[*] Consider the ordered set $M(N)$ of roots of the characteristic polynomial $p_N(x)$ of the bit-and matrix $(N \times N)$. The multiplicity of the roots is not counted.

A. Show that there exist such integer numbers $n = n(N)$ and $K = K(N)$ that the following is valid:

$$M(N) = \{p_{-n}, M(K) \setminus \{-1, 1\}, p_n\}.$$

B. Find the multiplicity $t = t(N)$ of roots p_{-n} and p_n.
C. Calculate the representation of $M(N)$ in A for $N = 9$.

Problem 4.10 Let H be the bit-and matrix $(N \times N)$ and let $\mathbf{x}$ be a given discrete-time signal of length N. Consider the matrix T of the cyclic shift transformation

$$T : \{x_0, x_1, x_2, ..., x_{N-1}\} \rightarrow \{x_{N-1}, x_0, x_1, x_2, ..., x_{N-2}\}.$$

For $N = 3, 4, ..., 9$, find the matrix R defined as $HT = RH$, and show that this matrix is the Nth root of the identity matrix, i.e., $R^N = I$.

Problem 4.11[*] Let $H = H_4$ be the Hadamard matrix (4×4).

A. Show that the following decomposition by the projection matrices holds: $H = 2P_1 - 2P_2$, where

$$P_1 = \frac{1}{4} \begin{bmatrix} 3 & 1 & 1 & 1 \\ 1 & 1 & 1 & -1 \\ 1 & 1 & 1 & -1 \\ 1 & -1 & -1 & 3 \end{bmatrix}, \quad P_2 = -\frac{1}{4} \begin{bmatrix} -1 & 1 & 1 & 1 \\ 1 & -3 & 1 & 1 \\ 1 & 1 & -3 & -1 \\ 1 & -1 & -1 & -1 \end{bmatrix}.$$

B. Show that the following properties hold:

$$P_1 P_2 = P_2 P_1 = 0, \quad P_1 + P_2 = I, \quad P_n = P_n', \ n = 1, 2.$$

C. Show that two matrices defined by

$$H^{\frac{1}{2}} = 2^{\frac{1}{2}} P_1 + (-2)^{\frac{1}{2}} P_2$$

are the square roots of H.

D. Show that the complex matrix

$$Q = P_1 + iP_2 = \frac{1}{4} \begin{bmatrix} 3+i & 1-i & 1-i & 1-i \\ 1-i & 1+3i & 1-i & -1+i \\ 1-i & 1-i & 1+3i & -1+i \\ 1-i & -1+i & -1+i & 3+i \end{bmatrix}$$

is the square root of the identity matrix, i.e., $|Q|^2 = Q\bar{Q}' = I$.

E. Show that the complex matrix Q can be expressed by the Hadamard matrix as $Q = \frac{1-i}{4}(H + 2iD)$, where the diagonal matrix

$$D = \begin{bmatrix} 1 & 0 & 0 & 0 \\ 0 & -1 & 0 & 0 \\ 0 & 0 & -1 & 0 \\ 0 & 0 & 0 & 1 \end{bmatrix}.$$

Problem 4.12 Let H_{12} and H_{20} be the known Hadamard matrices (12×12) and (20×20), respectively. Verify if these matrices are square roots of the identity matrix (12×12) and (20×20), respectively. Verify if for these matrices the following is valid: $H_n H_n' = nI$, where n is the order of the matrix and I is the identity matrix. (As H_{20} consider the matrix known as "had.20.toncheviv.")

Problem 4.13 Find the characteristic equations for the Hadamard matrices H_{12} and H_{20}.

Problem 4.14[*] Let H_{12} be the Hadamard (12×12). Analyze and answer if it is possible to find such a matrix A that the coefficients of the Hadamard matrix can be calculated by

$$h_{n,m} = (-1)^{(m,n)_{A,2}}, \quad n, m = 0 : (N-1),$$

Here we denote

$$(m,n)_{A,2} = \sum_{p=0}^{r-1} \sum_{s=0}^{r-1} m_p a_{p,s} n_s, \tag{4.53}$$

and $\{m_p\}$ and $\{n_p\}$ are coefficients of binary representation of numbers m and n, respectively. $a_{p,s}$ are coefficients of the matrix A.

Problem 4.15[*] Let H_{20} be the Hadamard (20×20). Analyze and answer if it is possible to find such a matrix A that the coefficients of the Hadamard matrix can be calculated by

$$h_{n,m} = (-1)^{(m,n)_{A,2}}, \quad n, m = 0 : (N-1).$$

Problem 4.16 Use the orthogonal basis of matrices $T_{m,n}$, and find the matrix D,

$$D = I^{[1/3]} = \sum_{n=0}^{1} \sum_{m=0}^{1} d_{m,n} T_{m,n},$$

which is the 3rd root of the identity matrix.

Problem 4.17[*] Prove that for any order $N > 1$, the Nth root of the identity matrix I can be defined in the T-decomposition as

$$I^{[1/N]} = d_{0,0} T_{0,0} + r \left(\cosh(\vartheta) \cos(\varphi) T_{0,1} + \cosh(\vartheta) \sin(\varphi) T_{1,0} + \sinh(\vartheta) T_{1,1} \right) \tag{4.54}$$

for some numbers r, ϑ, and φ.

Problem 4.18[*]
   A. Consider the projection operators P_1, P_2, and P_3 defined in Example 4.12. Find the N-roots of the identity matrix (3×3), for any $N > 1$.
   B. Show that the set of the N-th roots of the identity matrix (3×3) define a group $G(N)$ of matrices with cardinality N^3.
   C. Define the group operation in $G(N)$.

Problem 4.19 Determine if the matrix (3×3) of the Fourier transform belongs to the group $G(N)$.

Problem 4.20 Consider the 512-point discrete-time signal $x(n)$ by sampling uniformly the signal

$$x(t) = 2e^{-\frac{t}{32}} \sin\left(\frac{\pi t^2}{54}\right), \quad t \in [0, 50].$$

Compute the square root of the discrete identity transform, SR-DIT, of this signal by using the concept of the mixed transform with respect to the Fourier transform. Plot the signal and the real part of the square of the Fourier transform and the real part of the Fourier transform of this signal.

Problem 4.21 Calculate the determinant of the bit-and matrices $(N \times N)$ for $N = 1 : 9$. Derive the analytical formula for the determinant in the general $N \geq 1$ case.

Problem 4.22 We define another class of binary transformations, which we call bit-or transformations, whose matrices $O(N)$ are composed by elements calculated by this simple and elegant formula:

$$O_{n,m} = (-1)^{n \odot m}, \quad n, m = 0 : (N - 1),$$

where $n \odot m = n_0 \vee m_0 + n_1 \vee m_1 + \ldots + n_r \vee m_r$, and n_k and m_k are coefficients of the binary representation of numbers n and m, respectively. The operation $\vee$ is the binary "OR."

A. Calculate the matrix $O(2), O(3)$, and $O(4)$.

B. Calculate the matrix $R(3) = H(3)O^{-1}(3)$, where $H(3)$ is the bit-and matrix (3×3).

C. Derive the analytical form of the matrix $R^m(3)$, where $m \geq 1$.

Problem 4.23 Consider the following matrix:

$$F = \frac{1}{2} \begin{pmatrix} 1 & 1 & 1 & -1 \\ -1 & -1 & 1 & -1 \\ -1 & 1 & 1 & 1 \\ 1 & 1 & -1 & -1 \end{pmatrix}.$$

A. Verify that the matrix is orthogonal.

B. Verify that the matrix is the 8th root of the identity matrix, i.e., $F^8 = I$.

C. Show that the following is valid:

$$\sum_{k=1}^{8} F^k = 0.$$

(D) Let T be the operation of transposition, i.e., $TA = A^T$. Show that $F = P \otimes Q$, where the matrices

$$P = \begin{pmatrix} 1 & T \\ -T & 1 \end{pmatrix}, \quad Q = \frac{1}{2} \begin{pmatrix} 1 & 1 \\ -1 & -1 \end{pmatrix}.$$

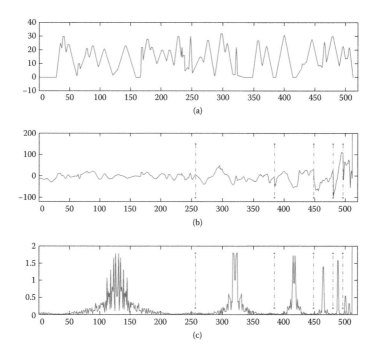

FIGURE 1.4
(a) The signal of length 512, (b) the paired transform of the signal, and (c) the splitting of the DFT of the signal (shown in absolute scale and cyclicly shifted to the centers). (The last value of the transform has been truncated.) These short DFTs together compose the 512-point DFT of the signal.

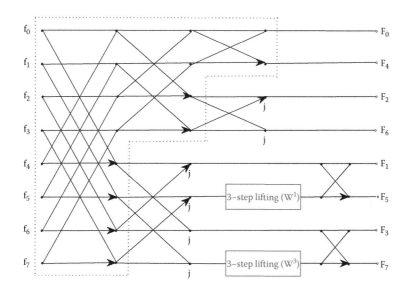

FIGURE 2.3
Signal-flow graph of calculation of the 8-point DFT by two lifting schemes.

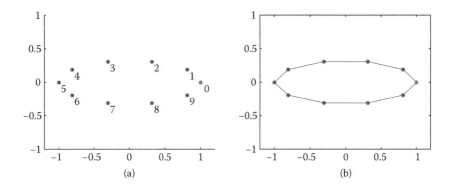

FIGURE 2.24
(a) Location of all points y_k and (b) the path of the point (1, 0).

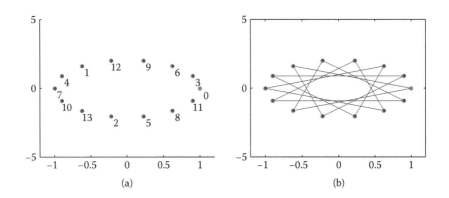

FIGURE 2.25
(a) Location of all points y_k and (b) scheme of the movement of the point (1, 0), when $N = 7$.

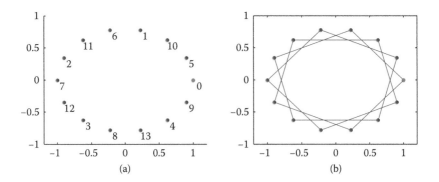

FIGURE 2.26
(a) Location of all points y_k and (b) scheme of the movement of the point (1, 0), when $N = 7$.

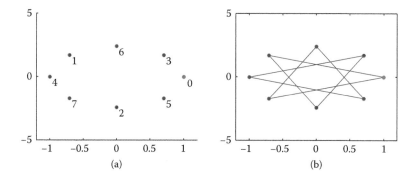

FIGURE 2.27
(a) Location of all points y_k and (b) scheme of the movement of the point $(1, 0)$, when $N = 8$.

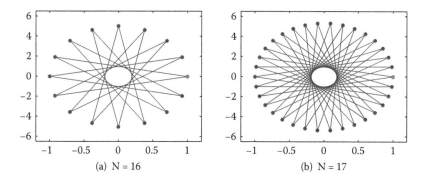

FIGURE 2.28
Schemes of movement of the point $(1, 0)$ for the cases, when (a) $N = 16$ and (b) $N = 17$.

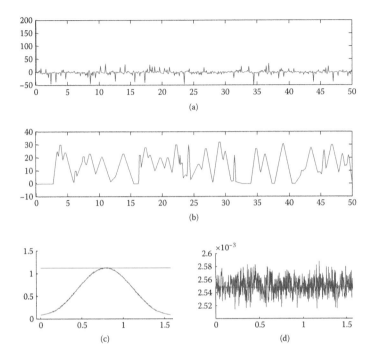

FIGURE 4.16

(a) Random realization of the degraded transform, (b) the inverse transform $\hat{y}$, (c) error $\varepsilon_{0;0,\varphi}(\hat{y}, y)$, and (d) error $\varepsilon_{2;0,\varphi}(\hat{y}, y)$, when $\varphi \in [0,\pi/2]$.

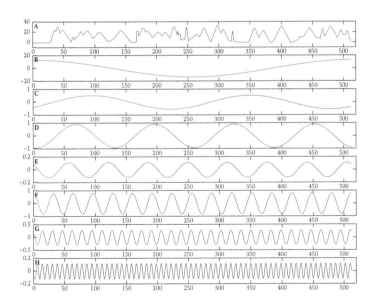

FIGURE 5.1

(a) Signal and seven (b)-(h) basis signals $r(p)_n$, for $p = 2^k$, $k = 0:6$.

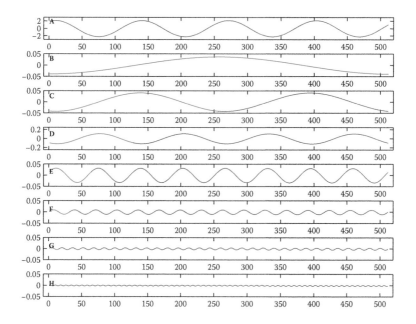

FIGURE 5.2
(a) Original signal of length 512, and seven (b)-(h) basis sinusoidal signals $r(p)_n$, for $p = 2^k$, $k = 0:6$.

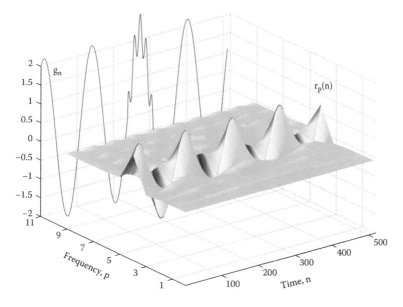

FIGURE 5.6
Original signal $g(t)$ and the surface of the part of the Fourier transform.

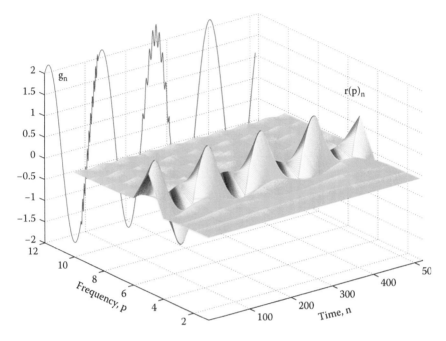

FIGURE 5.8
Signal $g(t)$ and the surface of the frequency-time image.

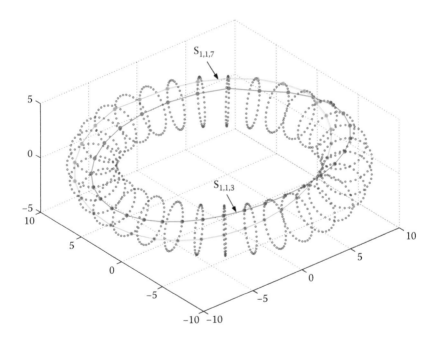

FIGURE 5.19
The net with knots of the grid 32×32 in the 3-D space with locus of two spirals $S_{1,1,3}$ and $S_{1,1,7}$.

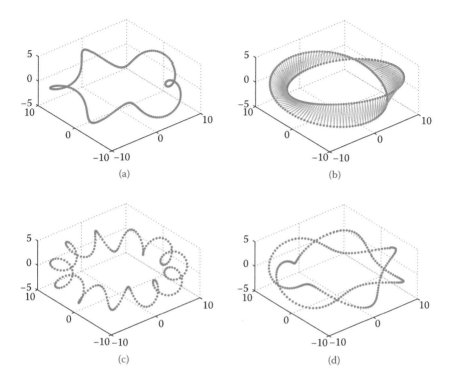

FIGURE 5.20
Locus of the spiral $\check{S}p,s$ in the grid 256×256 on the torus, when (p, s) equals (a) (5, 1), (b) (127, 1), (c) (11, 1), and (d) (125, 1).

FIGURE 5.27
(a) 766 generators of the set J'_{256}, and the girl image with amplified (b) 7th series image and (c) 6th series image. (All images are displayed in the same colormap.)

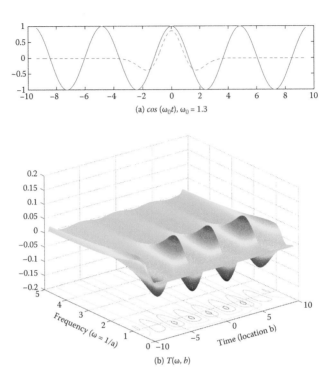

(a) $\cos(\omega_0 t)$, $\omega_0 = 1.3$

(b) $T(\omega, b)$

FIGURE 6.6
(a) Cosinusoidal signal and (b) the wavelet transform plot of the signal with the Mexican hat wavelet.

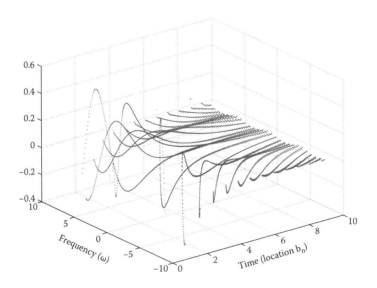

FIGURE 6.9
Wavelet transform of the signal $\cos(w_0 t)$, which is based on the cosine analyzing wavelet.

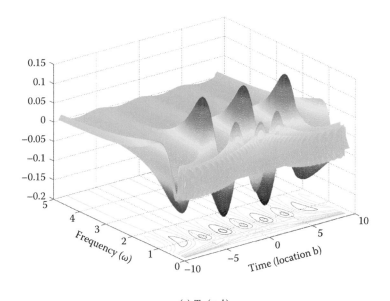

(a) $T_\psi (a, b)$

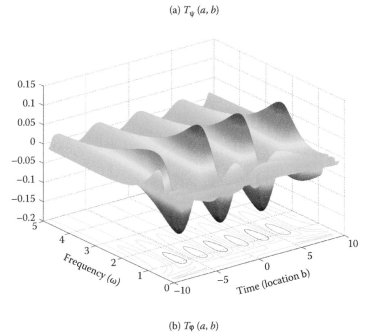

(b) $T_\phi (a, b)$

FIGURE 6.10
(a) Cosine and (b) sine wavelet transforms of the function $f(t) = \cos(w_0 t)$.

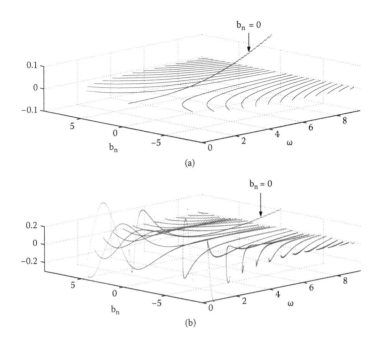

(a)

(b)

FIGURE 6.13
Wavelet transforms of the (a) signal $n(t)$ and (b) signal $y(t)$.

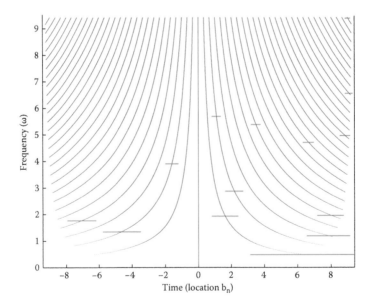

FIGURE 6.14
The locus of time-frequency points for the B-wavelet transform.

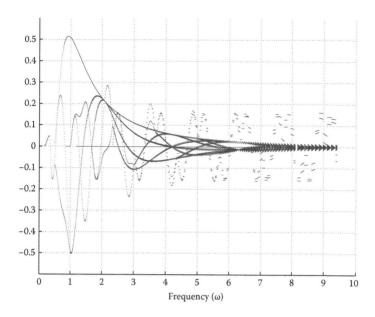

FIGURE 6.17
Wavelet transform plot of the signal $f(t)$ with the cosine analyzing function (projection on the frequency domain).

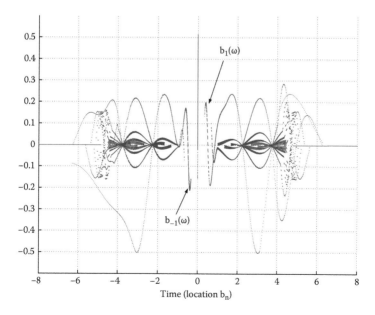

FIGURE 6.18
Wavelet transform plot of the signal $g(t)$ with the cosine analyzing function (projection on the time domain).

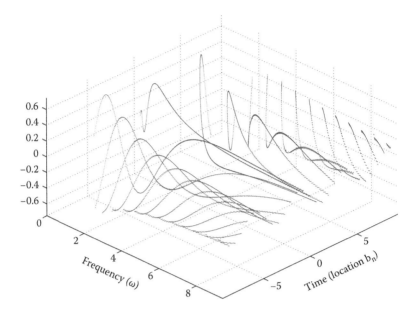

FIGURE 6.19
Hartley wavelet transform plot of the signal $\cos(w_0 t)$.

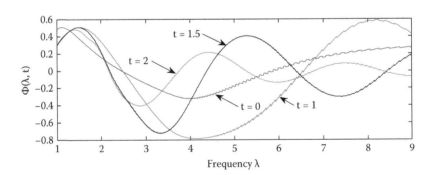

FIGURE 6.23
Transform (Ψ, t_n) calculated for $n = 0$ and time points t = 0, 1, 1.5, and 2.

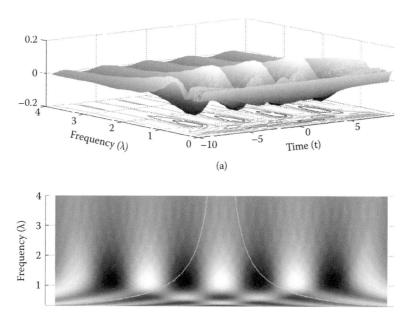

FIGURE 6.25
Cosine wavelet transform plot of function $f(t)$ in (a) 3-D view and (b) 2-D view with boundaries of set Δ_0.

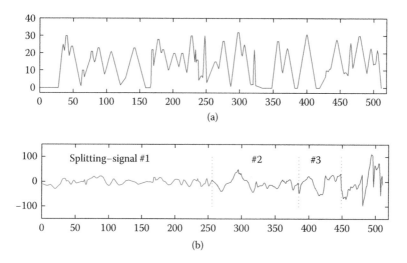

FIGURE 6.27
(a) Signal of length 128 and (b) the paired transform.

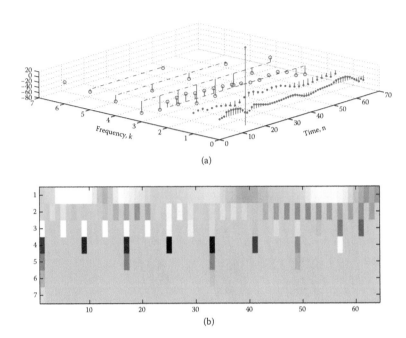

(a)

(b)

FIGURE 6.28
The paired transform of the signal of Fig. 6.27.

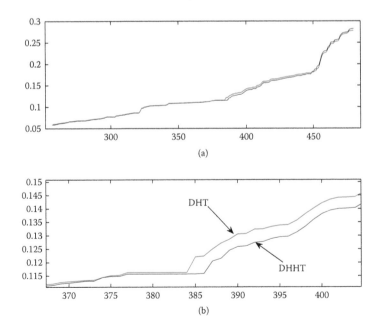

(a)

(b)

FIGURE 6.37
(a) The mean-square-root error curves for signal reconstruction after truncating L coefficients of the 512-point DGHT and DHT, where $L \in [256, 480]$, and (b) the same errors in the interval (365, 405).

5

Paired Transform-Based Decomposition

The application of the Fourier transform representation by series leads to the concept of a wavelet-type unitary transform, which we call the paired transform. This transform is considered as a part of the mathematical structure of the discrete Fourier transform, which allows for minimizing not only the computation cost of the fast Fourier and other transforms [9]-[13]. The paired transform represents the signal as a unique set of separate short and independent signals that can be processed separately for effectively solving different problems of signal and image processing. The splitting-signals have different lengths and carry the spectral information of the represented signal in disjoint sets of frequencies. The paired transform is fast and leads to an effective decomposition of signals. In this chapter, we consider the decomposition of 1-D and 2-D signals by introduced section basis signals and derive the inverse formula. Examples of the application of paired transforms for signal detection and noise filtration, and image enhancement are described.

5.1 Decomposition of 1-D signals

In this section, we describe the inverse discrete Fourier transformation, $\mathcal{F}_N^{-1}$, in a form related to the paired form of representation of the signal. The transform inverse to the N-point Fourier transform of the discrete signal f_n is defined by

$$f_n = \frac{1}{N} \sum_{p=0}^{N-1} F_p W^{-np}, \qquad n = 0 : (N-1).$$

The discrete signal f_n is decomposed by sinusoidal signals with frequencies $\omega_p = (2\pi/N)p$, $p = 0 : (N-1)$. Let f_n be a real signal. Then, for a given p, we define the following sinusoidal signal of frequency ω_p

$$r(p)_n = Re(F_p) \cos(\omega_p n) - Im(F_p) \sin(\omega_p n) = |F_p| \cos(\omega_p n + \theta_p),$$

where $n = 0 : (N-1)$ and the phase $\theta_p = \tan^{-1}(Im(F_p)/Re(F_p))$. For $p = 0$, $r(0) \equiv F_0/N$.

As an example, Figure 5.1 shows a signal of length 512 in part a, along with sinusoidal signals $r(p)_n$ for $p = 1, 2, 4, 8, 16, 32$, and 64 in parts b through h.

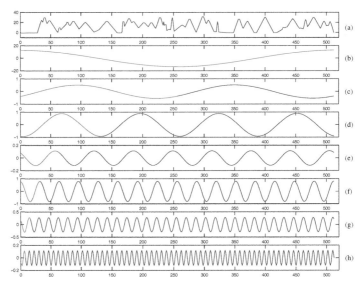

FIGURE 5.1 (See color insert following page 242.)
(a) Signal and seven (b)-(h) basis signals $r(p)_n$, for $p = 2^k$, $k = 0 : 6$.

A similar picture is shown in Figure 5.2 for the signal defined by

$$f(t) = 2\cos(\omega_0 t) + 1.5\sin(\omega_0 t - 0.25), \qquad \omega_0 = 1.3. \qquad (5.1)$$

512 sampled values f_n at points uniformly placed in the interval $[-3\pi, 3\pi]$ in part a, along with seven sinusoidal signals $r(p)_n$ for $p = 1, 2, 4, 8, 16, 32$, and 64, in parts b through h.

The real discrete signal is the averaged sum of N sinusoidal signals

$$f_n = [r(0)_n + r(1)_n + r(2)_n + \cdots + r(N-1)_n]/N, \ n = 0 : (N-1). \quad (5.2)$$

The signal $r(p)_n$ is referred to as *the ω_p-frequency basis sinusoidal signal* in the decomposition of the signal f_n.

The frequency-time image $R(p, n) = \{r_p(n)\}$ of all 512 sinusoidal signals $r(p)_n$ for $p = 0 : 511$ is given in Figure 5.3. The frequency, p, is measured along the vertical of the image, and the time, n, along the horizontal. This is the so-called frequency-time image of the original signal f_n. We consider this image in a convenient form for visualization. Namely, we perform a permutation of rows of the image that corresponds to the paired representation of the signal. The rows are sorted in the order that corresponds to the partition

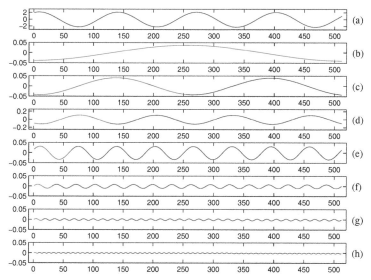

FIGURE 5.2 (See color insert following page 242.)
(a) Original signal of length 512, and seven (b)-(h) basis sinusoidal signals $r(p)_n$, for $p = 2^k$, $k = 0 : 6$.

$\sigma' = (T'_{2^k}; k = 0 : 9)$ of set $X_{512} = \{0, 1, 2, ..., 511\}$ of frequencies, i.e., as follows $X_{512} = \{1, 3, 5, ..., 511\} \cup \{2, 6, 10, ..., 510\} \cup \{4, 12, 20, ..., 508\} \cup ... \cup \{128, 384\} \cup \{256\} \cup \{0\}$.

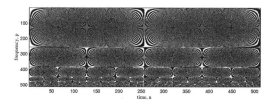

FIGURE 5.3
The image of 512 reordered basis signals.

This figure shows an interesting picture of dividing the frequency-time image by sections. In each section we observe a packing of homogeneous concentric type curves at centers uniformly placed along the horizontal and doubled along the vertical for each next section. Each section along the horizontal has a regular structure that replicates the next two sections. The number of sections increases and the size decreases by a factor of 2. Since ω_p-frequency sinusoids $r(p)_n$ are periodic, we consider only one period for each of such signals.

It should be noted that $r(N-p)_n = r(p)_n$, because of the following property of the DFT for real data: $F_{N-p} = \overline{F_p}$, when $p = 1 : (N/2 - 1)$. Indeed, $F_{N-p}W^{-n(N-p)} = \overline{F_p}W^{-np}$. Therefore, a real discrete signal can be written as the following sum of $(N/2 + 1)$ sinusoidal signals

$$f_n = \frac{1}{N}r(0)_n + \frac{2}{N}\sum_{p=1}^{N/2-1} r(p)_n + \frac{1}{N}r(N/2)_n, \quad n = 0 : (N-1). \quad (5.3)$$

It should be noted that $r(0)_n + r(0)_{N/2} = F_0 + (-1)^n F_{N/2}$, which equals the sum of f_{2k} or f_{2k+1}, depending on the evenness of n. Figure 5.4 shows the frequency-time image of the signal of (5.1) when only 257 required sinusoidal signals are considered.

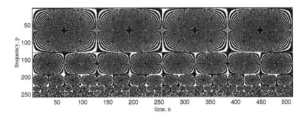

FIGURE 5.4

The image of the first 257 reordered basis signals.

Figure 5.5 illustrates a change in the frequency-time image when the impulse noise with amplitude 3 occurs at point $n = 210$. The picture has a regular structure, and we also observe a cyclic shift of the regular structure by $N/2 - n_0 = 256 - 210 = 46$ points to the left.

Example 5.1

Let $g(t)$ be the signal composed by the signal defined by (5.1) with the additional high-frequency cosine signal $x(t)$ of short time in the interval $[0.6, 2]$,

$$g(t) = f(t) + x(t) = \begin{cases} f(t) + 0.5\cos(8\omega_0 t), & \text{if } t \in [0.6, 2], \\ f(t), & \text{otherwise,} \end{cases} \quad \omega_0 = 1.3. \quad (5.4)$$

The signal $x(t)$ is referred to as a noise added to the signal $f(t)$. Figure 5.6 shows the 512-point sampled signal $g(t)$ in the XZ-plane with the part of the surface of the frequency-time transform $R(p, n)$ when $p = 1 : 9$. The rest of the transform $R(p, n)$ is very small in magnitude, has an almost flat surface, and is not shown in the figure. The reconstruction of the original signal by using only the first ten sinusoidal signals $r_p(n)$, $p = 1 : 10$, is shown in Figure 5.7 in part

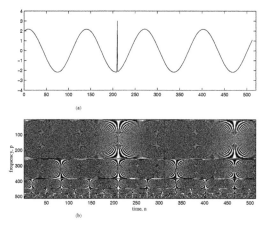

FIGURE 5.5

(a) The original signal with the impulse noise at point $n = 210$, and (b) the image of 512 basis signals.

a. The reconstruction, $\hat{f}_n$, is calculated by means of the following truncation of the sum in (5.3): $\hat{f}_n = 2/N[r(1)_n + r(2)_n + \cdots + r(10)_n]$, $n = 0 : (N-1)$.

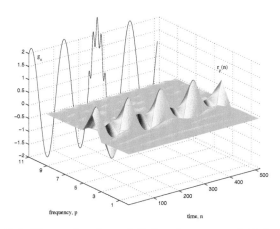

FIGURE 5.6 (See color insert following page 242.)

Original signal $g(t)$ and the surface of the Fourier transform.

A significant amount of the additional short-time signal x_n can be extracted from the composite signal $g(t)$ by choosing a band of frequencies to be deleted from the sum in (5.3). As an example, Figure 5.7(b) shows the result of signal composition by only 50 cosine basis signals $r(p)_n$, when $p = 11 : 60$, and $r(0)_n$, i.e., $\hat{x}_n = r(0)_n/N + 2/N[r(11)_n + r(12)_n + \cdots + r(60)_n]$, $n = 0 : (N - 1)$. We obtain a perfect reconstruction of the original signal with almost full

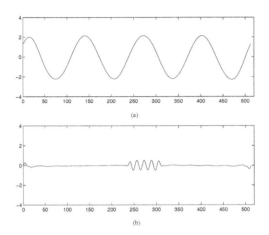

FIGURE 5.7
(a) Signal reconstructed by the first ten basis signals $r_p(n)$, $p = 1 : 10$. (b) Signal extracted from the composite signal when the next 50 basis functions are used.

extraction of the noise signal. The above formulas describe examples of low-pass and band-pass filters in terms of frequency basis signals $r_p(n)$. ▯

Example 5.2
Let $g(t) = f(t) + x(t)$ be the signal composed by the above mentioned signal $f(t)$ with the additional two-high-frequency cosine signal $x(t)$ of short duration in the time-intervals $[-0.6, 1.8]$ and $[-6.1, -4.7]$

$$g(t) = f(t) + x(t) = \begin{cases} f(t) + 0.25\cos(16\omega_0 t), & \text{if } t \in [-0.6, 1.8] \\ f(t) + 0.25\cos(32\omega_0 t), & \text{if } t \in [-6.1, -4.7] \\ f(t), & \text{otherwise} \end{cases}$$

where $\omega_0 = 1.3$. Figure 5.8 shows the 512-point sampled signal $g(t)$ in the XZ-plane with the part of the surface of the transform $R(p, n)$ when $p = 1 : 10$. The reconstruction of the original signal by using only the ten sinusoidal signals $r_p(n)$ from the first four sections

$$\hat{f}_n = \frac{1}{256}\left[\sum_{p=1,3,5,7,9} r(p)_n + \sum_{p=2,6,10} r(p)_n + r(4)_n + r(8)_n\right]$$

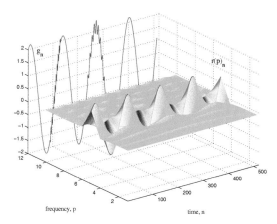

FIGURE 5.8 (See color insert following page 242.)
Signal $g(t)$ and the surface of the frequency-time image.

where $n = 0 : 511$ is shown in Figure 5.9(a). Namely, five basis functions have been taken from section 1, three from section 2, and one each from sections 3 and 4. We call this reconstruction a $(5, 3, 1, 1)$-filtration. In general, parameters of such a reconstruction can be specified for each section depending on the case under consideration. The five first basis signals are periodic with the period 256, the next three with period 128, and the last two with periods 64 and 32, respectively. Figure 5.9(b) shows the result of the signal composition by only 90 cosine basis signals $r(p)_n$, when $p = 61 : 150$, plus $r(0)_n$, which is referred to as $(0, 61\text{-}150)$-pass filtration.   ☐

5.1.1   Section basis signals

In this section, we describe the relation between different sections of the discrete Fourier transform and the paired transform of the signal.

For a generator $p \in J'$, we define the following incomplete Fourier transform:

$$A(p)_{p_1} = \begin{cases} F_{p_1}, & \text{if } p_1 = (2k+1)p \bmod N, \ k \in \{0, 1, 2, \ldots, L_p - 1\} \\ 0, & \text{otherwise,} \end{cases} \tag{5.5}$$

where $L_p = N/(2p)$, when $p \neq 0$, and $L_p = 1$, when $p = 0$. The inverse Fourier transform of these data results in the following real signal:

$$a(p)_n = \left(\mathcal{F}_N^{-1} \circ A(p) \right)_n = \frac{1}{N} \sum_{p_1 \in T'_p} A(p)_{p_1} W^{-np_1}$$

$$= \frac{1}{N} \sum_{p_1=0}^{N/(2p)-1} r(p_1)_n, \quad n = 0 : (N-1), \quad (p \neq 0), \tag{5.6}$$

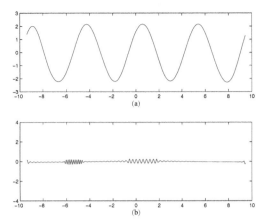

FIGURE 5.9
(a) Signal reconstructed by the first ten basis signals $r_p(n)$, $p = 1 : 10$. (b)
Signal extracted from the composite signal when 91 basis functions are used.

which we call the *p-section basis signal* of f_n.

Since sets T'_p compose a partition of X_N, the sum of all transforms $A(p)$
across all frequencies $p \in J'$ coincides with the N-point DFT of f_n,

$$F_p = A(1)_p + A(2)_p + A(4)_p + ... + A(N/2)_p + A(0)_p, \quad p = 0 : (N - 1).$$

The sum of p-section basis signals yields thus the decomposition of f_n

$$f_n = a(1)_n + a(2)_n + a(4)_n + ... + a(N/2)_n + a(0)_n, \quad n = 0 : (N - 1).$$

The following calculations follow from (5.5) and (5.6), when $p \neq 0$:

$$a(p)_n = \frac{1}{N} \sum_{p_1=0}^{N-1} A(p)_{p_1} W^{-np_1} = \frac{1}{N} \sum_{\{p_1=(2k+1)p\}} F_{(2k+1)p \bmod N} W^{-n \cdot (2k+1)p}$$

$$= \left[\frac{1}{N} \sum_{k=0}^{N/(2p)-1} F_{(2k+1)p \bmod N} W_{N/(2p)}^{-kn} \right] W^{-np}$$

$$= \left[\frac{1}{2p} f'_{p,np} W^{np} \right] W^{-np} = \frac{1}{2p} f'_{p,np} = \pm \frac{1}{2p} f'_{p,np \bmod N/2},$$

where we consider the following property of the paired representation:

$$f'_{p,t+N/2} = -f'_{p,t}, \quad t = 0 : (N/2 - 1). \tag{5.7}$$

When $p = 0$, the basis function is the constant, $a(0)_n \equiv \frac{1}{N} F_0 = \frac{1}{N} f'_{0,0}$.
Therefore, the decomposition of the signal f_n can be written as

$$f_n = \sum_{p \in J'} a(p)_n = \sum_{p \in J' \setminus \{0\}} \frac{1}{2p} f'_{p,np} + \frac{1}{N} f'_{0,0}, \quad n = 0 : (N - 1). \tag{5.8}$$

For instance, when $N = 8$ and $n = 1$ and $n = 4$, we obtain, respectively:

$$f_1 = \frac{1}{2}f'_{1,1} + \frac{1}{4}f'_{2,2} - \frac{1}{8}f'_{4,0} + \frac{1}{8}f'_{0,0}, \ f_4 = -\frac{1}{2}f'_{1,0} + \frac{1}{4}f'_{2,0} + \frac{1}{8}f'_{4,0} + \frac{1}{8}f'_{0,0}.$$

We obtain the simple analytical formula for the inverse discrete paired transform. The signal f_n is composed by the splitting signals. Namely, the signal of length $N = 2^r$ is composed by r anticyclic splitting-signals or functions with frequencies $p = 1, 2, 4, 8, ..., 2^{r-1}$, and 0. Indeed, the p-section function is periodic with period $2L_p$, when $p \neq 0$. For instance, for $p = 1$ and 4, $a(1)_{n+N} = a(1)_n$ when $n = 0 : (N-1)$, and $a(4)_{n+N/4} = a(4)_n$ when $n = 0 : (N/4-1)$. Moreover, according to (5.7), we obtain $a(p)_{n+L_p} = -a(p)_n$, when $n = 0 : (L_p-1)$. For instance, $a(1)_{n+N/2} = -a(1)_n$, for $n = 0 : (N/2-1)$.

The decomposition of a signal of length N by the discrete Fourier transform in the original presentation of (5.2) is fulfilled by N basis sinusoidal signals $r(p)_n$ with N different frequencies. The sinusoids $r(p)$ have frequencies $\omega_p = (2\pi/N)p$, and amplitudes and phases equal respectively to amplitudes and phases of coefficients F_p. Each sinusoidal signal $r(p)$ carries information of the Fourier transform only at point p. This is why the number of such sinusoids in the signal decomposition equals N. In contrast, the use of the paired transform yields the decomposition of the signal by $(r + 1)$ functions $a(p)_n$

$$f_n = \sum_{p=1,2,4,8,...,N/2,0} a(p)_n, \quad n = 0 : (N - 1).$$

Each function $a(p)$ carries information of the Fourier transform at $N/(2p)$ frequency-points of the set T'_p that contains the particular point p. The number of functions $a(p)$ in the signal decomposition equals $(r + 1)$. Each signal $a(p)$ in the decomposition is not a sinusoidal wave, but a sum of such.

5.2   2-D paired representation

In the paired representation, a two-dimensional (2-D) image is represented uniquely by a complete set of 1-D signals, so-called *splitting-signals*, that carry the spectral information of the image at frequency-points of specific sets that divide the whole domain of frequencies [9, 100]. The processing of the image, such as image enhancement, for example, can thus be reduced to processing splitting-signals and such a process requires a modification of only a few spectral components of the image, for each signal. For instance, the α-rooting method of image enhancement can be fulfilled through processing one or a few splitting-signals. We describe an effective formula for inverse 2-D $N \times N$-point paired transform, where N is a power of 2. New concepts of direction and series images are introduced that define the resolution and periodic structures of the image components, which can be packed in the form of the "resolution map" of the size of the image. A simple method of image enhancement by series images is described.

Let $f = \{f_{n,m}\}$ be an image of size $N \times N$, where $N = 2^r$ and $r > 1$. The $N \times N$-point 2-D DFT of the image, accurate to the normalizing factor $1/N$, is defined by

$$F_{p,s} = (\mathcal{F}_{N,N} \circ f)_{p,s} = \sum_{n=0}^{N-1} \sum_{m=0}^{N-1} f_{n,m} W^{np+ms}, \quad W = e^{-\frac{2\pi j}{N}}.$$

Frequency-points (p, s) are from the square lattice $X = X_{N,N} = \{(p, s); p, s = 0 : (N - 1)\}$. This lattice can be split into a set of disjoint subsets in such a way that the 2-D DFT of the image at frequency-points of each subset is represented by the 1-D DFT of a corresponding signal. Such a partition consists of $(3N - 2)$ subsets and can be defined as

$$\sigma' = \left(\left((T'_{2^k p, 2^k s})_{(p,s) \in J_{2^{r-k}}} \right)_{k=0:(r-1)}, \{(0,0)\} \right)$$

where the subsets equal $T'_{p,s} = \left\{ (\overline{(2m+1)p}, \overline{(2m+1)s}); m = 0 : (N/2 - 1) \right\}$, and $\bar{l} = l \bmod N$, and the subsets J_{2^k} of generators are defined by

$$J_{2^k} = \{(1, s), s = 0 : (2^k - 1)\} \cup \{(2p, 1), p = 0 : (2^{k-1} - 1)\}.$$

We denote by $J' = J'_N$ the set of $(3N - 2)$ generators of all subsets $T'_{p,s}$ of the partition σ', $J' = J'_N = \{(2^k p, 2^k s), (p, s) \in J_{2^{r-k}}, k = 0 : (r - 1)\} \cup \{(0,0)\}$. The number of frequency-points of the subset $T'_{p,s}$ equals $N/2^{k+1}$, where $2^k = g.c.d.(p, s)$, $k \geq 0$. With each such subset, we associate a 1-D short signal $f_{T'_{p,s}}$ of length $N/2^{k+1}$, which is called the *paired splitting-signal*,

$$f_{T'_{p,s}} = \{f'_{p,s,0}, f'_{p,s,2^k}, f'_{p,s,2\cdot 2^k}, \dots, f'_{p,s,N/2-2^k}\}.$$

The components of this signal are calculated by

$$f'_{p,s,t} = \sum_{V_{p,s,t}} f_{n,m} - \sum_{V_{p,s,t+N/2}} f_{n,m}, \quad t = 0, 2^k, \dots, N/2 - 2^k.$$

Disjoint sets $V_{p,s,t}$, $t = 0 : (N-1)$, are defined as $\{(n, m); np + ms = t \bmod N\}$ and compose a partition of the set X of points (n, m) in the image plane.

For a given frequency-point (p, s), the following statement directly follows from the definition of the paired representation:

$$F_{\overline{(2m+1)p}, \overline{(2m+1)s}} = \sum_{t=0}^{N/2^{k+1}-1} \left(f'_{p,s,2^k t} W^t_{N/2^k} \right) W^{mt}_{N/2^{k+1}}$$

where $m = 0 : (N/2^{k+1} - 1)$. The 2-D DFT at frequency-points of the subset $T'_{p,s}$ is thus defined by the $N/2^{k+1}$-point DFT of the splitting-signal $f_{T'_{p,s}}$ modified by the vector of twiddle coefficients $\{W^t_{N/2^k}; t = 0 : (N/2^{k+1} - 1)\}$. The transformation $\chi'_{N,N} : \{f_{n,m}\} \rightarrow \{f'_{p,s,t}; (p,s,t) \in U\}$ is called *the 2-D paired transformation* [8]. The paired transform does not require multiplications. U is the set of all number-triplets (p,s,t) of components of the splitting-signals generated by frequencies $(p,s) \in J'$, i.e.,

$$U = \{(p,s,t); (p,s) \in J', t = 0 : (2^{r-1}/g.c.d.(p,s) - 1)\},$$

where we consider $t = 0$ when $(p,s) = (0,0)$. The complete system of paired functions

$$\chi'_{p,s,t}(n,m) = \begin{cases} 1; & np + ms = t \bmod N, \\ -1; & np + ms = (t + N/2) \bmod N, \qquad (p,s,t) \in U, \\ 0; & \text{otherwise}, \end{cases}$$

is extracted from the structure of the 2-D discrete Fourier transformation. To illustrate splitting-signals, Figure 5.10 shows the tree image $f_{n,m}$ of size 256×256 and the 2-D DFT of the image in absolute mode and shifted to the center. Figure 5.11 shows the totality of all 766 paired splitting-signals of this

(a)   (b)

FIGURE 5.10
(a) Tree image (256×256) and (b) 2-D DFT of the image.

image in part a. The first 384 splitting-signals of length 128 each are shown in the form of the "image" 384×128. The next 192 splitting-signals of length 64 each are shown in the form of the image 192×64, the splitting-signals of length 32 are shown in the form of the image 96×32, and so on. The whole picture represents the paired transform of the tree image. The totality of the 1-D DFTs of all modified splitting-signals is shown in the form of a similar figure in b. These 1-D DFTs represent the splitting of the 2-D DFT by frequency-points which are distributed by disjoint subsets T' in the lattice 256×256. The number of elements in both images equals $(256)^2$.

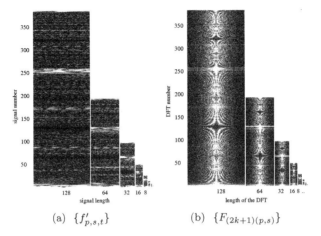

(a) $\{f'_{p,s,t}\}$ (b) $\{F_{(2k+1)(p,s)}\}$

FIGURE 5.11
(a) Splitting-signals of lengths $128, 64, 32, 16, 8, 4, 2, 1$, and (b) 1-D DFTs of
the modified splitting-signals (the transforms are shown in the absolute mode
and shifted to the center).

5.2.1 Set-frequency characteristics

Each subset $T'_{p,s}$ of frequency-points generates the splitting-signal $f_{T'_{p,s}}$, which
carries the spectral information of the image $f_{n,m}$ at these frequency-points.
The splitting-signal is thus the set-frequency characteristics of the image. This
signal defines the corresponding direction image-component of $f_{n,m}$. Indeed,
let D_{p_1,s_1} be the following 2-D DFT composed only from the components of
the 2-D DFT which lie on the given subset $T'_{p,s}$:

$$D_{p_1,s_1} = \begin{cases} F_{p_1,s_1}; & \text{if } (p_1, s_1) \in T'_{p,s}, \\ 0; & \text{otherwise.} \end{cases}$$

We first consider the case when g.c.d.$(p, s) = 1$, i.e., $(p, s) \in J_{2^r}$. The inverse
transform of the defined 2-D DFT can be calculated as follows:

$$d_{n,m} = \frac{1}{N^2} \sum_{p_1=0}^{N-1} \sum_{s_1=0}^{N-1} D_{p_1,s_1} W^{-(np_1+ms_1)} = \frac{1}{N^2} \sum_{(p_1,s_1)\in T'_{p,s}} F_{p_1,s_1} W^{-(np_1+ms_1)}$$

$$= \frac{1}{N^2} \sum_{k=0}^{N/2-1} F_{\overline{(2k+1)p},\overline{(2k+1)s}} W^{-(n\overline{(2k+1)p}+m\overline{(2k+1)s})}$$

$$= \frac{1}{N^2} \sum_{k=0}^{N/2-1} F_{\overline{(2k+1)p},\overline{(2k+1)s}} W^{-(2k+1)\overline{(np+ms)}} \qquad [t = (np + ms) \bmod N]$$

$$= \frac{1}{2N}\left(\frac{2}{N} \sum_{k=0}^{N/2-1} F_{\overline{(2k+1)p},\overline{(2k+1)s}} W_{N/2}^{-kt}\right) W^{-t} = \frac{1}{2N}\left(f'_{p,s,t} W^t\right) W^{-t} = \frac{1}{2N} f'_{p,s,t}$$

where we consider $f'_{p,s,t+N/2} = -f'_{p,s,t}$, $t = 0 : (N/2 - 1)$. Thus the direction image is defined as

$$d_{n,m} = \frac{1}{2N} f'_{p,s,(np+ms) \bmod N}, \quad (n, m) \in X_{N,N}.$$

In other words, the direction image $N \times N$ is filled by the $N/2$ values of the splitting-signal $f_{T'_{p,s}}$, which are placed along the parallel lines $np + ms = t \bmod N$, $t = 0 : (N-1)$. The direction of these lines is defined by coordinates of the frequency-point (p, s).

In the case when g.c.d.$(p, s) = 2^k$, $k \geq 1$, the inverse 2-D DFT of the transform D_{p_1,s_1} results in the following direction image:

$$d_{n,m} = \frac{1}{2^{k+1}N} f'_{p,s,(np+ms) \bmod N}, \quad (n, m) \in X_{N,N}.$$

As an example, Figure 5.12 shows the splitting-signal $f_{T'_{1,1}}$ of the tree image in part a, along with the 128-point DFT (in absolute scale) of the modified splitting-signal in b, the set of frequency-points $T'_{1,1}$ on the square grid 256×256 in c, and the direction image $d_{n,m}$ of the size 256×256 in d. The frequency-

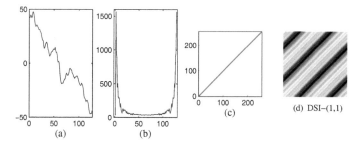

FIGURE 5.12
(a) Splitting-signal with number $(p, s) = (1, 1)$, (b) 128-point DFT of the modified splitting-signal, (c) set of frequency-points $T'_{1,1}$, and (d) direction image 256×256.

points of the set $T'_{1,1}$ are located on the main diagonal of the grid; they occupy each second point on the diagonal. Figure 5.13 shows similar characteristics of the tree image when the splitting-signal $f_{T'_{1,4}}$ is taken. The subset $T'_{1,4}$ consists of 128 frequency-points which lie on four parallel lines at angle $\arctan(4)$ to the horizontal.

All $(3N - 2)$ subsets $T'_{p,s}$, with generators (p, s) from the set J' compose the partition of the grid $X_{N,N}$. Therefore, the sum of all direction images $d_{n,m} = d_{n,m;p,s}$ results in the original image $f_{n,m}$. In other words, we obtain

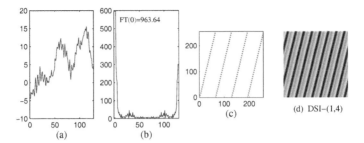

FIGURE 5.13
(a) Splitting-signal with number $(p, s) = (1, 4)$, (b) 128-point DFT of the modified splitting-signal, (c) set of frequency-points $T'_{1,4}$, and (d) direction image 256×256.

the following decomposition of the image by the direction images:

$$f_{n,m} = \sum_{(p,s)\in J'} d_{n,m;p,s} = \frac{1}{2N} \sum_{k=0}^{r} \frac{1}{2^k} \sum_{(p,s)\in 2^k J_{2^{r-k}}} f'_{p,s,(np+ms) \bmod N} \quad (5.9)$$

where the $k = r$ case corresponds to the set $J_0 = (0, 0)$ and normalize coefficient $1/2^{k-1}$ instead of $1/2^k$. Equation 5.9 is the formula of reconstruction of the image by its paired transform, by using operations of addition and division by powers of two.

The processing of the splitting-signal $f_{T'_{p,s}}$ leads to the change of the 2-D DFT of the image in the frequency-points of the subset $T'_{p,s}$, or the change of the image in the spatial domain along the corresponding parallel lines. The direction of these lines is defined by coordinates of the generator (p, s). When processing only one splitting-signal with number (p_0, s_0), $f_{T'_{p_0,s_0}} \to \hat{f}_{T'_{p_0,s_0}}$, the image is changed as

$$f_{n,m} \to f_{n,m} - \frac{1}{2^{k+1}N} \Delta f'_{p_0,s_0(np_0+ms_0) \bmod N}$$

where $\Delta f'_{p_0,s_0,t} = \hat{f}'_{p_0,s_0,t} - f'_{p_0,s_0,t}$, $t = 0 : (N-1)$. Figure 5.14 shows the tree image after amplifying the splitting-signal $f_{T'_{1,1}}$ by the factor of 4 in part a, along with the image with amplified splitting-signal $f_{T'_{1,4}}$ in b, and $f_{T'_{1,17}}$ in c. One can observe how the direction images corresponding to these splitting-signals superpose on the image. Splitting-signals can also be used for modeling images with noise along different directions. One can note that the picture in c shows the hexagonal net on the tree image.

The processing of the image through its splitting-signals is the process along parallel lines in certain directions, which are referred to as projection data. Some splitting-signals or direction images are highly expressed and others little, by their characteristics (see Figure 5.11). For instance, the splitting-signals have different energy, $E_{p,s} = \sqrt{(f'_{p,s,0})^2 + (f'_{p,s,1})^2 + \cdots + (f'_{p,s,N/2-1})^2}/N$,

(a) DSI–(1,1)   (b) DSI–(1,4)   (c) DSI–(1,17)

FIGURE 5.14
Tree image after amplifying the splitting-signal with number (a) $(1,1)$, (b) $(1,4)$, and (c) $(1,17)$.

which can be used for selecting the signal with high energy. The energy curve

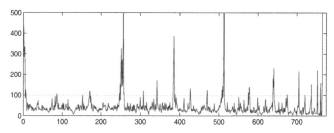

FIGURE 5.15
Energy of all splitting-signals of the tree image.

of all 766 splitting-signals of the tree image 256×256 is given in Figure 5.15. The signals are numbered by the order of generators (p, s) in the set J', which is divided by the subsets $J_{256}, 2J_{128}, ..., 128J_2$, plus $(0, 0)$. Taking the threshold for energy equal $E_0 = 100$, we obtain 50 splitting-signals with high energies $E_{p,s} > E_0$. All together, the images of these splitting-signals or their direction images compose the image shown in Figure 5.16a, and the rest of the direction images compose the image in b. The sum of these two images equals the original tree image. The image in a provides no details but a very smooth and "hot" picture of the image, and opposite, the image in b provides the details of the tree image but lacks brightness.

5.2.2 Image reconstruction by projections

The derived formula in (5.9) can be used for effective reconstruction of images by their projections. To show that, we briefly consider the simple discrete model of image reconstruction that is used in finite series-expansion reconstruction methods [69]. This discrete model of image reconstruction for the

   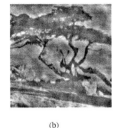

(a)   (b)

FIGURE 5.16
(a) The sum of 50 direction images defined by splitting-signals of high energy, and (b) the sum of the remaining direction images.

typical parallel-beam scanning scheme can be used for calculating all components $f'_{p,s,t}$ of the paired representation of the image. In other words, the paired transform of the image to be reconstructed is completely defined by the finite number of projections [99]. The number of such projections equals $3N/2$ if N is a power of two, and $(N+1)$ if N is a prime number.

Suppose that a reconstruction image $f(x, y)$ occupies the quadratic domain $L \times L$, on which the quadratic lattice $N \times N$ of image elements (IEs) is marked. We assume that the absorption function of the (n, m)th image element, where $n, m = 0 : (N - 1)$, takes a constant value $f_{n,m}$. We also assume that the radiation source and detector represent themselves the points, and the rays spreading between them are straight. The measured value of the total attenuation energy along the lth ray, denoted via y_l, $l \in \{1, ..., M\}$, can be represented in the form of the finite series of the unknown image $\{f_{n,m}\}$ along this ray:

$$y_l = \sum_{n=0}^{N-1} \sum_{m=0}^{N-1} a_{n,m}^l f_{n,m}. \tag{5.10}$$

The attenuation measurements y_l are also called the summation coefficients with the lth ray. We assume that the size of the image elements is small and $a_{n,m}^l = 1$ if the lth ray intersects the (n, m)th IE, and 0 otherwise, for all $l = 1 : M$ and $n, m = 0 : (N - 1)$. The rays pass along knots of the discrete lattice, and one can consider that values $f_{n,m}$ correspond to the samples of the discrete image at points with the coordinates (n, m). The set of the measurements y_l taken at a fixed direction is called a *projection*. Since it was difficult to find the direct solution f of the complete system (5.10), iterative procedures of approaches of the reconstructed image were proposed [69]-[72].

We now describe the direct solution. Given a triplet (p, s, t), where $p, s, t = 0 : (N-1)$, we consider in the lattice $X_{N,N}$ the set $V_{p,s,t}$ that was used in the tensor and paired representations, and its characteristic function $\chi_{p,s,t}(n, m)$. The set $V_{p,s,t}$, if it is not empty, is the set of points (n, m) along a maximum of $p + s$ parallel straight lines at an angle of $\psi = arctg(s/p)$ to the horizontal axis. The equations for the lines are

$$xp + ys = t, \; xp + ys = t + N, \; \ldots, xp + ys = t + (p + s - 1)N \qquad (5.11)$$

in the square domain $Y = [0, N] \times [0, N]$. We will assume for simplicity that $L = N$ and define the above set of lines by $\mathcal{L}_t = \mathcal{L}_{p,s,t}$. As an example, Figure 5.17 shows the elements of the set $V_{1,2,2}$ on the lattice $X_{8,8}$ that lie on three lines. Two points of the set $V_{1,2,2}$ are on the line $x + 2y = 2$, four points are on the line $x + 2y = 10$, and two points are on the line $x + 2y = 18$. All samples of the set $V_{p,s,t}$ lie on the parallel rays passing along samples

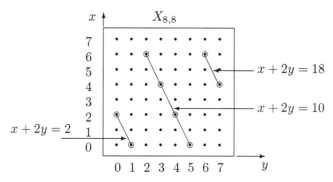

FIGURE 5.17
The elements of the set $V_{1,2,2}$ lie on the three straight lines: $x \cdot 1 + y \cdot 2 = 2$, $x \cdot 1 + y \cdot 2 = 10$, and $x \cdot 1 + y \cdot 2 = 18$. Therefore, $f_{1,2,2} = (f_{0,1} + f_{2,0}) + (f_{0,5} + f_{2,4} + f_{4,3} + f_{6,2}) + (f_{4,7} + f_{6,6})$.

of the discrete net $X_{N,N}$ traced on the initial image. It means that all '1's in the mask of the 2-D function $\chi_{p,s,t}(n, m)$ lie on parallel lines. We denote them by $r(p, s, t)_1, r(p, s, t)_2, \ldots, r(p, s, t)_q, \; q \geq 1$. The relation between the components of image-signals and summation coefficients is described by the following statement.

Statement 5.1 Given a group $T_{p,s}$, the components of the corresponding image-signal $f_{T_{p,s}}$ of the reconstructed image $\{f_{n,m}\}$ equal the sum of summation coefficients $f_{p,s,t} = y(p, s, t)_1 + y(p, s, t)_2 + \ldots + y(p, s, t)_q$, where

$$y(p, s, t)_k = \sum_{(n,m) \in r(p,s,t)_k} f_{n,m}, \quad k = 1 : q, \; q = p + s.$$

The number of summation coefficients q depends on the frequency (p, s).

As an example, Figure 5.18 illustrates the image 256×256 in part a, along with the 1-D DFT over the image-signal $f_{T_{1,3}}$ of length 256 in c, and the spectrum of the image in d. Three bright parallel lines on the spectrum show the samples at points of the group $T_{1,3}$ at which the 2-DFT of the image is the 1-DFT of the image-signal. The 2-D DFT in points in this group has been amplified in order to see location of the group and directions of the projection along which the components of the tensor are calculated as linear integrals. The image after amplifying the 2-DFT at frequency-points of the group $T_{1,3}$ is shown in b. The projection is calculated at the angle $\psi = 18.4349°$ and the 1-D DFT is filled the 2-D DFT along three lines at angle $\theta = 90 - \psi = 71.5651°$. The elements of the group $T_{p,s}$ lie on parallel lines at an angle of $\theta = \text{arctg}(p/s)$

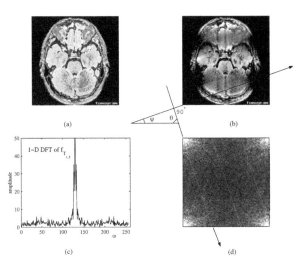

FIGURE 5.18
(a) Image 256×256. (b) Image after amplifying the 2-DFT at samples of the group $T_{1,3}$. (c) Absolute value of the 1-D DFT of the image-signal $f_{T_{1,3}}$ (zero component is shifted to the center and truncated). (d) 2-D DFT of the image with amplified samples at the group $T_{1,3}$. Angles $\psi = 18.4349°$ and $\theta = 71.5651°$.

to the horizontal axis x.

In the 3-D space, one can identify the opposite sides of boundaries of the square Y and consider it as a torus and the lattice $X_{N,N}$ as a net traced on the torus. Given (p, s), the straight lines of $\mathcal{L}_t$ of (5.11) will compose one discrete spiral $S_t = S_{p,s,t}$ on the torus, because of the equality $\overline{xp + ys} = \overline{t + kN} = t$ for integers k. As an example for $N = 32$, Figure 5.19 shows the locus of two spirals S_3 and S_7 on the net, for $(p, s) = (1, 1)$. They correspond respectively to the parallel straight lines of families $\mathcal{L}_{1,1,3}$ and $\mathcal{L}_{1,1,7}$. The sums $\{f_{s,p,t}\}$, that are calculated on N spirals S_t, $t = 0 : (N - 1)$, determine the 2-D DFT of the image $f_{n,m}$ at net points that are situated on a spiral $\tilde{S}_{p,s}$ that passes through the initial point of the net and makes an angle $\pi/2$ with the spirals

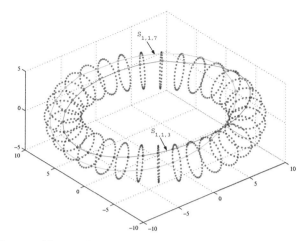

FIGURE 5.19 (See color insert following page 242.)
The net with knots of the grid 32×32 in the 3-D space with locus of two spirals $S_{1,1,3}$ and $S_{1,1,7}$.

S_t. As an example, Figure 5.20 illustrates the locus of spiral $\tilde{S}_{5,1}$ on the 3-D

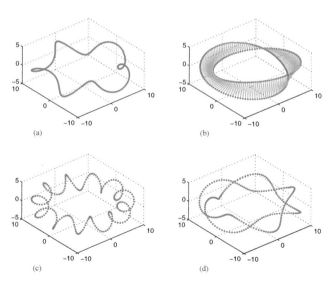

FIGURE 5.20 (See color insert following page 242.)
Locus of the spiral $\tilde{S}_{p,s}$ in the grid 256×256 on the torus, when (p, s) equals (a) $(5, 1)$, (b) $(127, 1)$, (c) $(11, 1)$, and (d) $(125, 1)$.

net with the knots of the grid 256×256 in part a, along with spiral $\tilde{S}_{127,1}$ in b, spiral $\tilde{S}_{11,1}$ in c, and spiral $\tilde{S}_{125,1}$ in d. When a point runs along the spiral $S_{5,1}$, it rotates around the torus seven times. At points of this spiral the 2-D DFT of the image is calculated by the 1-D 256-point DFT over the image-signal $f_{T_{5,1}}$ of length 256.

The property similar to the splitting in the tensor representation,

$$F_{\overline{kp,ks}} = (\mathcal{F}_N \circ f_{T_{p,s}})_k = \sum_{t=0}^{N-1} f_{p,s,t} W^{kt}, \quad k = 0 : (N-1), \qquad (5.12)$$

is well known in computerized tomography as the theorem of projections [9]. We may think that the components $f_{p,s,t}$ represent values of the Radon transform of the 2-D sequence $f_{n,m}$ written on the discrete net on the torus. But it seems that it is not appropriate to name $f_{p,s,t}$ to be a discrete version of the Radon transform, because this transform is defined on the plane with the polar system of coordinates, and the tensor is defined on the original plane with the Cartesian system of coordinates. Moreover, the tensor representation is universal; the splitting in (5.12) holds for the Hartley and Hadamard transforms. Thus, the projections can be used, for instance, for computing the 2-D Hadamard transform of the image, which will save a large number of operations when compared with the Fourier method.

We now consider the paired representation of the image, when the image is described by a set of $(3N - 2)$ image-signals

$$f_{T'_{p,s}} = \{f'_{p,s,0}, f'_{p,s,2^n}, f'_{p,s,2\cdot2^n}, \dots, f'_{p,s,N/2-2^n}\} \qquad (5.13)$$

where the integer $n \geq 0$ is such that $2^n = g.c.d.(p,s)$. These paired image-signals have lengths $N/2^{n+1}$ and define the 2-D DFT of the image at samples of the corresponding subsets $T'_{p,s}$ by

$$F_{\overline{(2k+1)p},\overline{(2k+1)s}} = \sum_{t=0}^{N/2^{n+1}-1} \left(f'_{p,s,2^n t} W^t_{N/2^n} \right) W^{kt}_{N/2^{n+1}}, \ k = 0 : (N/2^{n+1} - 1).$$

According to the definition, the masks of paired functions are composed by '1's and '−1's that lie on separable parallel lines. For instance, the following mask corresponds to the function $\chi'_{1,1,1}(n,m)$:

$$M_{1,1,1} = \begin{bmatrix} 0 & 0 & 1 & 0 & 0 & 0 & -1 & 0 \\ 0 & 0 & 0 & 1 & 0 & 0 & 0 & -1 \\ -1 & 0 & 0 & 0 & 1 & 0 & 0 & 0 \\ 0 & -1 & 0 & 0 & 0 & 1 & 0 & 0 \\ 0 & 0 & -1 & 0 & 0 & 0 & 1 & 0 \\ 0 & 0 & 0 & -1 & 0 & 0 & 0 & 1 \\ 1 & 0 & 0 & 0 & -1 & 0 & 0 & 0 \\ 0 & 1 & 0 & 0 & 0 & -1 & 0 & 0 \end{bmatrix}$$

wherein all '1's and all '-1's lie on two parallel lines each.

The set of N^2 components $f'_{p,s,t} = (\chi'_{p,s,0} \circ f) = f_{p,s,t} - f_{p,s,t+N/2}, (p,s,t) \in U$ is the paired transform of the discrete image $\{f_{n,m}\}$. According to Statement 5.1, the components $f'_{p,s,t}$ can be calculated by the summation coefficients as

$$f'_{p,s,t} = y(p,s,t) + \ldots + y(p,s,t)_{m_1} - y(p,s,t+\frac{N}{2})_1 - \ldots - y(p,s,t+\frac{N}{2})_{m_2}$$

where numbers $m_1 = m(p,s,t)$, $m_2 = m(p,s,t+N/2)$. Thus, the initial system (5.10) of linear equations can be used for calculating the paired transform $\{f'_{p,s,t}\}$ of the discrete reconstructed image. Then, it can be used for calculating the 2-D DFT of the image, processing it in the frequency domain if needed, and, then, performing the inverse 2-D DFT. We can also obtain the image reconstruction from its projections by direct calculation of the image by means of direction images as was discussed above,

$$f_{n,m} = \frac{1}{2N} \sum_{k=0}^{r} \frac{1}{2^k} \sum_{(p,s)\in 2^k J_{2^{r-k}}} f'_{p,s,(np+ms) \bmod N} \cdot \quad (5.14)$$

This is the formula of image reconstruction by using the projection data through the paired representation. One can notice that only operations of addition and subtraction, as well as multiplication by negative powers of two, are used in this reconstruction.

5.2.3 Series images

We now consider the decomposition of the image by its direction images in more detail. It follows from the definition of the paired representation that from each image specific periodic structures can be extracted, which all together compose the image. These structures do not have the smooth forms of cosine or sine waves, but the forms which are defined by binary paired basis functions united by subsets. To illustrate this property, we call the sum of direction images corresponding to the subset of generators $2^k J_{2^{r-k}}$,

$$S_{n,m}^{(k)} = \sum_{(p,s)\in 2^k J_{2^{r-k}}} d_{n,m;p,s}, \quad k = 0 : r-1, \quad S_{n,m}^{(r)} = d_{n,m;0,0} \equiv \frac{1}{N^2} F_{0,0}$$

the kth series image. Figure 5.21 shows the first five series images for the tree image in parts a through e. One can see that each series image, starting from the second one, has a periodic structure with a resolution which increases exponentially with the number of the series. We call the number 2^k the resolution of the kth series image. This is an interesting fact: each resolution is referred to as a periodic structure of one part of the image. The first series image is the component of the image with the lowest resolution, and the $(r-1)$th series image is the component of the image with the highest

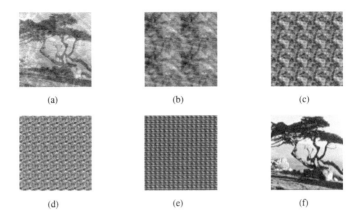

FIGURE 5.21

(a)-(e) Five series images of the tree image, (f) and the sum of these images.

resolution. The constant image $S^{(r)}$ has 0 resolution. The sum of the series images equals the original image, as shown in f, where the image is the sum of only the first five series images.

The consequent sum of the four first series images of the tree image is given in Figure 5.22; we can see that series images with resolution 1, 2, and 4 result in a tree image of good quality. The other four resolutions add more detail to the image which are difficult to notice. Periodic structure of the

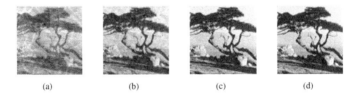

FIGURE 5.22

(a) The 1st series image, (b) the 1st plus 2nd series images, (c) the sum of the first three series images, and (c) the sum of the first four series images.

series-images takes place for other images as well. As an example, Figure 5.23 shows the first series image of the girl image in part a, along with the next six series images in (b)-(g), and the sum of these series images in h. Note that series images have different ranges of intensities, which decrease when the resolution increases. For instance, the first four series images have values that vary in range $255, 101, 45$, and 15, respectively. For better illustration, all series images in this figure and Figs. 5.21 and 5.47 are scaled by using the

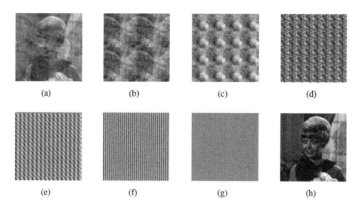

(a)   (b)   (c)   (d)

(e)   (f)   (g)   (h)

FIGURE 5.23
(a)-(g) Seven series-images of the girl image, and (h) the sum of these images.

MATLAB-based command "imagesc(.)".

5.2.4 Resolution map

It is important to mention that the first series image is also composed by periodic structures $N/2 \times N/2$. In this image, as well as the rest of the series images, we can separate subsets of direction images in the following way. The set of generators J_{2^r} is divided into three parts as

$$J_{2^r}^{(1)} = \{(1, 2s); \ s = 0 : (N/2 - 1)\}, \quad J_{2^r}^{(2)} = \{(2p, 1); \ p = 0 : (N/2 - 1)\},$$
$$J_{2^r}^{(3)} = \{(1, 2s + 1); \ s = 0 : (N/2 - 1)\}.$$

In the first two sets, the coordinates of the generators are replaced, i.e., these two sets are symmetric to each other. The directions of the direction images which correspond to the first set of generators are positive, and negative for the second set. The directions defined by the third set of generators are unique. The division of the first series image $S^{(0)}$ by these subsets we denote as

$$P_{n,m}^{(0)} = \sum_{(p,s) \in J_{2^r}^{(1)}} d_{n,m;p,s}\,, \quad N_{n,m}^{(0)} = \sum_{(p,s) \in J_{2^r}^{(2)}} d_{n,m;p,s}\,,$$
$$U_{n,m}^{(0)} = \sum_{(p,s) \in J_{2^r}^{(3)}} d_{n,m;p,s},$$

so that $S^{(0)} = P^{(0)} + N^{(0)} + U^{(0)}$.

Figure 5.24 shows the image $P^{(0)}$ for the girl image in part a, along with the images $N^{(0)}$ and $U^{(0)}$ in b and c, respectively. In these images, one can notice different parts of the girl image with their negative versions periodically shifted by 128 along the horizontal, vertical, and diagonal directions. Each image is divided by four parts $N/2 \times N/2$ with similar structures, which can

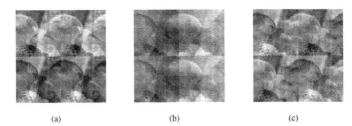

(a)   (b)   (c)

FIGURE 5.24

Three components of the first series images of the girl image.

be used for composing the entire series image $S^{(0)}$. Indeed, it follows directly from the definition of the paired functions that the following equations are valid:

$$
\begin{cases}
f'_{1,2s,(n+N/2)+2ms \bmod N} = -f'_{1,2s,n+2ms \bmod N} \\
f'_{1,2s,n+2(m+N/2)s \bmod N} = f'_{1,2s,n+2ms \bmod N} \\
f'_{1,2s,(n+N/2)+2(m+N/2)s \bmod N} = -f'_{1,2s,n+2ms \bmod N}
\end{cases}
$$

$$
\begin{cases}
f'_{1,2s+1,(n+N/2)+m(2s+1) \bmod N} = -f'_{1,2s+1,n+m(2s+1) \bmod N} \\
f'_{1,2s+1,n+(m+N/2)(2s+1) \bmod N} = -f'_{1,2s+1,n+m(2s+1) \bmod N} \\
f'_{1,2s+1,(n+N/2)+(m+N/2)(2s+1) \bmod N} = f'_{1,2s+1,n+m(2s+1) \bmod N}
\end{cases}
$$

for $s = 0 : (N/2 - 1)$, and

$$
\begin{cases}
f'_{2p,1,2(n+N/2)p+m \bmod N} = f'_{2p,1,2np+m \bmod N} \\
f'_{2p,1,2np+(m+N/2) \bmod N} = -f'_{2p,1,2np+m \bmod N} \\
f'_{2p,1,2(n+N/2)+(m+N/2) \bmod N} = -f'_{2p,1,2np+m \bmod N}
\end{cases}
$$

for $p = 0 : (N/2 - 1)$. Therefore, the series image components $P^{(0)}, N^{(0)}$, and $U^{(0)}$ can be defined from their first quarters which we denote by P_1, N_1, and U_1, respectively, as follows:

$$
P^{(0)} = \begin{bmatrix} P_1 & P_1 \\ -P_1 & -P_1 \end{bmatrix}, \quad
N^{(0)} = \begin{bmatrix} N_1 & -N_1 \\ N_1 & -N_1 \end{bmatrix}, \quad
U^{(0)} = \begin{bmatrix} U_1 & -U_1 \\ -U_1 & U_1 \end{bmatrix}.
$$

Figure 5.25 shows the decomposition of the next series image $S^{(1)}$ for the girl image. The decomposition of the third series image $S^{(2)}$ is shown in Figure 5.26. For this series image, as well as the rest of the series images $S^{(k)}$, $k = 2 : (r - 1)$, similar decompositions hold. Each such image can be defined by the three quarters P_{k+1}, N_{k+1}, and U_{k+1} of their periods $N/2^{k+1} \times N/2^{k+1}$ in a way similar to the first series image. As a result, the following resolution map (RM) associates with the image f:

$$
RM[f] =
\begin{array}{|c|c c|}
\hline
P_1 & \multicolumn{2}{c|}{U_1} \\
\hline
 & P_2 & U_2 \\
N_1 & \multicolumn{2}{c|}{} \\
 & P_3 & U_3 \\
 & N_2 & \\
 & N_3 & \cdots \\
\hline
\end{array}
. \tag{5.15}
$$

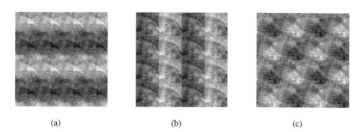

(a) (b) (c)

FIGURE 5.25
Three components of the 2nd series image of the girl image.

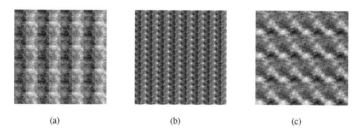

(a) (b) (c)

FIGURE 5.26
Three components of the 3rd series image of the girl image.

This resolution map has the same size as the image and contains all periodic parts of the series images, i.e., all periods by means of which the original image can be reconstructed. Each periodic part is extracted from the direction images, whose directions are given by subsets of generators of J'. In other words, the RM represents itself the image packed by its periodic structures that correspond to a specific set of projections. The resolution map can be used to change the resolution of the entire image, by processing direction images for desired directions. We now consider examples of using the resolution map for image enhancement.

5.2.5 *A*-series linear transformation

Our preliminary experimental results show that good results of image processing, including the enhancement, can be achieved when working with one or a few high energy splitting-signals, as well as the sets of splitting-signals which are combined by series and correspond to different resolutions written in the image RM. Figure 5.27 shows all 766 generators $(p, s) \in J'_{256}$ in part a, where the twelve generators for the 6th series image and six generators for the 7th series image are marked by "•" and "+", respectively. The girl image with amplified series image of number 7 by the factor of 2 is shown in b, and images with the amplified series images of numbers 6 and 7 respectively by the factors of 1.2 and 1.5 in c. These two images are enhanced by resolutions 64 and

FIGURE 5.27 (See color insert following page 242.)
(a) 766 generators of the set J'_{256}, and the girl image with amplified (b) 7th series image and (c) 6th series image. (All images are displayed in the same colormap.)

64 with 32, respectively. We now consider the splitting method of α-rooting enhancement, when the image is enhanced by only one splitting-signal. The effective method of image enhancement, when splitting-signals in the tensor representation are used, is described in detail in [9],[100]-[102]. The paired representation is a more advanced form, since in the tensor representation, all required $3N/2$ splitting-signals are of length N and do not provide a partition of the 2-D DFT, but a covering which leads to the redundancy of calculation of the 2-D DFT.

5.2.6 Method of splitting-signals for image enhancement

The purpose of image enhancement is to improve the quality of the digital image when the critical details are not seen clearly enough. For instance, in medical imaging, such as computer tomography and magnetic resonance, three-dimensional images (or a stack of 2-D images) of different organs and tissues are produced. There are many sources of interference in the production of medical images, such as the movement of a patient, insufficient performance and noise of imaging devices.

The basic idea behind the frequency domain methods consists in computing a discrete unitary transform of the image, for instance, the 2-D discrete Fourier transform (2-D DFT), manipulating the transform coefficients, and then performing the inverse transform, as shown in Figure 5.28.

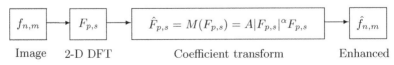

Image   2-D DFT   Coefficient transform   Enhanced

FIGURE 5.28
Block-diagram of the transform-based image enhancement (with application for the α-rooting method for given parameters A and α).

As an example, Figure 5.29 shows the original image of size 256×256 in part a, along with the 2-D DFT of the image (in absolute scale) in b, the coefficients to be multiplied pointwise by the 2-D DFT in c, the modified 2-D DFT in d, and the inverse 2-D DFT that yields the enhanced image in e.

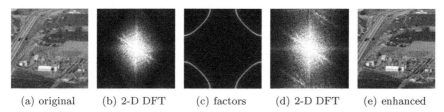

(a) original   (b) 2-D DFT   (c) factors   (d) 2-D DFT   (e) enhanced

FIGURE 5.29
The amplitudes of the 2-D DFT (b) of the chemical plant image (a) are multiplied by coefficients (c), and the new 2-D DFT (d) results in the enhanced image (e).

Transform-based image enhancement methods include techniques such as alpha-rooting, weighted α-rooting, modified unsharp masking, and filtering, which are all motivated by the human visual response [3],[73]-[76]. The main advantages of transform-based image enhancement techniques are a low complexity of computations, high quality of enhancement, and the critical role of unitary transforms in digital image processing.

The α-rooting method of image enhancement by the Fourier transform is performed by the following three steps.

Step 1: Calculate the 2-D DFT $\mathcal{F}[f]$ of the image $f_{n,m}$.

Step 2: Multiply the transform coefficients, $F_{p,s}$, by real factors $A|F_{p,s}|^{\alpha-1}$, where A is a positive constant and $\alpha \in (0,1)$.

Step 3: Calculate the inverse 2-D discrete Fourier transform over the modified spectral coefficients $F_{p,s}$.

The phase of the Fourier transform of the image does not change. The selection of the best or optimal value of parameter α is image dependent and can be adjusted interactively by the user, or by an automated method when using a quantitative measure of image enhancement. In general, the optimality refers also to the 2-D unitary transform which may differ from the Fourier transform. For instance, the Hadamard transform in many cases leads to image enhancement comparable to the Fourier transform, but its implementation requires fewer arithmetical operations.

The quantitative measure of enhancement of the image $\hat{f}_{n,m}$ of size $N \times N$ by a transform Φ can be defined in the following way [100]. The image is divided by k^2 blocks of a preassigned size $L \times L$, where $k = \lfloor N/L \rfloor$ and the operation $\lfloor \cdot \rfloor$ denotes the floor function. We denote the (m, l)th block of the image as $\hat{f}[m, l]$. The measure of image enhancement is defined by [100]

$$EME(\hat{f}) = EME_\Phi(\hat{f}) = \frac{1}{k^2} \sum_{m=1}^{k} \sum_{l=1}^{k} 20 \log_2 \frac{\max(\hat{f}[m, l])}{\min(\hat{f}[m, l])}.$$

Instead of the max and min operations, respectively, the 2nd and $(L^2 - 1)$th order statistics of the image within the (m, l)th blocks can also be used. The value of $EME_I(f)$ when there is no processing of the image, in other words when the transform F is considered equal to the identity transform I, is called *enhancement measure* of the original image f and is denoted by $EME(f)$.

For images of Figure 5.29 in parts a and e, the enhancement measure EME takes values equal to 9.38 and 12.48, respectively, when blocks are of size 5×5. The quality of the image has been improved, and this improvement is estimated by the measure as $EME_F(\hat{f}) - EME(f) = 12.48 - 9.48 = 3.10$. Figure 5.30 illustrates image enhancement by applying the α-rooting method. The operation of a Fourier-transform-based image enhancement has been parameterized by α varying in the interval $(0.2, 1]$. The curve of $EME(\alpha) = EME(\hat{f}_\alpha)$ as a function of α has a maximum at the point $\alpha_0 = 0.80$ as shown in part a. The original truck image f is shown in b, and the image $\hat{f}_\alpha$ enhanced by the α-rooting method when $\alpha = 0.80$ in c. The enhancement of the image increases by the value of $EME(\hat{f}_{0.8}) - EME(f) = 20.59 - 7.83 = 12.76$. This figure also shows the function $EME(\alpha)$ in d calculated for the tank image in e, and the image enhancement by the α-rooting when $\alpha = 0.85$ in f. The enhancement measure of the image equals $EME(\hat{f}_{0.85}) = 20.58$.

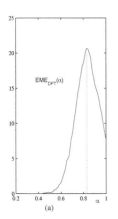

(a)

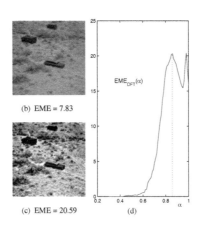

(b) EME = 7.83

(c) EME = 20.59

(d)

(e) EME = 16.09

(f) EME = 20.58

FIGURE 5.30
(a) Measure EME for α-rooting enhancement, (b) the truck image, and (c) the enhancement by 0.8-rooting method. (d) EME graph for α-rooting enhancement of (e) the tank image and (f) the enhancement by 0.84-rooting.

5.2.7 Fast methods of α-rooting

We now analyze image enhancement by using the concept of splitting-signals. For that, the original tensor representation of the image is modified into the paired form. The proposed block-scheme of enhancement of an image $f_{n,m}$ of size $N \times N$ is given in Figure 5.31. The image is transfered to a totality of splitting-signals of length N each that represent the image in tensor representation. Two cases of most interest are considered when N is a prime number and a power of two; in these cases such a totality consists of $(N+1)$ or $3N/2$ signals, respectively. Rather than process an image by the traditional α-rooting method by the Fourier transform, $\mathcal{F}_{N,N}$, we can separately process all (or only a few) splitting-signals and then calculate and compose the 2-D DFT of the processed image by means of new 1-D DFTs of the processed splitting-signals. The enhanced image can be calculated by the 2-D inverse DFT, or directly from the totality of these splitting-signals.

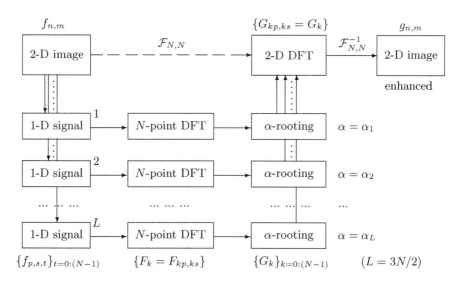

FIGURE 5.31

Image processing by splitting-signals in the tensor representation.

The main idea of the tensor representation can be described in the following way. The set of all frequency-points $X = \{(p, s); p, s = 0 : (N-1)\}$ is covered by a family of subsets $\sigma = (T_k)_{k=1:l}$, where $l > 1$, in a way that the 2-D DFT of the image $f_{n,m}$ at a subset T_k becomes an image of the 1-D N-point DFT, $\mathcal{F}_N$, of a 1-D signal, $f^{(k)}$. This supposition means the following. The set of splitting-signals $f^{(k)}$, $k = 1 : l$, completely defines the image, f, and at the same time, the 1-D DFTs over these signals determine the entire 2-D DFT of the image and compose a splitting of the transform:

$$f \to \begin{bmatrix} f^{(1)} \\ f^{(2)} \\ \cdots \\ f^{(l)} \end{bmatrix} \longleftrightarrow \begin{bmatrix} \mathcal{F}_N[f^{(1)}] \\ \mathcal{F}_N[f^{(2)}] \\ \cdots \\ \mathcal{F}_N[f^{(l)}] \end{bmatrix} \to \mathcal{F}_{N,N}[f]. \tag{5.16}$$

The set T of the covering σ is defined as the cyclic group with generator (p, s), $T_{p,s} = \{(\overline{kp}, \overline{ks}); \ k = 0 : (N - 1)\}$, $(T_{0,0} = \{(0,0)\})$, the 2-D DFT is defined by the 1-D DFT over some N-point signal. We can thus compose an irreducible covering of $X_{N,N}$ by groups $T_{p,s}$ and then define the 2-D DFT by 1-D DFTs. As an example, we consider the arrangement of frequency of groups $T_{1,1}, T_{1,0}, T_{1,2}$, and $T_{0,1}$ that compose a covering, σ, of the lattice 3×3 :

$$\underbrace{\begin{bmatrix} \bullet & \bullet & \bullet \\ \bullet & \bullet & \bullet \\ \bullet & \bullet & \bullet \end{bmatrix}}_{X_{3,3}} = \underbrace{\begin{bmatrix} \bullet & \bullet & \bullet \\ \circ & \circ & \circ \\ \circ & \circ & \circ \end{bmatrix}}_{T_{0,1}} \cup \underbrace{\begin{bmatrix} \bullet & \circ & \circ \\ \circ & \bullet & \circ \\ \circ & \circ & \bullet \end{bmatrix}}_{T_{1,1}} \cup \underbrace{\begin{bmatrix} \bullet & \circ & \circ \\ \circ & \circ & \bullet \\ \circ & \bullet & \circ \end{bmatrix}}_{T_{2,1}} \cup \underbrace{\begin{bmatrix} \bullet & \circ & \circ \\ \bullet & \circ & \circ \\ \bullet & \circ & \circ \end{bmatrix}}_{T_{1,0}}.$$

In the general case when N is a prime, the covering σ consists of $N+1$ groups with generators $(0, 1), (1, 1), (2, 1), \ldots, (N - 1, 1)$, and $(1, 0)$. The irreducible covering σ of the set X, which is composed by groups $T_{p,s}$ is unique. The $N = 2^r$ case is considered similarly. The following important relation holds in the frequency domain. The 2-D DFT of the image $\{f_{n,m}\}$ at frequency-points of the group $T_{p,s}$ is the N-point DFT, $\mathcal{F}_N$, of the image-signal $f_T = f_{T_{p,s}} = \{f_{p,s,0}, f_{p,s,1}, \ldots, f_{p,s,N-1}\}$.

Example 5.3

Consider $N = 8$ and $(p, s) = (1, 2)$. All values of t in equations $np + ms = t \bmod 8$ can be written in the form of the following matrix:

$$||t = (n \cdot 1 + m \cdot 2) \bmod 8||_{n,m=0:7} = \begin{vmatrix} 0 & 1 & 2 & 3 & 4 & 5 & 6 & 7 \\ 2 & 3 & 4 & 5 & 6 & 7 & 0 & 1 \\ 4 & 5 & 6 & 7 & 0 & 1 & 2 & 3 \\ 6 & 7 & 0 & 1 & 2 & 3 & 4 & 5 \\ 0 & 1 & 2 & 3 & 4 & 5 & 6 & 7 \\ 2 & 3 & 4 & 5 & 6 & 7 & 0 & 1 \\ 4 & 5 & 6 & 7 & 0 & 1 & 2 & 3 \\ 6 & 7 & 0 & 1 & 2 & 3 & 4 & 5 \end{vmatrix}.$$

Therefore, the image-signal $f_{T_{1,2}}$ is defined as follows

$$f_{T_{1,2}} = \begin{cases} f_{1,2,0} = f_{0,0} + f_{6,1} + f_{4,2} + f_{2,3} + f_{0,4} + f_{6,5} + f_{4,6} + f_{2,7} \\ f_{1,2,1} = f_{1,0} + f_{7,1} + f_{5,2} + f_{3,3} + f_{1,4} + f_{7,5} + f_{5,6} + f_{3,7} \\ f_{1,2,2} = f_{2,0} + f_{0,1} + f_{6,2} + f_{4,3} + f_{2,4} + f_{0,5} + f_{6,6} + f_{4,7} \\ f_{1,2,3} = f_{3,0} + f_{1,1} + f_{7,2} + f_{5,3} + f_{3,4} + f_{1,5} + f_{7,6} + f_{5,7} \\ f_{1,2,4} = f_{4,0} + f_{2,1} + f_{0,2} + f_{6,3} + f_{4,4} + f_{2,5} + f_{0,6} + f_{6,7} \\ f_{1,2,5} = f_{5,0} + f_{3,1} + f_{1,2} + f_{7,3} + f_{5,4} + f_{3,5} + f_{1,6} + f_{7,7} \\ f_{1,2,6} = f_{6,0} + f_{4,1} + f_{2,2} + f_{0,3} + f_{6,4} + f_{4,5} + f_{2,6} + f_{0,7} \\ f_{1,2,7} = f_{7,0} + f_{5,1} + f_{3,2} + f_{1,3} + f_{7,4} + f_{5,5} + f_{3,6} + f_{1,7} \end{cases} \quad (5.17)$$

The subset $J_{N,N}$ of $3N/2$ frequency-generators (p, s) that are required to calculate the complete 2-D DFT of the image $\{f_{n,m}\}$ by image-signals can be taken as

$$J_{N,N} = \{(1,0), (1,1), \dots, (1, N-1), (0,1), (2,1), (4,1), \dots, (N-2,1)\}. \quad (5.18)$$

The totality of sets $\sigma_{N,N} = (T_{p,s}; (p, s) \in J_{N,N})$ is the irreducible covering of the lattice $X_{N,N}$. We can select splitting-signals $f_{T_{p,s}}$ by maximums of the energy $E_{p,s}$ they carry. It is interesting to analyze the image by processing only one splitting-signal. As an example, Figure 5.32 shows the process of image enhancement by the splitting-signal generated by $(p, s) = (4, 1)$. For

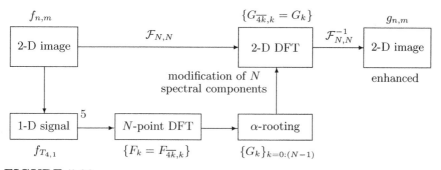

FIGURE 5.32

Block-diagram of enhancement of an image 512×512 by one splitting-signal.

image enhancement, we can select one or a few splitting-signals. For the truck image, Figure 5.33 shows the graph of function $E_{p,s}$ for all generators (p, s) of groups $T_{p,s}$ in the order given in (5.18). The splitting-signal with the maximum energy 28.1 is $f_{T_{0,1}}$. The next three signals of high energy are $f_{T_{128,1}}$, $f_{T_{1,0}}$, and $f_{T_{1,258}}$. These splitting-signals can be used for image enhancement. Splitting-signals with high energy can be selected for the tank image, as shown in b.

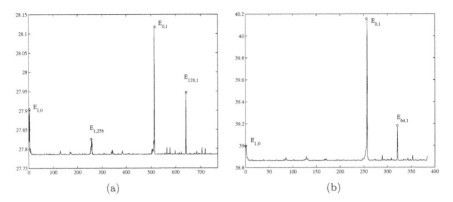

FIGURE 5.33

The energy curves of 768 and 384 splitting-signals of (a) the truck and (b) tank images, respectively.

Image enhancement can be achieved by processing only one splitting-signal with the corresponding optimal value of α (the optimality is with respect to EME). The 2-D DFT of the image changes by the 1-D DFT of this splitting-signal only at N frequency-points of the group $T_{p,s}$. As an example, Figure 5.34 shows the 513th splitting-signal $f_{T_{0,1}}$ for the truck image in part a, along with the result $g_{n,m}$ of image enhancement by this signal in b. The achieved enhancement equals $EME(\hat{f}_\alpha) = 17.28$, when $\alpha = 0.93$. The traditional α-rooting by the 2-D DFT yields the optimal value 0.85 with image enhancement 20.59. The 513th splitting-signal leads to the highest enhancement by EME, when considering parameter α to be equal to 0.93. For the tank image, the enhancement by such a signal equals $EME(\hat{f}_\alpha) = 19.11$ when $\alpha = 0.99$.

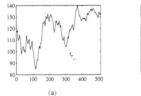

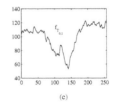

FIGURE 5.34

(a) Splitting-signal $f_{T_{0,1}}$ and (b) truck image enhanced by this signal, (c) splitting-signal $f_{T_{0,1}}$ and (d) tank image enhanced by this signal with the method of α-rooting.

Figure 5.35(a) shows the graph of the enhancement measure $EME(n; \alpha_o)$ of the truck image that was calculated after processing only one, the nth splitting-signal for $\alpha_o = 0.98$, where $n = 0 : 767$. The splitting-signal $f_{T_{1,256}}$ is shown in b, along with coefficients C_k, $k = 0 : 511$ in c, and the enhanced image $\hat{f}$ in d. The enhancement $EME(\hat{f}_{0.98}) = 11.19$; it can be improved if we use an optimal value of α for this splitting-signal. Results of processing the tank image are given in e-h when the achieved enhancement equals 14.18.

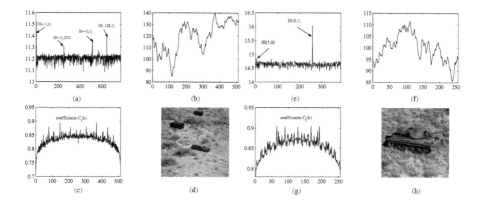

(a) (b) (e) (f)

(c) (d) (g) (h)

FIGURE 5.35

(a) Enhancement measure $EME(n, \alpha_o)$ for $\alpha_o = 0.98$, when $n = 0 : 767$. (b) Splitting-signal $f_{T_{1,256}}$. (c) Coefficients $C_1(k)$, $k = 0 : 511$, of the 1-D α-rooting enhancement. (d) Truck image enhanced by the splitting-signal. (e)-(g) Results of processing the tank image, and (h) the image enhanced by the splitting-signal.

5.2.7.1 Fast paired method of α-rooting

In the paired representation, the α-rooting enhancement can be achieved by processing one or a few splitting-signals by the following scheme:

$$f'_{p_0,s_0,t} \rightarrow f'_{p_0,s_0,t} W^t \overset{\text{1-D DFT}}{\longrightarrow} F_m \overset{\alpha\text{-rot}}{\longrightarrow} \{\hat{F}_m = c(m)F_m\} \overset{\text{1-D IDFT}}{\longrightarrow} \times W^{-t} = \hat{f}'_{p_0,s_0,t}.$$

The following are the main steps of the paired splitting α-rooting algorithm, when processing splitting-signal with number (p_0, s_0) with g.c.d.$(p_0, s_0) = 2^k$, $k \geq 0$.

Step 1: Calculate the splitting-signal $f_{T'_{p_0,s_0}}$.

Step 2: Calculate the 1-D DFT, F_m, of the modified splitting signal.

Step 3: Calculate coefficients $c(m) = |F_m|^{\alpha-1}$, $m = 0 : (N/2^{k+1} - 1)$.

Step 4: Change values of the 1-D DFT by

$$F_m \rightarrow \hat{F}_m = c(m)F_m, \quad m = 0 : (N/2^{k+1} - 1).$$

Step 5: Calculate the enhanced splitting-signal $\hat{f}_{T'_{p_0,s_0}}$ by the inverse 1-D DFT as follows:

$$\hat{f}_{p_0,s_0,t} = W_{N/2^k}^{-t} \sum_{m=0}^{N/2^{k+1}-1} \hat{F}_m W_{N/2^{k+1}}^{-mt}.$$

Step 6: Calculate the new direction image by

$$\hat{d}_{n,m} = \frac{1}{2^{k+1}N}\hat{f}'_{p_0,s_0,(np_0+ms_0) \bmod N}, \quad n,m = 0:(N-1).$$

As an example, Figure 5.36 shows the enhancement of the truck image of size 512×512. The curve of EME of the traditional α-rooting, when α runs in

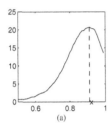  

      (a)   (b)   (c)

FIGURE 5.36
(a) EME curve, (b) α-rooting of the truck image when $\alpha = 0.91$, and (c) 1-D α-rooting by the paired splitting-signal with generator $(1,1)$.

the interval $(0.5, 1)$, is given in part a. The maximum of the EME is at point $\alpha = 0.91$. The result of the 0.91-rooting is shown in b. For comparison, the result of the 0.91-rooting method of enhancement by using only one splitting-signal $f_{T'_{1,1}}$ of length 256 is given in c. This method can be generalized by using different values of the α parameter for different splitting-signals.

The set of $3N - 2$ image-signals $f_{T'}$, $T' \in \sigma'$ corresponds to the *paired* representation of f with respect to the 2-D DFT. The summary length of all image-signals equals N^2 which coincides with the size of the image (and 2-D DFT). These 1-D signals describe uniquely the original image and at the same time they split the mathematical structure of the 2-D DFT. There is no redundancy in the spectral information carried by different image-signals. In this sense, the image-signals representing the 2-D image are independent and can be processed separately.

It should be noted that the splitting-signal $f_{T_{p,s}}$ defines $r + 1$ signals $f_{T'_{p_1,s_1}}$ in the paired representation, where $(p_1, s_1) = (p, s), (\overline{2p}, \overline{2s}), (\overline{4p}, \overline{4s}), \ldots, (0, 0)$. All together these $(r + 1)$ signals represent the 1-D paired transform of the signal $f_{T_{p_1,s_1}}$. Thus, the splitting-signal in tensor decomposition can be split, and each of its parts can be processed separately with a different (optimal) value of α, to achieve image enhancement, as shown in Figure 5.37.

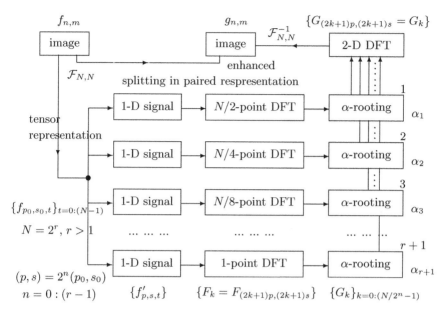

FIGURE 5.37
Image processing by short splitting-signals in the paired representation.

Figure 5.38(a) shows ten image-signals $f_{T'_{2^n,0}}$, $n = 0 : 9$, representing the 512-point image-signal $f_{1,0}$ of the truck image in the tensor representation. The signals have been processed by the 1-D α-rooting with optimal values of $\alpha_n = 0.95$ for $n = 0 : 6$, and $\alpha_n = 0.99$ for $n = 7 : 9$. The result of image enhancement with $EME = 21.28$ is shown in b. For comparison of all the above discussed methods of image enhancement, Figure 5.39 shows the original image in part a, along with the image enhanced by one image-signal $f_{T_{1,0}}$ in the tensor representation in b, the image enhanced by ten short image-signals $f_{T_{2^n,0}}$ in paired representation in c, and the image enhanced by the traditional α-rooting. One can note that the processing of short image-signals in the paired representation allows enhancement higher than the tensor representation. The result of enhancement equals $EME = 21.28$ and other paired signals can be processed to achieve enhancement greater than $EME = 22.70$ provided by the traditional α-rooting.

5.2.7.2 Directional denoising

Oceonagraphic aerial images are commonly used for studying ocean current flow, seabed structures, rock locations, sediment formation, etc. Usually, these

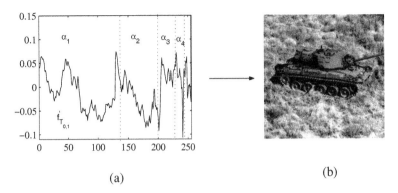

(a)   (b)

FIGURE 5.38
(a) 1-D paired transform of the splitting-signal $f_{T_{1,0}}$ and (b) the image enhanced by ten short paired image-signals.

(a)   (b)   (c)   (d)

FIGURE 5.39
(a) Image with $EME = 11.61$, and enhancement (b) by splitting-signal $f_{T_{1,0}}$ with $\alpha = 0.95$ and $EME = 20.90$, (c) by ten short paired image-signals with $EME(\alpha) = 21.28$, and (d) by the traditional α-rooting with $\alpha_{opt} = 0.87$ and $EME = 22.70$.

aerial images are captured with wave clutters because of ocean wave generating sources. The clutters behave like additional noise and interfere with useful information over the surface. The clutters are classified into two types: ripple wave (long-waves) and spark wave (short-waves). These waves are modeled and generated according to [77]. These waves are treated as noise, thus the process of removing them will be referred to as denoising. Common denoising techniques such as bandpass filtering cannot be applied to the oceonagraphic images since the background information should be preserved. Therefore direction image-signals are employed for this purpose.

Recently a few algorithms were proposed by using wavelets [78, 79] to improve these images. A new approach was proposed in [77], where a hybrid technique combines X-ray wavelet transform (XWT) and Markov random field. In the denoising by wavelets the soft thresholding of wavelet coefficients is used. In the wavelet transform of multiresolution decomposition, a 1-D signal $f(x)$ may be decomposed into detail signals at various scales 2^j and various locations n with corresponding coefficients d_{jn} and a coarse component with coefficients S_{jn}, i.e.,

$$f(x) = \sum_{j=1}^{J} \sum_{n} d_{jn} \psi_{jn}(x) + \sum_{n} S_{jn} \phi_{jn}(x), \qquad (5.19)$$

where $d_{jn} = \int f(x)\psi_{jn}(x)dx$, $S_{jn} = \int f(x)\phi_{jn}(x)dx$, and $\psi(x)$ and $\phi(x)$ are the mother wavelet function and scaling function, respectively. j is the level decomposition. The simple algorithm to denoise a 1-D signal can be described as follows: 1) Perform the forward wavelet transform with j level; 2) Discard certain coefficients that have noise; 3) Reconstruct the denoised signal by the inverse wavelet transform. The coefficients that are to be discarded are all high frequency coefficients at $j-1$ levels (for more details see [80]).

Both long-waves and short-waves are assumed to be additive to the original image. Long-waves are modeled as sinusoidal waves with varying frequency and expressed as: $c_{ripple}(x, y_o) = A\sin(2\pi/T(x))$, $T(x) = \alpha + \sqrt{x/\beta}$, where the constant α controls the initial period of long-waves, the constant β controls the variation of the ripple frequencies, and A is the amplitude of the wave. y_o is the location at which the long-wave occurs. The c_{ripple} function has a higher frequency near the locations where $x = 0$ and lower frequency far from there. Short-waves are modeled and generated as follows:

$$c_{spark}(x, y) = \frac{A}{1 + (\sqrt{2} - 1) \cdot \left(\frac{D(x,y)}{D_o}\right)}, \quad D(x, y) = \sqrt{\frac{(x - x_o)^2}{a^2} + \frac{(y - y_o)^2}{b^2}},$$

where (x_o, y_o) is the location of the short-wave. Constants a, b, and D_o determine the width and length of each wave, and A is the amplitude of the short-wave peak.

The generated long-waves are shown in Figure 5.40 in part b and short-waves in c. The image with directional clutters is defined by $I(x, y) = f(x, y) + c_\theta(x, y)$, where $f(x, y)$ stands for original image and $c_\theta(x, y)$ stands for directional clutter with angle θ. The synthesized image $I(x, y)$ for long-waves is shown in d and for short-waves in e. The generating long-waves have been synthesized with the test image shown in a, and as a result new wavy images have been obtained. We can denoise these wavy images by using splitting-signals, or image-signals. Image-signals have a directional effect on image structure. The clutters in test images are vertical; thus the horizontal image-signal is the most corrupted one in the wavy image. In Figure 5.41, some of these image-signals of long-waved image are given. One can observe the change in all image-signals of the original image after adding long-waves by looking at the energy of the image-signals. The signal $f_{T_{0,2}}$ is mostly corrupted.

The 2-D denoising problem can be reduced to a 1-D signal denoising problem, and we demonstrate it in the example. Since the image-signal $f_{T(0,1)}$ is the most corrupted signal, this signal has been processed. The Daubechies

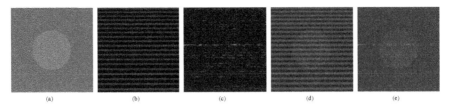

FIGURE 5.40

Images tested with waveforms: (a) original image, (b) generated long waves, (c) generated short waves, (d) the synthesized image with long waves, and (e) the synthesized image with short waves.

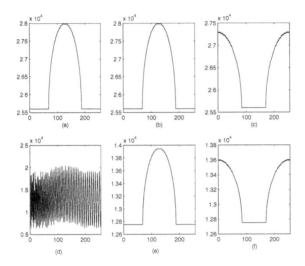

FIGURE 5.41

Splitting-signals (a,d) $f_{T_{0,1}}$, (b,e) $f_{T_{1,0}}$, (c,f) $f_{T_{1,1}}$ of the original and noised image, respectively.

wavelet five has been used and high pass coefficients were discarded. The denoised signal is shown in Figure 5.42. After combining the denoised image-signal with the noisy image, we got a wave free image shown in Figure 5.43.

In the process of denoising short-waves, the image-signals can be used as well. In Figure 5.44 three image-signals of the noisy image are given. The most corrupted image-signal $f_{T(0,1)}$ is used in the first step of the design of the SMEME filter that is a refined version of the SMEM filter defined in [77]. In order to use the SMEME filter first, clutters should be located, then the abnormal pixels at those points are removed by SMEME filtering. The image-signal $f_{T(0,1)}$ is used in locating the clutters. The points of the peaks in the image-signal $f_{T(0,1)}$ are the locations of short-waves in the spatial domain.

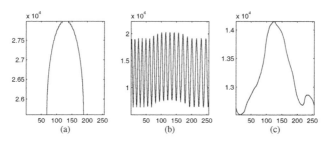

FIGURE 5.42
(a) Original image-signal $f_{T_{0,1}}$, (b) noisy image-signal, and (c) denoised image-signal.

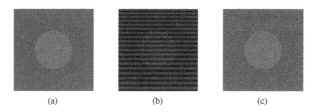

FIGURE 5.43
Denoising long waves by image-signals: (a) the original, (b) noisy, and (c) denoised images.

After locating the clutters, the image is filtered at these points with the SMEME filter. The idea motivation behind the SMEME filter is to remove the abnormal pixels of the noisy image but keep the background information data as much as possible. Therefore the size of the filter window is chosen such that part of it would cover abnormal pixels and part of it would cover background. Then the average value of the last K smallest values will be a value from background and the first P largest values within the filter will belong to wave peaks. Let $W(i,j)$ be a window size 5×5 centered at position (i,j) and $X(m,n)$ be the pixel value of image X at position (m,n); the SMEME algorithm is the following: 1) Find the peaks in image-signal $f_{T_{0,1}}$ by differentiating the signal. 2) Find the last K pixels $X_i(m,n)$ which have the smallest gray values within the filter window and calculate the average by $\text{Avg} = 1/K[X_1(m,n) + X_2(m,n) + ... + X_K(m,n)]$. 3) Replace the pixel at the center of the filter window $X(m,n)$ by Avg obtained in step 2. 4) Apply this procedure again if needed to the denoised image. The final algorithm of denoising short waves is described below: 1) Compute the tensor representation of the noisy image and find the image-signal that has the information in the clutter direction. 2) Find the peaks of the 1-D image-signal by differentiating. 3) Center the SMEME filter at the clutter and smooth the waves. After SMEME filtering all waves are cleared up except a few residual pixels,

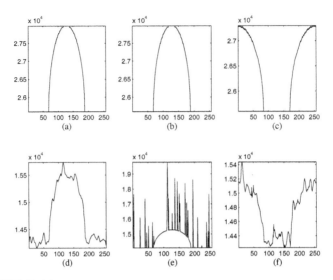

FIGURE 5.44

Image-signals: original (a) $f_{T_{0,1}}$, (b) $f_{T_{1,0}}$, (c) $f_{T_{1,1}}$, and noised (d) $f_{T_{0,1}}$, (e) $f_{T_{1,0}}$, (f) $f_{T_{1,1}}$.

then we apply SMEME a second time and observe that all waves are cleaned without any undesired effects.

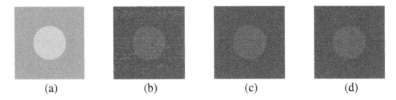

FIGURE 5.45

(a) The circle image, (b) shortwaves added to the image, (c) SMEME filtered one time, and (d) SMEME filtered twice.

For short-waves, the wavy images have been denoised by using tensor transform and the SMEME filter and we applied contrast enhancement for visual purposes; this result is shown in Figure 5.45. For quantitative comparison SNR is used and defined as follows:

$$SNR = 10 \log_{10} \frac{\max(s)^2 - \min(s)^2}{\frac{1}{N^2} \sum_{n=1}^{N} (s-n)^2}$$

where s is the original signal and n is the noisy signal. The SNR for the long waved image is 16.28 and after denoising the SNR improved to 34.40. Likewise the SNR for the short waved image is 15.26 and after denoising the SNR increased to 25.88. The proposed denoising algorithm was applied to real time satellite images and the visual results are presented in Figure 5.46. One of the main advantages of image-signals is the fact that they can be estimated according to direction of clutter and do not require image rotation or interpolation.

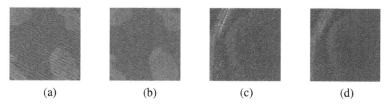

(a)   (b)   (c)   (d)

FIGURE 5.46
(a) Image 1, (b) image 2, (c) denoised image 1, and (d) denoised image 2.

5.2.8   Method of series images

We now define the following simple method of image enhancement. Let A be a set of r nonnegative parameters, $A = \{a_0, a_1, ..., a_{r-1}\}$, which are considered to be the weighted coefficients for the series images. The image $f_{n,m}$ enhanced by the set A is defined as

$$\hat{f}_{n,m} = \sum_{k=0}^{r} a_k S_{n,m}^{(k)}, \qquad (n,m) \in X_{N,N}.$$

In the case $a_k = 1$, $k = 0 : (r-1)$, the image $\hat{f}_{n,m} = f_{n,m}$. The operation $\mathcal{A} : f_{n,m} \rightarrow \hat{f}_{n,m}$ we call *the A-series linear transformation (A-SLT)*. The selection of the coefficient a_k to be greater than 1 means that the resolution 2^k of the image increases, and in the $a_k < 1$ case, this resolution decreases in the image. As an example, Figure 5.47 shows the first four series images of the truck image of size 512×512, with resolutions $1, 2, 4$, and 8, respectively.

One can improve the quality of this image by manipulating the resolution of series images in the desired way. As an example, for the truck image, Figure 5.48 shows the result of the A-SLT, when the set of parameters A equals $\{1.5, 2, 1.5, 1, 1, 1, 1.5, 1.5, 1\}$ in part a, and $\{1, 2, 1, 1, 1, 2, 4, 2, 1\}$ and $\{1, 3, 2, 1, 1, 1, 1, 1.5, 1\}$ in b and c, respectively. In the truck image in part a, the resolutions $1, 2, 4, 64$, and 128 of series images have been increased, and for images in b and c the resolutions $2, 32, 64, 128$ and $2, 4, 128$ have been increased, respectively.

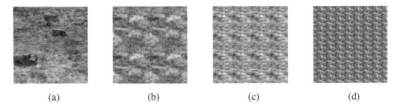

(a)   (b)   (c)   (d)

FIGURE 5.47

The first four series images of the truck image. (All images are displayed by "imagesc(.)" in "colormap(gray(256)".)

(a)   (b)   (c)

FIGURE 5.48

A-SLT of the truck image by the set of parameters (a) $\{1.5, 2, 1.5, 1, 1, 1, 1.5, 1.5, 1\}$, (b) $\{1, 2, 1, 1, 1, 2, 4, 2, 1\}$, and (c) $\{1, 3, 2, 1, 1, 1, 1, 1.5, 1\}$.

Problems

Problem 5.1 For the discrete-time signal defined by $f(t) = t/16 \cos(\omega_0 t^2)$, $\omega_0 = 1.3$, with 512 sampled values f_n at points uniformly placed in the interval $[-3\pi, 3\pi]$,

A. Compute and plot the first six sinusoidal signals $r(p)$, for $p = 2^r$, $r = 0 : 5$.

B. Compute the frequency-time image by sections, by using the first 257 reordered basis signals (see as an example Figure 5.3c).

Problem 5.2 Compute the complete set of p-section basis signals for the discrete-time signal f_n of length 512 sampled from $f(t) = t/16 \cos(3\omega_0 t^2)$, $\omega_0 = 1.3$, in the interval $[-3\pi, 3\pi]$. Use the formula of reconstruction (5.8) to compute the signal f_n.

Problem 5.3 For the tree image, compute the direction images with numbers $(p, s) = (1, 2), (2, 1)$, and $(1, 8)$.

Problem 5.4 Enhance the truck image by the method of α-rooting with only one splitting-signal with number $(p, s) = (128, 1)$. For that, use the enhancement measure EME to find the optimal value of α for this signal.

Problem 5.5 Compute and display the first five series images of the truck image.

Problem 5.6 Compute the resolution map (see (5.15)) of the tree image.

Problem 5.7* Reconstruct the tree image from its resolution map.

6

Fourier Transform and Multiresolution

Wavelet analysis has been developed as multiresolution signal processing, which is used effectively for signal and image processing, compression, computer vision, medical imaging, etc. [81]-[85]. In wavelet analysis, a fully scalable modulated window is used for frequency localization [88]-[91]. The window is sliding, and the wavelet transform of a part of the signal is calculated for every position. The result of the wavelet transform is a collection of time-scaling representations of the continuous-time signal with different resolutions. In other words, wavelet methods are referred to as methods of cross correlations of the signal with a given family of scaled waves. In contrast, the Fourier transform is considered as a transform without time resolution, since the basis cosine and sine functions are defined everywhere on the real line. Each Fourier component depends on the global behavior of the signal.

In this chapter, we describe different methods of representation of the continuous-time Fourier transform by the cosine and sine type wavelet transforms, namely, wavelet-like transforms, with fully scalable modulated windows. The Fourier transform provides the multiresolution signal processing because cosine and sine type waveforms of every frequency participate in Fourier analysis. We also dwell here on the fundamental problem of defining the unitary discrete transformations providing the time-frequency representation. The paired transformation and its modification, or the Haar transformation, are examples of such transformations. The class of such transformations is much wider, and as an example we present the class of Givens-Haar type transformations which are unitary and generated by discrete signals.

6.1   Fourier transform

In this section, we describe properties of the integral Fourier transform of a function which will be denoted by $f(t)$ or $f(x)$, where t and x are one dimension variables as the time or coordinate of the point. The letters ω, λ, and f are used for frequency. The basic functions of the Fourier transform are composed by pairs of the cosine and sine waves, $(\cos(\omega t), \sin(\omega t))$, of any frequencies ω. The Fourier transform $F(\omega)$ of a function $f(t)$ at frequency-point ω is defined as the complex integral of the sinusoidal waves of frequency

ω superposed on the function

$$F(\omega) = \int_{-\infty}^{\infty} f(t)e^{-j\omega t}\,dt = \int_{-\infty}^{\infty} f(t)\cos(\omega t)dt - j\int_{-\infty}^{\infty} f(t)\sin(\omega t)dt, \qquad (6.1)$$

where $\omega \in (-\infty, +\infty)$. We assume that the Fourier transform of $f(t)$ exists, which is true, for instance, when the function is square-absolute integrable (or has a finite energy) $\int_{-\infty}^{+\infty} |f(t)|^2 dt < \infty$, but this is not a necessary condition for existence of the transform.

Example 6.1

Let $f(t)$ be an exponential function $e^{-a|t|}$, where $a > 0$. This function decays in both sides when t tends to infinity. Then $F(\omega) = 2a/(a^2 + \omega^2)$, $\omega \in (-\infty, +\infty)$. As an example, Figure 6.1 shows the exponential function $e^{-2|t|}$ in the time interval $[-20, 20]$ in part a, along with the Fourier transform in the frequency interval $[-20, 20]$ in b. ▯

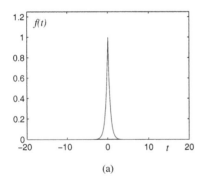
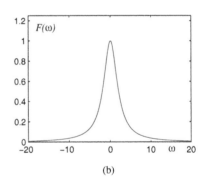

(a)   (b)

FIGURE 6.1

(a) The exponential function and (b) the Fourier transform.

Example 6.2

Let $u(t)$ be the unit step function being 1 for $t > 0$, and 0 for $t < 0$. We consider an exponential function $e^{-at}u(t)$ defined on the right part of the real line. This function describes many linear time-invariant systems in practice, and can be defined as the impulse response function of such systems. For instance, we can consider the elementary electric network, RC circuit with the impulse response $h(t) = \frac{1}{RC}e^{-\frac{t}{RC}}u(t)$, where R and C are respectively two limped parameters of the resistance and capacitance. The Fourier transform of $h(t)$, which is called *the transfer function*, or *frequency response* of the system, is calculated as

$$H(\omega) = \int\limits_{0}^{+\infty} \frac{1}{RC} e^{-\frac{t}{RC}} e^{-j\omega t} dt = \frac{1}{RC} \int\limits_{0}^{+\infty} e^{-[j\omega + \frac{1}{RC}]t} dt = \frac{\frac{1}{RC}}{j\omega + \frac{1}{RC}}.$$

The magnitude and phase frequency responses are defined as

$$|H(\omega)| = \sqrt{\frac{\left(\frac{1}{RC}\right)^2}{\omega^2 + \left(\frac{1}{RC}\right)^2}} = \sqrt{\frac{1}{1 + (RC\omega)^2}}, \quad \arg H(\omega) = -\tan^{-1}[RC\omega].$$

As an example, Figure 6.2 shows the magnitude and phase responses of the RC circuit, for the cases when the time constant $RC = 1$ and 4. ⬚

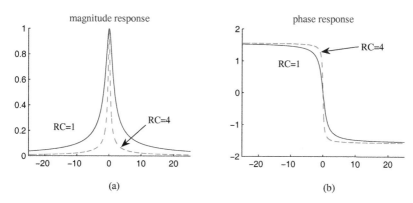

(a)   (b)

FIGURE 6.2
(a) Magnitude frequency responses and (b) phase responses.

Example 6.3
Let $h(t)$ be the following rectangle function on the interval $(-T, T)$, where $T > 0$,

$$h(t) = rect\left(\frac{t}{2T}\right) = \begin{cases} 1, & \omega \in (-T, T); \\ 0, & \text{otherwise.} \end{cases}$$

Fourier transform of this function equals* $H(\omega) = 2T\text{sinc}(\omega T) = 2\sin(\omega T)/\omega$. As an example, Figure 6.3 shows the rectangle function on the time interval $[-16, 16]$ in part a, along with the amplitude of the Fourier transform in (b). The $\text{sinc}(\omega)$ function is shown only in the frequency interval $[-25, 25]$. We

---

*Another definition is also used: $\text{sinc}(\omega) = \sin(\pi\omega)/(\pi\omega)$.

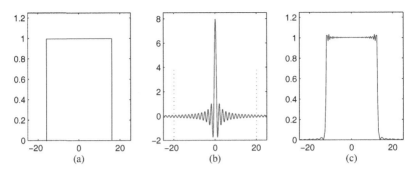

FIGURE 6.3
(a) The rectangle unit signal, (b) the Fourier transform of the signal, and (c)
inverse transform of the truncated sinc function $H_{20}(\omega)$ which equals $H(\omega)$,
if $|\omega| \le 20$, and 0 otherwise.

note that the $sinc(\omega)$ function is not an absolute integrable function. The
inverse Fourier transform of this function truncated by the cutoff frequency
$\omega_{cut} = 20$ is given in c.   ◻

The inverse Fourier transform is defined as follows:

$$f(t) = \frac{1}{2\pi} \int_{-\infty}^{\infty} F(\omega)e^{j\omega t}d\omega = \int_{-\infty}^{\infty} F(f)e^{j2\pi ft}df, \quad t \in (-\infty, +\infty), \qquad (6.2)$$

if the function is continuous at point t. In the case of discontinuity at t, these
integrals result in the value $[f(t+)+f(t-)]/2$. The Fourier transform is linear
and has many interesting properties, among which we mention the following:

1 (*duality*). It follows from (6.2) that the inverse Fourier transform is defined as the Fourier transform of $F(\omega)$ at point $-t$ and magnified by the
number $1/(2\pi)$. Therefore, the Fourier transform of the function $F(t)$ equals
$2\pi f(-\omega)$. For instance, the graphics of Figure 6.3, which describe the time-
frequency-time representation and approximation of the rectangle function,
can be considered as the frequency-time-frequency representation and approximation of an ideal low pass filter. The only exception is the magnitude of
the sinc function in b to be multiplied by 2π.

2 (*real case*). If $f(t)$ is a real function, then the Fourier transform at the
pair frequencies ω and $-\omega$ are complex conjugate, i.e., $\bar{F}(\omega) = F(-\omega)$. The
Fourier transform in absolute mode is thus symmetric with respect to the
vertical axis, as shown in Figure 6.3(b).

3 (*oddness*). The transform is an odd function when $f(t)$ is odd, i.e., $F(\omega) = -F(-\omega)$, if $f(t) = -f(-t)$.

4 (*convolution*). The product $F(\omega) = F_1(\omega)F_2(\omega)$ of Fourier transforms
corresponds in the time domain to the linear convolution of functions $f_1(t)$

and $f_2(t)$, i.e.,

$$F(\omega) \rightarrow f(t) = (f_1 * f_2)(t) = \int\limits_{-\infty}^{\infty} f_1(x)f_2(t-x)dx.$$

5 (*Parseval's equality*). The transform preserves the average energy of the function $f(t)$, i.e.,

$$||f||_2^2 = \int\limits_{-\infty}^{\infty} |f(t)|^2 dt = ||F||_2^2 = \frac{1}{2\pi} \int\limits_{-\infty}^{\infty} |F(\omega)|^2 d\omega.$$

As a conclusion it follows that the distance between two functions $f_1(t)$ and $f_2(t)$ in time and frequency domains is the same, $||f_1 - f_2||_2 = ||F_1 - F_2||_2$.

6 (*power*). Given a nonnegative function $f(t)$, the power of the function is concentrated in the zero-frequency $\int\limits_{-\infty}^{\infty} f(t)dt = F(0) \geq |F(\omega)|$, for all $\omega \in (-\infty, +\infty)$. Therefore, the symmetric graph of the positive function has a pike at the original point (see Figure 6.3(b) for the rectangle unit function).

6.1.1 Powers of the Fourier transform

We consider the interesting property of the Fourier transform which relates to the roots of the identical transform. The Fourier transform coincides with one of the roots of the identical transform. Indeed, let us define the following four operators in the space of continuous functions $f(t)$

$$\hat{F}^0[f(x)] = \int\limits_{-\infty}^{\infty} f(y)\delta(y-x)dy = f(x)$$

where $\delta(x)$ is the generalized delta function, and

$$\hat{F}^1[f(x)] = \tfrac{1}{\sqrt{2\pi}} \int\limits_{-\infty}^{\infty} f(y)e^{-jxy}dy = F(x),$$

$$\hat{F}^2[f(x)] = \tfrac{1}{\sqrt{2\pi}} \int\limits_{-\infty}^{\infty} f(y) \int\limits_{-\infty}^{\infty} e^{-jyz}e^{-jxz}dzdy = \int\limits_{-\infty}^{\infty} f(y)\delta(x+y)dy = f(-x)$$

$$\hat{F}^3[f(x)] = \tfrac{1}{\sqrt{2\pi}} \left[\int\limits_{-\infty}^{\infty} f(-y)e^{-jxy}dy = \int\limits_{-\infty}^{\infty} f(y)e^{jxy}dy \right] = (\hat{F}^1)^*[f(x)] = F(-x)$$

where * denotes the complex conjugate. Since $\hat{F}^2[f(x)] = f(-x)$, we obtain $\hat{F}^4[f(x)] = \hat{F}^0[f(x)] = f(x)$. In other words, the fourth power of the Fourier transform is the identical operator which we denote by I, i.e., $\hat{F}^4 = I$.

We now consider an operator being a linear combination of the powers of the Fourier transforms $\hat{G} = a_0\hat{F}^0 + a_1\hat{F}^1 + a_2\hat{F}^2 + a_3\hat{F}^3$ and define a condition when the inverse operator $(\hat{G})^{-1}$ exists. Defining the inverse operator by $(\hat{G})^{-1} = b_0\hat{F}^0 + b_1\hat{F}^1 + b_2\hat{F}^2 + b_3\hat{F}^3$, we obtain the following equations to be solved

$$\sum_{n=0}^{3} a_n b_{m-n} = \delta_{m,0}, \quad m = 0 : 3,$$

where $\delta_{m,0}$ is the delta symbol equal to 1 if $m = 0$, and 0 otherwise. These linear equations can be written in the matrix form as

$$\mathbf{Ab} = \begin{pmatrix} a_0 & a_3 & a_2 & a_1 \\ a_1 & a_0 & a_3 & a_2 \\ a_2 & a_1 & a_0 & a_3 \\ a_3 & a_2 & a_1 & a_0 \end{pmatrix} \begin{pmatrix} b_0 \\ b_1 \\ b_2 \\ b_3 \end{pmatrix} = \begin{pmatrix} 1 \\ 0 \\ 0 \\ 0 \end{pmatrix}, \tag{6.3}$$

and the vector $\mathbf{b}$ of coefficients b_n can be found if the determinant of the matrix $\mathbf{A}$ is not zero.

Example 6.4
Consider the operator $\hat{G}$ with coefficients $1, 2, 3$, and 4, i.e., $\hat{G} = \hat{F}^0 + 2\hat{F}^1 + 3\hat{F}^2 + 4\hat{F}^3$. Then, the mixed transform of a function $f(t)$ can be written as $g(t) = \hat{G}f(t) = f(t) + 2F(t) + 3f(-t) + 4F(-t)$, and the matrix $\mathbf{A}$ in (6.3) and its inverse are equal, respectively,

$$\mathbf{A} = \begin{pmatrix} 1 & 4 & 3 & 2 \\ 2 & 1 & 4 & 3 \\ 3 & 2 & 1 & 4 \\ 4 & 3 & 2 & 1 \end{pmatrix} \quad \text{and} \quad \mathbf{A}^{-1} = \frac{1}{4}\begin{pmatrix} -0.9 & 0.1 & 0.1 & 1.1 \\ 1.1 & -0.9 & 0.1 & 0.1 \\ 0.1 & 1.1 & -0.9 & 0.1 \\ 0.1 & 0.1 & 0.1 & -0.9 \end{pmatrix}.$$

The vector $\mathbf{b}$ equals $(-0.9, 1.1, 0.1, 0.1)'/4$. The inverse formula can thus be written as

$$f(t) = \hat{G}^{-1}g(t) = \frac{1}{4}\left(-0.9g(t) + 1.1G(t) + 0.1g(-t) + 0.1G(-t)\right), \tag{6.4}$$

where $G(t)$ denotes the Fourier transform of the function $g(t)$.

Let us consider the cosinusoidal wave $f(t) = \cos(\omega_0 t)$ with a frequency $\omega_0 > 0$. The mixed transform of this wave equals

$$\begin{aligned} g(t) &= \cos(\omega_0 t) + \left(\delta(t - \omega_0) + \delta(t + \omega_0)\right) \\ &\quad + 3\cos(-\omega_0 t) + 2\left(\delta(-t - \omega_0) + \delta(-t + \omega_0)\right) \\ &= 4\cos(\omega_0 t) + 3\left(\delta(t - \omega_0) + \delta(t + \omega_0)\right). \end{aligned} \tag{6.5}$$

Thus, the mixed transform is the cosinusoidal signal of amplitude 4 plus two infinite impulses at frequencies $\pm\omega_0$. As example, Figure 6.4 shows the cosinusoidal wave $\cos(\pi t/3)$ in part a, along with the mixed transform in b.

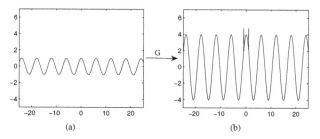

(a) (b)

FIGURE 6.4

(a) Cosinusoidal wave $\cos(\pi t/3)$ and (b) the mixed transform.

According to the inverse formula of (6.4), the cosinusoidal wave can be reconstructed from the mixed transform by

$$\begin{aligned}
\cos(\omega_0 t) &= \frac{1}{4}\left(-0.9g(t) + 1.1G(t) + 0.1g(-t) + 0.1G(-t)\right) \\
&= \frac{1}{4}\left(-0.8g(t) + 1.1G(t) + 0.1G(-t)\right),
\end{aligned}$$
(6.6)

since $g(-t) = g(t)$. Thus, two equidistant impulses on the cosinusoidal wave can be removed by means of the inverse mixed transform. Note that two impulses at time+frequency points $\pm\omega_0$ appear as a result of superposition of the transforms on the original signal. They are not time-impulse signals that appear at time-points $\pm\omega_0$, but because of the frequency ω_0. □

We also can define the square root of the Fourier transform as a linear combination of the powers of the transform. Indeed, let us consider the operator $\hat{G} = a_0\hat{F}^0 + a_1\hat{F}^1 + a_2\hat{F}^2 + a_3\hat{F}^3$, such that $(\hat{G})^2 = \hat{F}$. The solution of this task is reduced to solving the following matrix equation

$$\mathbf{Aa} = \begin{pmatrix} a_0 & a_3 & a_2 & a_1 \\ a_1 & a_0 & a_3 & a_2 \\ a_2 & a_1 & a_0 & a_3 \\ a_3 & a_2 & a_1 & a_0 \end{pmatrix} \begin{pmatrix} a_0 \\ a_1 \\ a_2 \\ a_3 \end{pmatrix} = \begin{pmatrix} 0 \\ 1 \\ 0 \\ 0 \end{pmatrix},$$

which leads to the coefficients

$$a_0 = \frac{1 + \sqrt{2} - j}{4}, \quad a_1 = \frac{1 - \sqrt{2} - j}{4}, \quad a_2 = \frac{1 + \sqrt{2} + j}{4}, \quad a_3 = \frac{1 - \sqrt{2} + j}{4}.$$

For instance, for the wave $f(t) = \cos(\omega_0 t)$, the square root of the Fourier transform results in the mixed signal

$$g(t) = \frac{1 + \sqrt{2}}{2}\cos(\omega_0 t) + \frac{1 - \sqrt{2}}{4}\left(\delta(t - \omega_0) + \delta(t + \omega_0)\right).$$
(6.7)

The inverse to this square root transform is defined by complex vector-coefficient

$$\mathbf{b} = \frac{1}{2}(0.8536 + j, 0.1464 - j, 0.8536 - j, 0.1464 + j)'.$$

6.2   Representation by frequency-time wavelets

Fourier analysis includes the short-time or windowed Fourier transform used in speech signal processing when a non-stationary signal is analyzed [81, 86],[92]-[97]. For this modified version of the Fourier transform, a signal is cut by a usually compactly supported window function into parts, and the Fourier transform is analyzed for every cut. The window is translated by a chosen step along the time axis to cover the entire time domain. The short-time Fourier transform uses a single window for all frequency components, and that does not allow one to determine the locations where the frequencies are present. In wavelet analysis, a fully scalable modulated window is used for frequency localization [26, 98]. The window is sliding, and the wavelet transform of a part of the signal is calculated for every position. Then a slightly longer or shorter window is used for each new stage of calculation. The scale variable is inversely proportional to frequency. Short windows at high frequencies and long windows at low frequencies are used in the wavelet transform. The result of the wavelet transform is a collection of time-scaling representations of the signal with different resolutions.

In this section, a concept of the A-wavelet transform is introduced and the representation of the Fourier transform by the A-wavelet transform is described. The A-wavelet transform is defined on a specific set of points in the frequency-time plane. This transform uses a fully scalable modulated window, but not all possible shifts. A geometrical locus of frequency-time points for the A-wavelet transform is considered "optimal" for the Fourier transform when a signal can be recovered by using only values of its wavelet transform defined on the locus. The concept of the A-wavelet transform can be extended for representation of other unitary transforms, and such an example for the Hartley transform is described and the reconstruction formula is given.

6.2.1   Wavelet transforms

We consider briefly the representation of a function $f(t)$ in the form of the wavelet transform. To perform the wavelet transform, we take the Mexican hat function as an analyzing wavelet $\psi(t) = (1 - t^2)e^{-t^2/2}$, $-\infty < t < +\infty$. This function and its few time-transformations are shown in Figure 6.5.

The Mexican hat-based wavelet transform of a function $f(t)$ is defined as the cross-correlation of the function with a family of wavelets scaled by the time transformation $t \to t/a$,

$$T(a, b) = w(a) \int_{-\infty}^{\infty} f(t)\psi\left(\frac{t-b}{a}\right) dt, \quad a > 0, \ b \in (-\infty, +\infty), \qquad (6.8)$$

where the weighting function $w(a)$ is considered to be $1/\sqrt{a}$. The parameter

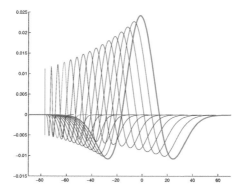

FIGURE 6.5
Mexican hats.

a is referred to as a *dilation* parameter and b as a *location* parameter.

As an example, the cosine waveform $f(t) = \cos(\omega_0 t)$ with frequency $\omega_0 = 1.3\text{rad/s}$ defined in the time interval $(-3\pi, 3\pi)$ is shown in Figure 6.6 in part a, along with the wavelet transform surface plot versus $\omega = 1/a$ and b in part b. The parameters ω and b vary respectively in intervals $(0, 5)$ and $[-3\pi, 3\pi]$. One can see that the wavelet transform $T(a, b) = T(\omega, b)$ leads to zero as frequency parameter $\omega = 1/a$ becomes greater than 4 (or dilation parameter a smaller than $1/4$). The pikes of the surface are located at points where the Mexican hat wavelet $\psi_{a,b}(t)$ has a form comparative with the cosine waveform move in phase and out of phase with the signal $f(t)$. The maxima and minima occur at the following seven points: $(\omega_m, b_m) = (a_m^{-1}, b_m) = (0.74, mP/2)$, $m = -3 : 3$, where $P = 2\pi/\omega_0$ is the period of $f(t)$. The Mexican hat function scaled by $a = 0.74$ is shown in Figure 6.6(a) by the dashed line. If we consider the weighted function $w(a)$ to be equal to $1/\sqrt{a}$ then the maximum frequency a_m will be shifted to 0.82rad/s, i.e., to the scale of approximately $0.63P$.

The function $f(t)$ can be recovered from its continuous wavelet transform $T(a, b)$ by integrating over all locations and dilations

$$f(t) = \frac{1}{\pi} \int_0^\infty w(a) \left[\int_{-\infty}^\infty T(a, b) \psi\left(\frac{t-b}{a}\right) db \right] \frac{da}{a^2}. \tag{6.9}$$

This complex formula includes two integrals, one of which is the cross-correlation of the wavelet transform with the analyzing wavelet.

6.2.2 Fourier transform wavelet

We here describe the integral Fourier transform in a way that differs from the method of calculation of the continuous wavelet transform. The description differs also from the well-known concept of the continuous short-time Fourier

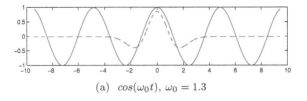

(a) $cos(\omega_0 t)$, $\omega_0 = 1.3$

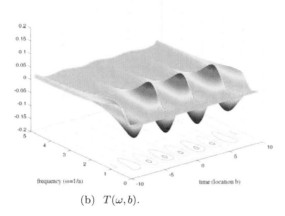

frequency ($\omega=1/a$) time (location b)

(b) $T(\omega, b)$.

FIGURE 6.6 (See color insert following page 242.)

(a) Cosinusoidal signal and (b) the wavelet transform plot of the signal with the Mexican hat wavelet.

transform, STFT. Such a transform is based on a joint time-frequency signal representation and defined by

$$F(t, \omega) = \int\limits_{-\infty}^{\infty} f(\tau) g(\tau - t) e^{-j\omega\tau} d\tau, \quad t, \omega \in (-\infty, +\infty), \qquad (6.10)$$

when a time-sliding window function $g(t)$ is used, which emphasizes "local" frequency properties. The window function is typically considered to be real, symmetric, non-zero only in a region of interest, and with unit norm in $L_2(R)$, the space of square-integrable functions. For instance, $g(t)$ can be taken equal to the rectangular function $L^{-1} rect(t/L)$ of length $L > 0$, or the Gaussian function $(\sqrt{\pi\sigma})^{-1} \exp(-t^2/\sigma)$ with a symmetric finite support, where $\sigma > 0$ is a fixed number defining a "width" of the window. The signal $f(t)$ can be reconstructed from the short-time Fourier transform by

$$f(t) = \frac{1}{2\pi} \int\limits_{-\infty}^{\infty} \int\limits_{-\infty}^{\infty} F(\tau, \omega) g(t - \tau) e^{j\omega t} d\omega d\tau.$$

Let $\psi(t)$ and $\varphi(t)$ be functions that are zero outside the interval $[-\pi, \pi)$ and coincide respectively with the cosine and sine functions inside this interval, i.e., $\psi(t) = \cos(t)$ and $\varphi(t) = \sin(t)$ when $t \in [-\pi, \pi)$, and $\psi(t) = \varphi(t) = 0$ otherwise. We consider the family $\{\psi_{\omega;b_n}(t), \varphi_{\omega;b_n}(t)\}$ of the following time-scale and shift transformations of these functions $t \to \omega t$, $t \to t - b_n$, where ω frequency varies along the real line (or a finite interval) and b_n takes values of a finite or infinite set that will be defined later on. Thus, we define the functions $\psi_{\omega;b_n}(t) = \psi(\omega[t - b_n])$, and $\varphi_{\omega;b_n}(t) = \varphi(\omega[t - b_n])$, where $t \in (-\infty, +\infty)$. These functions are periods of the cosine and sine waveforms $\cos(\omega t)$ and $\sin(\omega t)$ shifted by b_n; periods begin respectively at π/ω and 0.

Let $f(t)$ be a function for which the Fourier transform exists

$$F(\omega) = \int_{-\infty}^{\infty} f(t)e^{-j\omega t}dt = \int_{-\infty}^{\infty} f(t)\cos(\omega t)dt - j\int_{-\infty}^{\infty} f(t)\sin(\omega t)dt.$$

For a given frequency $\omega \neq 0$, we describe the process of $F(\omega)$ component formation when the cosine and sine waveforms of frequency ω are interfering with $f(t)$. For that we divide the time line R by intervals of length $2\pi/\omega$ with the centers at integer multiples of $2\pi/\omega$. These intervals are denoted by $I_n(\omega)$,

$$..., \ I_{-1} = \left[-\frac{3\pi}{\omega}, -\frac{\pi}{\omega}\right), \ I_0 = \left[-\frac{\pi}{\omega}, \frac{\pi}{\omega}\right), \ I_1 = \left[\frac{\pi}{\omega}, \frac{3\pi}{\omega}\right), ... \quad (6.11)$$

The Fourier transform $F(\omega)$, $\omega \neq 0$, can be written as follows:

$$
\begin{aligned}
F(\omega) &= \sum_{n=-\infty}^{\infty} \int_{I_n} f(t)\cos(\omega t)dt - j\sum_{n=-\infty}^{\infty} \int_{I_n} f(t)\sin(\omega t)dt \\
&= \sum_{n=-\infty}^{\infty} \int_{I_0} f\left(t + \frac{2\pi}{\omega}n\right)\cos(\omega t)dt - j\sum_{n=-\infty}^{\infty} \int_{I_0} f\left(t + \frac{2\pi}{\omega}n\right)\sin(\omega t)dt \\
&= \sum_{n=-\infty}^{\infty} \int_{-\infty}^{\infty} f(t)\psi\left(\omega\left[t - \frac{2\pi}{\omega}n\right]\right)dt - j\sum_{n=-\infty}^{\infty} \int_{-\infty}^{\infty} f(t)\varphi\left(\omega\left[t - \frac{2\pi}{\omega}n\right]\right)dt \\
&= \sum_{n=-\infty}^{\infty} \int_{-\infty}^{\infty} f(t)\psi(\omega t - 2\pi n)dt - j\sum_{n=-\infty}^{\infty} \int_{-\infty}^{\infty} f(t)\varphi(\omega t - 2\pi n)dt.
\end{aligned}
$$

We introduce the following transforms of the function $f(t)$:

$$T_\psi(\omega, b_n) = \int_{-\infty}^{\infty} f(t)\psi_{\omega,b_n}(t)\,dt, \quad \left(T_\psi(0,0) = \int_{-\infty}^{\infty} f(t)dt\right)$$

$$T_\varphi(\omega, b_n) = \int_{-\infty}^{\infty} f(t)\varphi_{\omega,b_n}(t)\,dt, \quad (T_\varphi(0,0) = 0) \quad\quad (6.12)$$

where $b_n = b_{n,\omega} = (2\pi/\omega)n$, n is an integer, and $b_n = 0$ if $\omega = 0$. $T_\psi(\omega, b_n)$ is the integral of the cosinusoidal signal $\psi(\omega t)$ of one period $2\pi/\omega$ that is superimposed on the $f(t)$ waveform at location b_n. The signal $\psi(\omega t)$ is moving along the $f(t)$ waveform at locations b_n which are integer multiples of $2\pi/\omega$. The location depends on the frequency ω of the signal. In a similar way, the transform $T_\varphi(\omega, b_n)$ is defined by the product of the $f(t)$ waveform with the sinusoidal signal $\varphi(\omega t)$ moving at locations b_n. As an example, Figure 6.7 shows the sinusoidal signal $f(t)$ and signal $\psi(\omega_1 t)$ located at point 0 in part a. The frequency of the signal is $\omega_1 = 1.75$. The integral of the function $f(t)$ being multiplied by the cosinusoidal signal defines the value of the transform at point $(\omega_1, b_0) = (1.75, 0)$, which equals $T_\psi(\omega_1, 0) = 0.0726$. In parts b and c, the cosinusoidal signal $\psi(\omega_1 t)$ has been moved respectively to locations $2\pi/\omega_1$ and $4\pi/\omega_1$. The values of transform $T_\psi(\omega, b_n)$ at these points equal to $T_\psi(\omega_1, 2\pi/\omega_1) = -0.0033$ and $T_\psi(\omega_1, 4\pi/\omega_1) = -0.0723$.

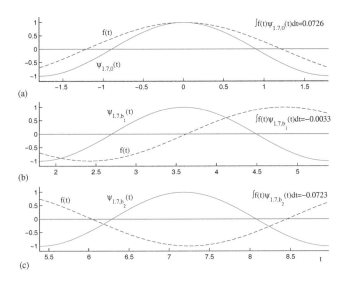

FIGURE 6.7
Integration of the signal with $\cos(1.7t)$ at locations 0, 3.70, and 7.39.

The Fourier transform as a complex transform is composed by the pair of transforms $T_\psi(\omega, b_n)$ and $T_\varphi(\omega, b_n)$

$$F(\omega) = \sum_{n=-\infty}^{\infty} T_\psi(\omega, b_n) - j \sum_{n=-\infty}^{\infty} T_\varphi(\omega, b_n). \qquad (6.13)$$

We now define the following set of points in the frequency-time plane

$$A = \left\{ (\omega, b_n); \ \omega \in (-\infty, +\infty), b_n = \frac{2\pi n}{\omega}, \ n = 0, \pm 1, \ldots, \pm N(\omega) \right\} \qquad (6.14)$$

where $N(0) = 0$, and $N(\omega) = \infty$ or is a finite number if the function $f(t)$ has a finite support. As was mentioned above, we consider $b_n = 0$ when $\omega = 0$.

Example 6.5

Let $f(t)$ be the $\cos(\omega_0 t)$ waveform of Figure 6.6(a), which is defined in the interval $(-3\pi, 3\pi)$. Figure 6.8 shows the following set of frequency-time points

$$A = \left\{ (\omega, b_n); \ \omega \in (0, 3\pi), \ b_n = \frac{2\pi n}{\omega}, \quad n = 0, \pm 1, \pm 2, \ldots, \pm N(\omega) \right\}$$

where $N(\omega)$ is defined as 0 when $\omega < 1/3$ and $N(\omega) = \left\lfloor \frac{3}{2}\omega - \frac{1}{2} \right\rfloor$, when $\omega \in (1/3, 3\pi)$ and the operation $\lfloor \cdot \rfloor$ denotes the floor function. Horizontal lines with centers located at a few points (ω, b_n) of set A, which show the widths of the corresponding cosine and sine basis signals $\varphi(\omega t)$ and $\psi(\omega t)$ of transforms $T_\psi(\omega, b_n)$ and $T_\varphi(\omega, b_n)$, are also given in the figure.

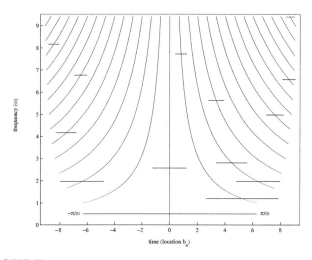

FIGURE 6.8
The locus of time-frequency points for the A-wavelet transform.

The function $N(\omega)$ shows how many shifted versions of signals $\psi(\omega t)$ and $\varphi(\omega t)$ are used for calculating values of transforms $T_\psi(\omega, b_n)$ and $T_\varphi(\omega, b_n)$ for frequency ω. For instance, when $\omega < 1$ only signals $\psi(\omega t)$ and $\varphi(\omega t)$ will be multiplied with $f(t)$ and then integrated to define $T_\psi(\omega, 0)$ and $T_\varphi(\omega, 0)$. No more translations of these signals are required for calculating the Fourier transform $F(\omega)$ at these frequencies. In other words, one can consider that $T_\psi(\omega, b_n) = T_\varphi(\omega, b_n) = 0$ if $n \neq 0$.

When $\omega \in (1, 5/3)$ the signals $\psi(\omega t)$ and $\varphi(\omega t)$ are required to be shifted to the right and left by $2\pi/\omega$. To calculate the Fourier transform $F(\omega)$ at such frequencies, only the following values are needed: $T_\psi(\omega, b)$ and $T_\varphi(\omega, b)$ for $b = 0, \pm 2\pi/\omega$. In the general case, the number of required translations $L(\omega) = 2N(\omega)+1$ can also be defined as the following step function: $L(\omega) = 1$ when $\omega \in (0, 1/3)$, and $L(\omega) = 2n+1$ when $\omega \in ((2n+1)/3, (2n+3)/3)$. We can consider that $T_\psi(\omega, b_n) = T_\varphi(\omega, b_n) = 0$ when $|n| > N(\omega)$.

According to (6.13), in order to calculate the Fourier transform of $f(t)$, the values of transforms $T_\psi(\omega, b_n)$ and $T_\varphi(\omega, b_n)$ at frequency-time points of the set A are required. Figure 6.9 shows the 3-D plot of required values of the transform $\{T_\psi(\omega, b_n); (\omega, b_n) \in A\}$ versus the frequency ω and location b.

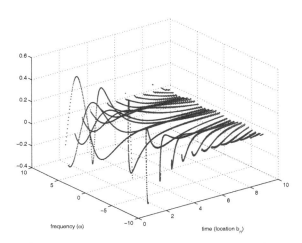

FIGURE 6.9 (See color insert following page 242.)
Wavelet transform of the signal $\cos(\omega_0 t)$, which is based on the cosine analyzing wavelet.

The representation of $f(t)$ by the pair of transforms $T_\psi(\omega, b_n)$ and $T_\varphi(\omega, b_n)$ (or by $T_\psi(\omega, b_n) - jT_\varphi(\omega, b_n)$) is called the *A-wavelet transform*. The transforms $\{T_\psi(\omega, b_n), (\omega, b_n) \in A\}$ and $\{T_\varphi(\omega, b_n), (\omega, b_n) \in A\}$ are called respectively the *C-wavelet* and *S-wavelet transforms*. ▯

6.2.3   Cosine- and sine-wavelet transforms

Similar to the Mexican hat-based wavelet transform, we define the wavelet transforms by using the cosine and sine signals $\psi(\omega, b)$ and $\varphi(\omega, b)$ instead of the Mexican hat wavelet. The transforms are the following

$$T_\psi(\omega, b) = \int\limits_{-\infty}^{\infty} f(t)\psi_{\omega,b}(t)dt, \quad T_\varphi(\omega, b) = \int\limits_{-\infty}^{\infty} f(t)\varphi_{\omega,b}(t)dt$$

where $\omega, b \in (-\infty, \infty)$. These transforms we call respectively *cosine-wavelet* and *sine-wavelet* transforms.

Figure 6.10 shows surface-plots of the cosine- and sine-wavelet transforms of the cosinusoidal waveform $f(t) = \cos(\omega_0 t)$ versus frequency ω and location b. Unlike the Mexican hat-based wavelet transform, these transforms are based on the analyzing wavelets with finite supports.

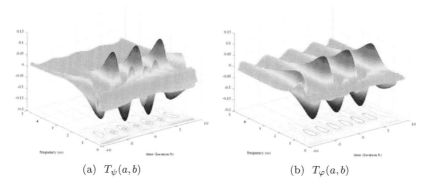

(a) $T_\psi(a, b)$ (b) $T_\varphi(a, b)$

FIGURE 6.10 (See color insert following page 242.)

(a) Cosine and (b) sine wavelet transforms of the function $f(t) = \cos(\omega_0 t)$.

The above defined A-wavelet transform takes information from the cosine- and sine-wavelet transforms at frequency-time points of set A. The A-wavelet transform differs from the wavelet-transform that uses values of transform at points (ω, b) (or (a, b)) of the whole range of ω and b. This is the main difference between the A-wavelet transform and existent continuous wavelet transforms. The size of window for the A-wavelet transform changes with ω frequency of the analyzing signals $\psi(\omega t)$ and $\varphi(\omega t)$. The window moves at only a finite number of specific locations depending on the frequency (as shown in set A of Figure 6.8), when the signal has a finite support.

To see the time-frequency resolution differences between the Fourier transform (or A-wavelet) and traditional wavelet transforms, Figure 6.11 shows the function coverage in the time-frequency plane with basis cosine and sine functions of the A-wavelet transform. The set A can be considered as an "optimal" geometrical locus of frequency-time points for the Fourier transform defined by the A-wavelet transform. This locus shows how to derive the minimum information from the cosine- and sine-wavelet transforms in order to determine the A-wavelet transform or the Fourier transform. Figure 6.9 shows that there is no need to calculate the wavelet transform across the whole range of frequency-time points, but to the points of set A. The locus of frequency-time points differs from grids used for discretization of the short-time Fourier transform and the wavelet transform. Indeed for discretization of

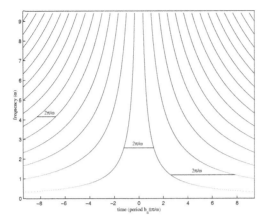

FIGURE 6.11
Coverage of the time-frequency plane by cosine and sine basis functions.

the short-time Fourier transform, a regular rectangular grid is used with time
and frequency steps t_0 and ω_0, $\{F_{n,m} = F(nt_0, m\omega_0); \ n, m = 0, \pm1, \pm2, \ldots\}$
that satisfy the frame bound condition $t_0\omega_0 \leq 2\pi$ [86]. For the wavelet
transform, the frames are constructed by sampling the dilation exponentially $a = a_0^n$ (with step $a_0 > 1$) and the translation proportionally a_0^n
$\{T_{n,m} = T\left(a_0^n, mb_0a_0^n\right); \ n, m = 0, \pm1, \pm2, \ldots\}$, where $b_0 > 0$ is a location parameter. For the Mexican hat wavelet transform, the sampling parameter is
$(a_0, b_0) = (\sqrt{2}, 0.5)$. To construct an orthonormal wavelet basis defining completely the signal, another simple grid is used with the effective discretization
of the wavelet transform by the parameter $(a_0, b_0) = (2, 1)$.

Example 6.6
We illustrate a simple application of the A-wavelet representation of the Fourier transform of a signal mixed with a short-in-time signal of a high frequency.
Consider the cosine signal $x(t) = \cos(\omega_0 t)$ with frequency $\omega_0 = 1.3\text{rad/s}$ in the
time-interval $(-3\pi, 3\pi)$. Suppose that the following high-frequency sinusoidal
signal with duration of 0.2s occurred in the signal $x(t) : n(t) = -3\sin(\omega_1 t)$
when $t \in [-0.1, 0.1]$, where the frequency $\omega_1 = 8\text{rad/s}$. The signal to be
analyzed is $y(t) = n(t)$ when $t \in [-0.1, 0.1]$, and $y(t) = x(t)$ when $t \in$
$[-3\pi, 3\pi] \setminus [-0.1, 0.1]$. Figure 6.12 shows the signal $y(t)$ sampled with the
rate of 100Hz in part a, along with the magnitude of the DFT of the signal
in the frequency-interval $[-0.5, 0.5]\text{rad/s}$ in part b. The appearance of the
short-time signal $n(t)$ causes ripples on the Fourier transform, which can be
seen in the figure. It is clear that those ripples can be removed by a low pass
filter with a short pass band, but the time of $n(t)$ signal appearance cannot
be seen in the Fourier spectrum.
    We now analyze the considered signals $n(t)$ and $y(t)$ by the A-wavelet

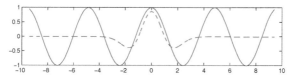

FIGURE 6.12

(a) Signal $y(t)$ and (b) the real part of the Fourier transform $Y(\omega)$.

representation of their Fourier transforms. The real part of the A-wavelet transform, i.e., C-wavelet transform of the additional signal $n(t)$, is shown in Figure 6.13 in part a, along with the C-wavelet transform of the signal $y(t)$ in part b. The C-wavelet transform of the signal $x(t)$ is given in Figure 6.9. One can see that the C-wavelet $T_\psi(\omega, b_n)$ of $n(t)$ is zero for all locations b_n except 0, i.e., $T_\psi(\omega, b) = 0$, if $b_n \neq 0$, for all frequencies $\omega \in [-3\pi, 3\pi]$. Therefore, the change in the C-wavelet transform occurs only at the location $b_n = 0$ which corresponds to the center of the support of the signal $n(t)$. Thus, the Fourier transform represented in the form of the A-wavelet transform allows for detecting the exact location of the higher-frequency signal $n(t)$. ▯

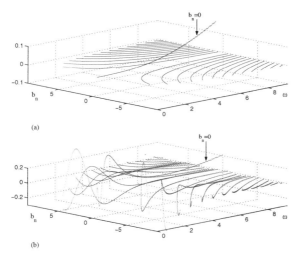

FIGURE 6.13 (See color insert following page 242.)

Wavelet transforms of the (a) signal $n(t)$ and (b) signal $y(t)$.

6.2.4   B-wavelet transforms

The above described representation of the Fourier transform in not unique. Different parts of one period for cosine and sine basic functions can be used to compose the Fourier transform representation as a sum of wavelet-type transforms. As an example, we can define the B-wavelet transform similarly to the B-wavelet transform, when using the cosine and sine functions of half periods. A geometrical locus of frequency-time points for the B-wavelet transform is derived similarly to the A-wavelet as well. Let $\psi(t)$ and $\varphi(t)$ be functions that coincide respectively with the cosine, $\cos(t)$, and sine, $\sin(t)$, functions inside the half of period interval $[-\pi/2, \pi/2]$ and equal to zero outside this interval. The half-periods begin at 0. We consider a family $\{\psi_{\omega;b_n}(t), \varphi_{\omega;b_n}(t)\}$ of time-scale and shift transformations of these functions $\psi_{\omega;b_n}(t) = \psi(\omega[t-b_n])$, $\varphi_{\omega;b_n}(t) = \varphi(\omega[t-b_n])$, where $t \in (-\infty, +\infty)$, frequency ω varies along the real line, and b_n takes values of a finite or infinite set to be defined below. The transform $F_\psi(\omega, b_n)$ as $T_\psi(\omega, b_n)$ is defined as the integral of the cosinusoidal signal $\psi(\omega t)$ of the half-period π/ω, which is multiplied on the $f(t)$ waveform at location b_n. The transform $F_\varphi(\omega, b_n)$ is defined by the inner product of the $f(t)$ waveform with the sinusoidal signal $\varphi(\omega t)$ of the half-period π/ω. The complex Fourier transform is composed by the pair of real transforms $F_\psi(\omega, b_n)$ and $F_\varphi(\omega, b_n)$ as

$$F(\omega) = \sum_{n=-\infty}^{\infty} (-1)^n F_\psi(\omega, b_n) - j \sum_{n=-\infty}^{\infty} (-1)^n F_\varphi(\omega, b_n). \qquad (6.15)$$

These transforms are calculated in the following set of points in the frequency-time plane

$$B = \left\{ (\omega, b_n); \ \omega \in (-\infty, +\infty), \ b_n = n\frac{\pi}{\omega}, n = 0, \pm 1, \pm 2, \ldots, \pm N(\omega) \right\}$$

where $N(0) = 0$, and $N(\omega)$ is the infinite in general, or a finite number if the function $f(t)$ has a finite support. Given an integer n, we call the set of points $B_n = \{(\omega, b_n); \ \omega \in (-\infty, +\infty)\}$ the nth *center-line* in the frequency-time plane. The locus B is the union of such center-lines, $B = \{\cup B_n; \ n = 0, \pm 1, \ldots, \pm N(\omega)\}$, as shown in Figure 6.14.

The representation of $f(t)$ by the pair of transforms $F_\psi(\omega, b_n)$ and $F_\varphi(\omega, b_n)$ (or by $F_\psi(\omega, b_n) - j F_\varphi(\omega, b_n)$) we name the *B-wavelet transform*. The transformations $f(t) \rightarrow \{F_\psi(\omega, b_n), (\omega, b_n) \in B\}$ and $f(t) \rightarrow \{F_\varphi(\omega, b_n), (\omega, b_n) \in B\}$ we call respectively the *cosine* and *sine* *B-wavelet* transformations.

Example 6.7

Let $f(t)$ be the $\cos(\omega_1 t)$ waveform of frequency $\omega_1 = \pi/3$ defined in the time interval $(-3\pi, 3\pi)$. We assume that in the interval $(-1.5, 1.5)$ a sinusoidal component $n(t) = 0.5 \sin(\omega_2 t)$ has been added to $f(t)$. This short-time signal has a frequency six times that of ω_1, i.e., $\omega_2 = 2\pi$. The signal to be analyzed

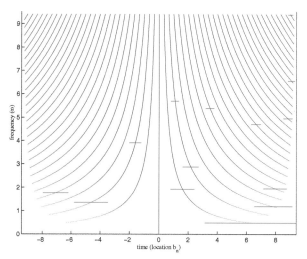

FIGURE 6.14 (See color insert following page 242.)
The locus of time-frequency points for the B-wavelet transform.

by the B-wavelet-transform is $g(t) = f(t) + n(t)$ when $|t| < 1.5$, and $g(t) = f(t)$ when $1.5 \leq |t| < 3\pi$ (see Figure 6.15). Figures 6.16 and 6.17 show

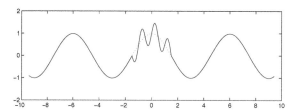

FIGURE 6.15
Cosine waveform with a high-frequency short-time signal.

two projections of the 3-D transform plot of the signal $f(t)$ in the time and frequency domains, respectively. Only the F_ψ transform part of the B-wavelet transform is illustrated. Figure 6.18 shows the projection of the 3-D B-wavelet transform plot of the signal $g(t)$ in the time domain.   The noise $n(t)$ yields a change of the transform plot at locations b_n lying in the interval $(-1.5, 1.5)$. The main changes occur in the time interval $(-1.5, 1.5)$ along the center-lines $B_1 = \{b_1(\omega) + \pi/\omega; \omega \in (0, 3\pi)\}$ and $B_{-1} = \{b_{-1}(\omega) + \pi/\omega; \omega \in (0, 3\pi)\}$.
▯

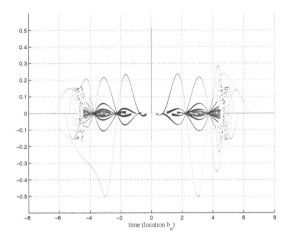

FIGURE 6.16

Wavelet transform plot of the signal $f(t)$ with the cosine analyzing function (projection on the time domain).

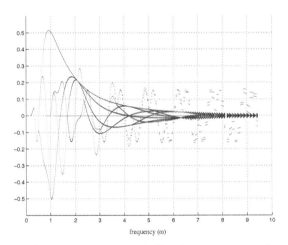

FIGURE 6.17 (See color insert following page 242.)

Wavelet transform plot of the signal $f(t)$ with the cosine analyzing function (projection on the frequency domain).

6.2.5 Hartley transform representation

Wavelet transforms similar to the A-wavelet transform can also be derived for other unitary transforms. For instance, for the Hartley transform

$$H(\omega) = \int\limits_{-\infty}^{\infty} f(t)cas(\omega t)dt = \int\limits_{-\infty}^{\infty} f(t)\left[\cos(\omega t) + \sin(\omega t)\right]dt$$

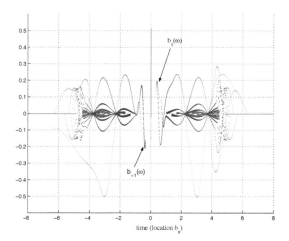

FIGURE 6.18 (See color insert following page 242.)
Wavelet transform plot of the signal $g(t)$ with the cosine analyzing function (projection on the time domain).

the following representation holds

$$H(\omega) = \sum_{n=-\infty}^{\infty} T_{\psi+\varphi}(\omega, b_n) = \sum_{n=-\infty}^{\infty} T_{\psi}(\omega, b_n) + \sum_{n=-\infty}^{\infty} T_{\varphi}(\omega, b_n). \qquad (6.16)$$

The locus of the frequency-time points (ω, b_n) is defined by the same set A defined in (6.14) for the Fourier transform. As an example, Figure 6.19 shows the A-wavelet transform for the Hartley transform

$$f(t) \to \{T_{\psi+\varphi}(\omega, b_n); \ (\omega, b_n) \in A\} \qquad (6.17)$$

of the waveform $f(t) = \cos(\omega_0 t)$ when $\omega_0 = 1.3\text{rad/s}$. The locus A for this transform is defined as in Example 6.5. The inverse Hartley transform can also be represented by a wavelet transform. Indeed, denoting by $\phi(\omega)$ the function $\psi(\omega) + \varphi(\omega)$, we obtain the following:

$$f(t) = \frac{1}{2\pi} \int_{-\infty}^{\infty} H(\omega)cas(\omega t)d\omega = \sum_{k=-\infty}^{\infty} W_\phi(t, \omega_k)$$

where the wavelet transform $W_\phi(t, \omega_k)$ is defined by

$$W_\phi(t, \omega_k) = \frac{1}{2\pi} \int_{-\infty}^{\infty} H(\omega)\phi_{t,\omega_k}(\omega)d\omega = \sum_{n=-\infty}^{\infty} \frac{1}{2\pi} \int_{-\infty}^{\infty} T_\phi(\omega, b_n)\phi_{t,\omega_k}(\omega)d\omega$$

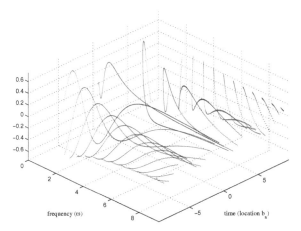

FIGURE 6.19 (See color insert following page 242.)
Hartley wavelet transform plot of the signal $\cos(\omega_0 t)$.

when $t \neq 0$, and $W_\phi(0, \omega_k) = 0$ for $\omega_k \neq 0$. Therefore the following reconstruction formula holds

$$f(t) = \sum_{k=-\infty}^{\infty} \sum_{n=-\infty}^{\infty} \frac{1}{2\pi} \int_{-\infty}^{\infty} T_\phi(\omega, b_n) \phi_{t, \omega_k}(\omega) d\omega.$$

6.3   Time-frequency correlation analysis

Let $f(t)$ be an absolute integrable function on the real line, R, that satisfies Dinni's condition at every point $t \in R$, $\int_{-\delta}^{\delta} |[f(t+x) - f(t)]x^{-1}| \, dx < \infty$ for some $\delta > 0$. For example, this condition is fulfilled for a continuous function that has a finite derivative. Then the function $f(t)$ can be represented as

$$f(t) = \frac{1}{\pi} \int_0^{\infty} d\lambda \int_{-\infty}^{\infty} f(x) \cos(\lambda(x-t)) \, dx. \qquad (6.18)$$

This is the well-known integral Fourier formula that in the complex form leads to the Fourier transform [87]. To break up (6.18), we consider the integral-function in the above decomposition of $f(t)$

$$F(\lambda, t) = \int_{-\infty}^{\infty} f(x) \cos(\lambda(x-t)) \, dx. \qquad (6.19)$$

The integral Fourier formula defines thus the pair of the Fourier transform with the first part in (6.19) and the second part written as

$$f(t) = \frac{1}{\pi} \int\limits_0^\infty F(\lambda, t) d\lambda. \tag{6.20}$$

Given a frequency λ, the function $F(\lambda, t)$ is periodic with period $2\pi/\lambda$. Function $F(\lambda, t)$ is a frequency-time, or time-frequency representation of the function $f(t)$, which is defined as the correlation of the function with the scaled cosine wave $\cos(\lambda t)$ when t runs from $-\infty$ to ∞. The value of function $f(t)$ at point t is defined as the sum of all these correlations calculated at this point.

The correlation in (6.19) leads to time-frequency analysis of functions by the Fourier transform. Indeed, let $\psi(t)$ be a function that is zero outside the interval $[-\pi, \pi)$ and coincides with the cosine function inside this interval, $\psi(x) = \cos(x)$ when $x \in [-\pi, \pi)$, and 0 otherwise. Let $\mathcal{A}$ be the family of the following time-scale and shift transformations $\{\psi_{\lambda;t}(x) = \psi(\lambda(x - t))\}$, where $\lambda > 0$ and $t \in (-\infty, \infty)$. These functions are referred to as one period of the cosine waveforms $\cos(\lambda x)$ with the centers at point t.

For a given frequency $\lambda > 0$ and time point t, we describe the process of $F(\lambda, t)$ component formation when one-period cosine waveforms of frequency λ correlate with $f(t)$. For that, we define the partition of the time line by specific intervals of length $2\pi/\lambda$ with the centers located at points being integer multiples of $2\pi/\lambda$ and shifted by t. The intervals are

$$I_n = I_n(\lambda, t) = \left[t + \frac{(2n - 1)\pi}{\lambda}, t + \frac{(2n + 1)\pi}{\lambda}\right)$$

where $n = 0, \pm 1, \pm 2, \ldots$. The partition of the time-line by such intervals depends on the frequency λ and time t. The direct calculations show the following

$$F(\lambda, t) = \sum_{n=-\infty}^\infty \int\limits_{I_n} f(x) \cos(\lambda(x - t)) \, dx = \sum_{n=-\infty}^\infty \int\limits_{-\infty}^\infty f(x) \psi_{\lambda, t_n}(x) dx$$

where $t_n = t_n(\lambda, t) = t + n(2\pi/\lambda)$, $n = 0, \pm 1, \pm 2, \ldots$.

The full Fourier correlation function can thus be represented as $F(\lambda, t) = \sum_{n=-\infty}^\infty \Psi(\lambda, t, n)$, where the transformation $\Psi(\lambda, t, n)$ is defined by

$$\Psi(\lambda, t, n) = \Psi(\lambda, t_n) = \int\limits_{-\infty}^\infty f(x) \psi_{\lambda, t_n}(x) dx. \tag{6.21}$$

If $\lambda = 0$, we consider that $I_0 = (-\infty, +\infty)$ and $\Psi(\lambda, 0) = F(\lambda, 0)$, and t_n is only defined for $n = 0$, when $t_0 = 0$.

The transform $\Psi(\lambda, t_n)$ is the cross-correlation of one period of the cosine waveform of frequency λ with the function $f(t)$ in the finite time interval $I_n(\lambda, t)$. The interval is located at time point t_n and has length equal to $2\pi/\lambda$. Thus $\Psi(\lambda, t_n)$ contains information if a cosine wave of frequency λ of short period within the interval of length $2\pi/\lambda$ is located (or has occurred) in the signal at time point t_n. The length of the interval within which the wave is analyzed is inversely proportional to the frequency. The intervals begin at t. The Fourier correlation function $F(\lambda, t)$ is the sum of correlations of one-period cosine wave $\psi(\lambda x)$ along the whole partition of the time line. As an example, Figure 6.20 shows the grid of 560 specified frequency-time points (λ_k, t_n) in the domain $[0, 200] \times [0, 1]$. The values of frequencies equal $\lambda_k = k/32$, $k = 1 : 32$. For each frequency λ_k, the time points t_n start from point $t = 1$, i.e., $t_n = 1 + 2\pi n/\lambda_k$.

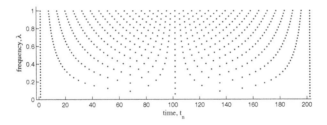

FIGURE 6.20
Grid with 560 frequency-time points (λ_k, t_n).

The reconstruction formula (6.20) can thus be written as

$$f(t) = \frac{1}{\pi} \int\limits_0^\infty \sum_{n=-\infty}^\infty \Psi(\lambda, t, n) \, d\lambda, \quad t \in (-\infty, \infty). \qquad (6.22)$$

The transform $f(t) \rightarrow \{\Psi(\lambda, t, n); \lambda \in [0, \infty), n = 0, \pm 1, \pm 2, ...\}$ describes the frequency-time analysis of the function $f(t)$. We call this transform the ψ-*resolution* of the function. It should be noted that the ψ-resolution is described by the totality of 2D functions $\{\Psi(\lambda, t, n); n = 0, \pm 1, \pm 2, ...\}$, all values of which are calculated in the 2D frequency-time plane. Indeed if $t = t_0 + 2\pi m/\lambda$, where $t_0 \in (0, 2\pi/\lambda)$ or $t_0 \in (-\pi/\lambda, \pi/\lambda)$ and m is an integer ≥ 0, then $\Psi(\lambda, t, n) = \Psi(\lambda, t_0, n + m)$, $n = 0, \pm 1, \ldots$, and $F(\lambda, t) = F(\lambda, t_0)$. Therefore $\Psi(\lambda, t, n)$ is required to be calculated only for triples (λ, t, n) from the set $\Delta = \{(\lambda, t, n); t \in [0, 2\pi/\lambda), \lambda > 0, n = 0, \pm 1, \ldots\}$, which is isomorphic to the 2D semi-plane $\{(\lambda, t'); \lambda > 0, t' \in (-\infty, \infty)\}$.

6.3.1   Wavelet transform and ψ-resolution

We dwell upon the reconstruction of the function $f(t)$ from its integral wavelet transform $T(\lambda, b)$ with respect to an analyzing function $\psi(t)$. Comparing formulas (6.22) with (6.8) and (6.21) with (6.9), we can note the following:

1. The wavelet transform is a redundant representation. All values of the wavelet transform $T(\lambda, b)$ are required in order to calculate the original function $f(t)$ at any point t.

2. The ψ-resolution $\Psi(\lambda, t, n)$ has one additional discrete parameter n. However, for any given triple (λ, t, n), the value of transform $\Psi(\lambda, t, n)$ is used only to calculate the original function at point t. This value is calculated as a cross-correlation of the function with $\psi(t)$ for the specified time-location $t_n = t_n(\lambda, t)$. The point itself can be calculated by $t = t_n - 2\pi n/\lambda$.

3. In the reconstruction of the original function from the wavelet transform, the cross-correlation of the transform with the analyzing function is used, as for the wavelet transform. The reconstruction of the function from the ψ-resolution does not require such a complex cross-correlation operation, but only the summation.

4. The cosine function is a simple trigonometric function, and, for many functions $f(t)$, the finite integrals $\Psi(\lambda, t_n)$ can be calculated and expressed in an analytical form. This fact may simplify the analysis of signals, for instance, when additional noisy signals are present.

Example 6.8
In the interval $(-3\pi, 3\pi)$, consider the signal $f(t) = \cos(\omega t) + 0.5\sin(2\omega t - 0.25)$, where $\omega = 1.3$rad/s. Suppose that the following high-frequency sinusoidal signal $n(t) = \sin(\lambda_n t)$ of frequency $\lambda_n = 8$rad/s with a duration of 0.4s has occurred in the signal $f(t)$ at points $T_1 = -4.425$s and $T_2 = 0.075$s with different amplitudes. The signal to be analyzed is

$$g(t) = \begin{cases} f(t) - 1.5n(t), & \text{if } t \in [-4.425, -4.025] \\ f(t) + 2n(t), & \text{if } t \in [0.075, 0.475] \\ f(t), & \text{otherwise.} \end{cases} \qquad (6.23)$$

Figure 6.21 shows the signal $g(t)$ sampled with the rate of 100Hz in part a, along with the values of the transform $\Psi(\lambda, t_n)$, when centers $t_n(t)$ of the intervals are defined by the center of the second peak $t = 0.2750$ of the noise in b. The values of the transform of the original signal change as a sinusoidal wave and the appearance of the noise signal causes the essential change in this wave at points which are close to both locations of the noise, T_1 and T_2.

Figure 6.22 illustrates another example, when the sinusoidal signal $\sin(\lambda_n t)$ of high frequency and of a duration of 0.8s has been superposed on the original signal $f(t)$ at the time points T_1 and T_2. The centers $t_n(t)$ of the intervals $I(\lambda, t_n)$ have been calculated for $t = 0.5$. Note that in both cases the transform detects the noisy signal at its two locations. Each pike of the noise signals can be determined by the transform $\Psi(\lambda, t_n)$.

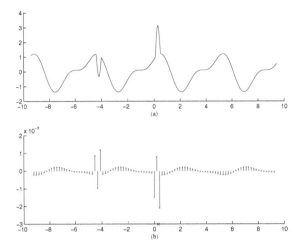

FIGURE 6.21
(a) Original signal $f(t)$ plus a noise signal of duration 0.4s and (b) the cosine wavelet transform $\Psi(\lambda, t_n)$, when $\lambda = 32\text{rad/s}$, $t_n \in (-3\pi, 3\pi)$, and $t = 0.275$.

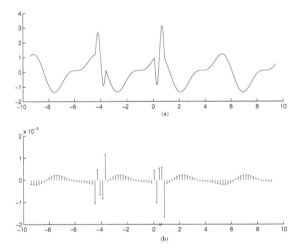

FIGURE 6.22
(a) Original signal plus two noisy signals of duration 0.8s and (b) the cosine wavelet transform $\Psi(\lambda, t_n)$ when $\lambda = 8\text{rad/s}$ and $t_n \in (-3\pi, 3\pi)$.

Figure 6.23 illustrates the transform $\Psi(\lambda, t)$ calculated for three different points $t = 0, 1, 1.5$ and 2, when the frequency λ varies in the interval $[1, 9]$. The transform is like a continuous function with respect to the frequency λ. As t increases, the wave of the transform takes the form of a sinusoidal signal with maximums that include the frequencies of the signal, 1.3 and 8rad/s.

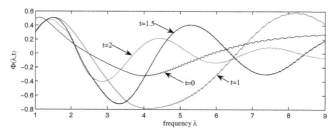

FIGURE 6.23 (See color insert following page 242.)
Transform $\Psi(\lambda, t_n)$ calculated for $n = 0$ and time points $t = 0, 1, 1.5$, and 2.

⬚

6.3.2 Cosine and sine correlation-type transforms

In this section, we consider another representation of the Fourier correlation function as well as the Fourier transform by two wavelet-like transforms. For that, we define the following transform based on the one-period cosine signal:

$$\Psi(\lambda, t) = \int_{-\infty}^{\infty} f(x)\psi_{\lambda,t}(x)dx,$$

where $(\lambda, t) \in [0, \infty) \times (-\infty, \infty)$. This is the $\psi(\lambda x)$-and-$f(x)$ correlation, or the cosine-wavelet transform, whose analyzing wavelet $\psi(x)$ has zero mean and vanishing moments of odd orders, i.e., $\int_{-\infty}^{\infty} x^n \psi(x)dx = \int_{-\pi}^{\pi} x^n \cos(x)dx = 0$, where $n = 0$ and $2k + 1$, $k \geq 0$. In the frequency domain, the function $\psi(t)$ is described by the sum of two shifted *sinc* functions

$$\hat{\psi}(\lambda) = \pi \left[\mathrm{sinc}(\pi(\lambda - 1)) + \mathrm{sinc}(\pi(\lambda + 1))\right] = \sin(\pi\lambda)\frac{2\lambda}{1 - \lambda^2}$$

which is shown in Figure 6.24. At frequency points $\lambda = \pm 1$, $\hat{\psi}(\lambda) = \pi$. It is not difficult to see that the admissibility condition holds: $C_{\hat{\psi}} < 2$.

Figure 6.25 shows surface-plot of the cosine-wavelet transform of the waveform $f(t) = \cos(1.5t) + 0.5\sin(12t - 0.25)$, versus frequency λ and location t. The frequency-time points (λ, t) are taken from the set $[0, 4] \times [-3\pi, 3\pi]$.

We now consider the partition of the time line $(-\infty, +\infty)$ by intervals I_n that begin at zero. In other words, for a given frequency $\lambda > 0$, let σ' be the following partition $\sigma' = \sigma'(\lambda) = (I_n(\lambda, 0); n = 0, \pm 1, \pm 2, \ldots)$ with centers of the intervals at the following set of time-points $c_n = c_n(\lambda) = n2\pi/\lambda$, $n = 0, \pm 1, \pm 2, \ldots$. If $\lambda = 0$, then we consider $I_0(\lambda, 0) = (-\infty, \infty)$.

The Fourier correlation function can be written as

$$F(\lambda, t) = \cos(\lambda t) \int_{-\infty}^{\infty} f(x)\cos(\lambda x)dx + \sin(\lambda t) \int_{-\infty}^{\infty} f(x)\sin(\lambda x)dx. \quad (6.24)$$

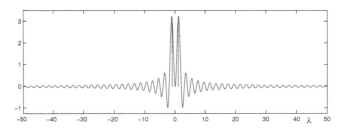

FIGURE 6.24

The Fourier transform $\hat{\psi}(\lambda)$ of the function $\psi(x)$.

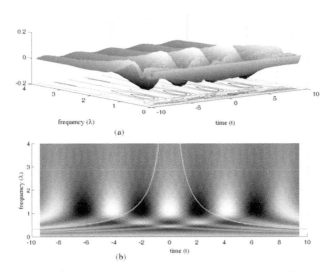

FIGURE 6.25 (See color insert following page 242.)

Cosine wavelet transform plot of function $f(t)$ in (a) 3-D view and (b) 2-D view with boundaries of set Δ_0.

Together with $\psi(x)$, we define by $\varphi(x)$ the period of the sine function ($\varphi(x) = \sin(x), x \in [-\pi, \pi]$ and 0 otherwise), and consider the family $\mathcal{B} = \{\varphi_{\lambda;t}(x) = \varphi(\lambda(x-t))\}$ of time-scale and shift transformations of this function $\varphi_{\lambda;t}(x) = \varphi(\lambda[x-t])$, where $\lambda > 0$, and $t \in (-\infty, \infty)$. Then, the correlation function $F(\lambda, t)$ can be written as follows

$$F(\lambda, t) = \cos(\lambda t) \sum_{n=-\infty}^{\infty} \Psi(\lambda, c_n) + \sin(\lambda t) \sum_{n=-\infty}^{\infty} \Phi(\lambda, c_n). \qquad (6.25)$$

$\Phi(\lambda, t)$ is the $\varphi(\lambda x)$-and-$f(x)$ correlation, or sine-wavelet transform of $f(x)$, which is defined by

$$\Phi(\lambda, t) = \int_{-\infty}^{\infty} f(x)\varphi_{\lambda,t}(x)dx, \qquad (\Phi(\lambda, \cdot) = 0).$$

In the above representation (6.25) of the Fourier correlation function, the cosine and sine wavelet transforms are calculated at specified time-points $c_n(\lambda)$. The higher the frequency λ, the denser the set of time-points $c_n(\lambda)$ used for calculations of these transforms. It is not difficult to see that the Fourier transform of the function $f(x)$ can be defined as the sum of the pair of the cosine- and sine-wavelets calculated at time points c_n,

$$\hat{f}(\lambda) = \sum_{n=-\infty}^{\infty} \left[\Psi(\lambda, c_n) - j\Phi(\lambda, c_n) \right].$$

The following simple relation holds between the Fourier correlation function and transform $F(\lambda, t) = |\hat{f}(\lambda)| \cos(\lambda t - \vartheta(\lambda))$, where $\vartheta(\lambda)$ is the phase of the Fourier transform at frequency λ. Thus, for a fixed frequency λ, the Fourier correlation function represents a cosine wave of that frequency and the amplitude is equal to the amplitude of the Fourier transform. Therefore, a large extremum of this correlation function may occur most probably at one point (or a few points) of equation $\lambda T - \vartheta(\lambda) = \pi k$, when k is an integer, or at point $T = \pi/\lambda k + \vartheta(\lambda)/\lambda$. For a high frequency, we can assume that k is even, $k = 2n$. The transform $\Psi(\lambda, t_n)$ will have thus an extremum at point $t_n = T$, as was observed in both examples shown in Figures 6.21 and 6.22.

6.3.3   Paired transform and Fourier function

In conclusion, we compare the Fourier integral function $F(\lambda, t)$ with paired representation of the Fourier transform. Let $f(x)$ be the continuous-time representation of the discrete-time signal f_n, i.e., $f(x) = \sum_{n=0}^{N-1} f_n \delta(x-n)$, where $\delta(x)$ is the delta function. Then components $f'_{p,pt}$ of the paired transform can be defined as coefficients of decomposition of function $f(x)$ by quantized cosine functions $\cos(2\pi/Np(x-t))$ considered in the set of integer time-points x of the set $\{0, 1, 2, ..., N-1\}$. Therefore, for $t = 0 : (N/(2p) - 1)$, we have

$$f'_{p,pt} = (f, q\cos_{p,pt}) = \cos_{p,pt} \circ f = \int_{-\infty}^{\infty} f(x)q\cos\left(\frac{2\pi}{N}p(x-t)\right) dx. \quad (6.26)$$

Comparing integrals in (6.26) and (6.19), one can see that the coefficients of the paired transform can be considered as a discrete integral-function $F(\lambda, t)$ calculated at frequency-time points $(2\pi p/N, t)$, where $p = 2^k$, $k = 0 : (r-1)$, and $t = 0 : (N/(2p) - 1)$. We can write $f'_{p,pt} = f'_{2^k, 2^k t} = F(\omega_k, t)$, when $t = 0 : (N/2^{k+1} - 1)$. Thus, the digitization of the Fourier formula on the discrete grid with knots at frequency-time points $(2^{k+1-r}\pi, t)$ yields the discrete paired transform. The numbering of coefficients of the paired transform is performed at frequency-time points $(2^k, 2^k t)$ of another dual grid. There is a one-to-one mapping of one grid to another. As an example, Figure 6.26 shows the grid $\mathbf{G}_{128} = \{\{(k, 2^k t); \ k = 0 : 6, \ t = 0 : 2^{6-k} - 1\}, (0, 0)\}$, for $N = 128$. Coordinates of frequencies $p = 2^k$, where $k = 0 : 6$, are plotted in the logarithmic scale as $k + 1$. One can note that this grid structure is similar to the scale indexing system which is used in the discrete wavelet transform.

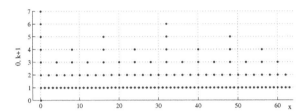

FIGURE 6.26
The grid $\mathbf{G}$ of the frequency-time points $(0, 0)$ and $(2^{k+1-r}\pi, 2^k t)$ plotted as $(k, 2^k t)$, for $N = 128$.

Figure 6.27 shows the signal f_n of length $N = 128$ in part a, along with the paired transform in b. The first three short splitting-signals are separated by dashed vertical lines. The paired transform of the signal at points of the grid

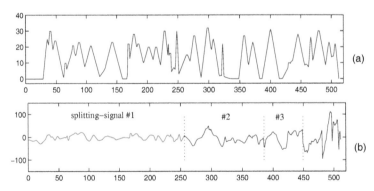

FIGURE 6.27 (See color insert following page 242.)
(a) Signal of length 128 and (b) the paired transform.

$\mathbf{G}_{128}$ is shown as a 3-D figure in Figure 6.28, and the matrix $||F(\omega_k, t)||_{k,t}$ in

b. The large value 2118.33 of the last coefficient of the transform with number $(0,0)$ has been truncated by 150.

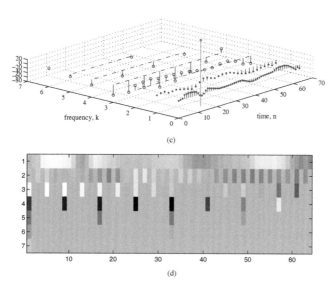

(c)

(d)

FIGURE 6.28 (See color insert following page 242.)
The paired transform of the signal of Figure 6.27.

Another example is given in Figure 6.29, where the paired transform of the cosine signal $f(t) = 2\cos(\omega_0 t) + 1.5\sin(\omega_0 t - 0.25)$, $\omega_0 = 1.3$, with 512 sampled values f_n at points uniformly placed in the interval $[-3\pi, 3\pi]$ is shown in part a, along with the 3-D view of this transform on the grid $\mathbf{G}_{512}$ in b. The paired transform of the signal at points of the grid $\mathbf{G}_{512}$ is also shown as the 3-D mesh in Figure 6.30 in part a, along with the matrix $\|F(\omega_k, t)\|_{k,t}$ in b. All splitting-signals are considered as 256-point signals, $f_{T'_{2^k}} = \{f'_{2^k,0}, f'_{2^k,1}, f'_{2^k,2}, \ldots f'_{2^k,255}\}$, $k = 0 : 8$, where components $f'_{2^k,t} = 0$, if t is not an integer multiple of 2^k. The largest value 2118.33 of the last coefficient of the transform with number $(0,0)$ has been truncated by 100.

6.4   Givens-Haar transformations

To complete our discussion of unitary transforms in discrete cases, we describe briefly a class of discrete unitary transformations which are defined by given signals and generate a motion of variable waves. These transforma-

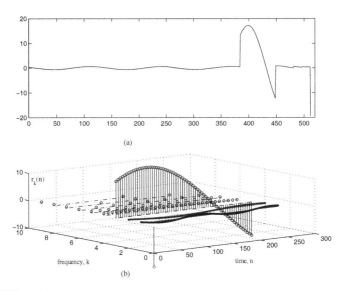

FIGURE 6.29
(a) The 512-point discrete paired transform of the sampled signal f_n. (b) The paired transform on the frequency-time grid $\mathbf{G}_{512}$.

tions, which we call *the discrete signal induced heap transformations*, DsiHT, are described in [103, 104], when the basic transformations are defined by the Givens transformations, or elementary rotations. For the DsiHT, a process of motion and transformation of one basic function into another, when starting from the wave-generator, is complicated. There are three stages which can be separated during this process. In the first stage, *the statical stage*, the generator itself is lying as the basic function. The second stage, *the evolution stage*, is related to the formation of a new wave. The last stage is *the dynamical stage*, when the newly established wave is moving to the end of the path. This wave is composed of two parts; the first part resembles the generator and the second part, or a splash, is a static wave increasing by amplitude. The vector-generator $\mathbf{x} = (x_0, x_1, x_2, ..., x_{N-1})'$ of the DsiHT and vectors, or discrete signals on which the transform is applied, are processed in the same way. For instance, the components of the generator can be proceeded by pairs in sequence $(0, 1)$, $(0, 2)$, $(0, 3)$, ..., and then $(0, N-1)$. This is a natural path P which is used also for input vectors $\mathbf{z} = (z_0, z_1, z_2, ..., z_{N-1})'$ as shown in the diagram of Figure 6.31. The transform is composed from basic transformations T_{φ_k} under some conditions. It is assumed that all parameters φ_k of these transformations are calculated by the given vector $\mathbf{x}$ and the set of constants $A = \{a_1, a_2, ..., a_{N-1}\}$. In general, the number of parameters in A is not necessarily equal to $(N-1)$ and the path of the transformation can be complex. We here describe a special class of the DsiHTs which are defined by the path

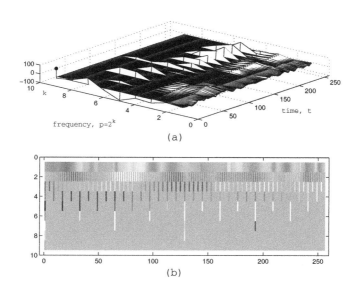

(a)

(b)

FIGURE 6.30

(a) 3-D view of the paired transform of the signal f_n at points of the dual grid.
(b) Gray scale discrete matrix plot of the discrete paired transform. Minimal
values of transform coefficients are shown in black and maximum values in
white.

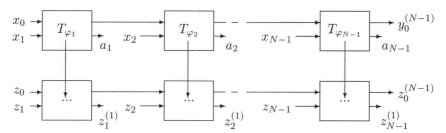

FIGURE 6.31

Signal-flow graph of composition and calculation of the transform with one
generator and a natural path.

borrowed from the Haar transformation [105]. These transformations are fast
and performed by simple rotations, can be composed for any order N, and
their complete systems of basic functions represent themselves variable waves
that are generated by signals. The 2^r-point discrete Haar transformation is
the particular case of the proposed transformations, when the generator is the
constant sequence $\{1, 1, 1, ..., 1\}$.

The composition of the DsiHT is based on the special selection of a set of
parameters which are initiated by the vector-generator through the so-called

decision equations. We dwell upon the case with two decision equations. Let $f(x, y, \varphi)$ and $g(x, y, \varphi)$ be functions of three variables; φ is referred to as the rotation parameter such as the angle, and x and y as the coordinates of a point (x, y) on the plane. These variables may have other meanings as well. It is assumed that, for a specified set of numbers a, the equation $g(x, y, \varphi) = a$ has a unique solution with respect to φ, for each point (x, y) on the plane or its chosen subset. We denote the solution of this equation by $\varphi = r(x, y, a)$. The system of equations

$$f(x, y, \varphi) = y_0, \quad g(x, y, \varphi) = a$$

is called *the system of decision equations*. The value of φ is calculated from the second equation, which we call *the angular equation*. Then, the value of y_0 is calculated from the given input (x, y) and angle φ.

Example 6.9
Given a real number a, we consider the functions $f(x, y, \varphi) = x \cos \varphi - y \sin \varphi$ and $g(x, y, \varphi) = x \sin \varphi + y \cos \varphi$. The basic transformation is defined as a rotation of the point (x, y) to the horizontal $Y = a$, $H_\varphi : (x, y) \rightarrow (x \cos \varphi - y \sin \varphi, a)$, where the rotation angle φ is calculated by

$$\varphi = \arccos \left(\frac{a}{\sqrt{x^2 + y^2}} \right) - \arctan \left(\frac{x}{y} \right),$$

$$(\varphi = \arcsin \left(\frac{a}{x} \right), \text{ if } y = 0). \tag{6.27}$$

The angular equation $g(x, y, \varphi) = x \sin \varphi + y \cos \varphi = a$ puts a constraint on the parameter a, since it is required that $a^2 \leq x^2 + y^2$. ☐

We now consider the basic transformation T defined by the simple binary orthogonal matrix 2×2,

$$\begin{bmatrix} y_0 \\ y_1 \end{bmatrix} = \frac{1}{\sqrt{2}} \begin{bmatrix} 1 & -1 \\ 1 & 1 \end{bmatrix} \begin{bmatrix} x_0 \\ x_1 \end{bmatrix}.$$

This transform corresponds to the case when the angle of rotation is $\varphi = \pi/4$, i.e., $T = T_{\pi/4}$. During the transformation, the energy of the input $\mathbf{v} = (x_0, x_1)$ is distributed between the components of the output vector $\mathbf{w} = (y_0, y_1)$ as

$$\begin{bmatrix} E[x_0] = x_0^2 \\ E[x_1] = x_1^2 \end{bmatrix} \rightarrow \begin{bmatrix} E[y_0] = \frac{x_0^2 + x_1^2}{2} - x_0 x_1 \\ E[y_1] = \frac{x_0^2 + x_1^2}{2} + x_0 x_1 \end{bmatrix}.$$

Thus, less or more than half of the energy of the input is distributed between y_0 and y_1, depending on the signs of the input components. If the input is such that $x_0 x_1 > 0$, then the first component of the output will receive less energy than the second component, and vice versa. We now consider the application of the basic transformations $T_{\pi/4}$ for computing a unitary transform with a few heaps. For that, we first analyze the recurrent algorithm of the discrete Haar transform in terms of the basic transformations. Let $\mathbf{x} = (x_0, x_1, x_2, ..., x_{N-1})$ be a vector-signal to be processed sequentially by transformations $T_{\pi/4}$. The process starts with the first two components x_0 and x_1

$$\begin{bmatrix} y_0 \\ y_1 \end{bmatrix} = \frac{1}{\sqrt{2}} \begin{bmatrix} 1 & -1 \\ 1 & 1 \end{bmatrix} \begin{bmatrix} x_0 \\ x_1 \end{bmatrix}.$$

The value of y_0 is considered to be the first heap. The next stage is not connected with this heap, and calculations are continued as

$$\begin{bmatrix} y_2 \\ y_3 \end{bmatrix} = \frac{1}{\sqrt{2}} \begin{bmatrix} 1 & -1 \\ 1 & 1 \end{bmatrix} \begin{bmatrix} x_2 \\ x_3 \end{bmatrix}.$$

The value of y_2 is considered to be the second heap. The next stage is not connected with these two heaps, as well as the following steps, when the calculations are performed as follows:

$$\begin{bmatrix} y_{2k} \\ y_{2k+1} \end{bmatrix} = \frac{1}{\sqrt{2}} \begin{bmatrix} 1 & -1 \\ 1 & 1 \end{bmatrix} \begin{bmatrix} x_{2k} \\ x_{2k+1} \end{bmatrix}, \quad k = 2, ..., (N/2 - 1),$$

and y_{2k} is recorded as the kth heap. As a result, we obtain the following N-dimensional vector with $N/2$ heaps:

$$\mathbf{y}^{(1)} = (\mathbf{y}_1^{(1)}, \mathbf{y}_2^{(1)}) = \Big(\underbrace{y_0, y_2, y_4, ..., y_{N-2}}_{N/2 \text{ heaps}}, y_1, y_3, y_5, ..., y_{N-1} \Big).$$

The process of calculation of the Haar transform is continued, but the second part $\mathbf{y}_2^{(1)}$ is not used for further calculations. $N/4$ transformations $T_{\pi/4}$ are applied over the first half of the obtained vector, in other words, over the heaps $\mathbf{y}_1^{(1)} = (y_0, y_2, y_4, ..., y_{N-2})$. As a result, other $N/4$ heaps are calculated in the output

$$\mathbf{y}_1^{(1)} \to \mathbf{y}^{(2)} = (\mathbf{y}_1^{(2)}, \mathbf{y}_2^{(2)}) = \Big(\underbrace{y_0^{(1)}, y_2^{(1)}, ..., y_{N/2-2}^{(1)}}_{N/4 \text{ heaps}}, y_1^{(1)}, y_3^{(1)}, ..., y_{N/2-1}^{(1)} \Big).$$

Continuing this process $(\log_2 N - 2)$ times more, until only one heap remains, we obtain the traditional N-point discrete Haar transform of the vector $\mathbf{x}$

$$\mathbf{x} \to H[\mathbf{x}] = \Big(y_1^{(r)}, y_2^{(r)}, y_2^{(r-1)}, ..., y_2^{(3)}, y_2^{(2)}, y_2^{(1)} \Big).$$

6.4.1   Fast transforms with Haar path

By using the above described approach for calculating different heaps, we can define a unitary transformation $\mathcal{H}$ induced by a given vector $\mathbf{x} = (x_0, x_1, x_2, ..., x_{N-1})$, when the basic transformation T_{φ_k} is generated and then used instead of $T_{\pi/4}$. The path for $\mathcal{H}$ is the same as for the Haar transformation. The process starts with the first two components (x_0, x_1)

$$\begin{bmatrix} y_0 \\ 0 \end{bmatrix} = \begin{bmatrix} \cos \varphi_1 & -\sin \varphi_1 \\ \sin \varphi_1 & \cos \varphi_1 \end{bmatrix} \begin{bmatrix} x_0 \\ x_1 \end{bmatrix}.$$

The angle φ_1 is calculated by $\varphi_1 = -\arctan(x_0/x_1)$ (or $\varphi_1 = \pi/2$, if $x_1 = 0$), and the value of y_0 is referred to as the first heap. The next heap is calculated by

$$\begin{bmatrix} y_2 \\ 0 \end{bmatrix} = \begin{bmatrix} \cos \varphi_2 & -\sin \varphi_2 \\ \sin \varphi_2 & \cos \varphi_2 \end{bmatrix} \begin{bmatrix} x_2 \\ x_3 \end{bmatrix}$$

where $\varphi_2 = -\arctan(x_2/x_3)$. The process is continued similarly and the kth heap y_{2k} is calculated by

$$\begin{bmatrix} y_{2k} \\ 0 \end{bmatrix} = \begin{bmatrix} \cos \varphi_{k+1} & -\sin \varphi_{k+1} \\ \sin \varphi_{k+1} & \cos \varphi_{k+1} \end{bmatrix} \begin{bmatrix} x_{2k} \\ x_{2k+1} \end{bmatrix}, \quad k = 2, ..., (N/2 - 1),$$

where $\varphi_2 = -\arctan(x_{2k}/x_{2k+1})$. As a result, we obtain the following new vector with $N/2$ heaps,

$$\mathbf{y}^{(1)} = \Big(\underbrace{y_0, y_2, y_4, ..., y_{N-2}}_{N/2 \text{ heaps}}, 0, 0, 0, ..., 0 \Big).$$

The process of calculation of the transform is continued, and new transformations T_{ψ_n}, $n = 1 : (N/2 - 1)$, are generated by the obtained heaps, $\mathbf{x}^{(1)} = (y_0, y_2, y_4, ..., y_{N-2})$. The calculation results in $N/4$ new heaps

$$\mathbf{y}^{(2)} = \Big(\underbrace{y_0^{(1)}, y_2^{(1)}, ..., y_{N/2-2}^{(1)}}_{N/4 \text{ heaps}}, 0, 0, ..., 0 \Big)$$

and more $N/8$ new heaps on the following stage and so on, until we obtain only one heap. The transformation thus is generated by $N/2 + N/4 + ... + 2 + 1 = N - 1$ basic transformations $\{\{T_{\varphi_k}\}, \{T_{\psi_n}\}, ...\}$. This transformation is called *the discrete Givens-Haar transformation* (DGHT) generated by the vector $\mathbf{x}$.

When applying this transformation to a vector $\mathbf{z} = (z_0, z_1, z_2, ..., z_{N-1})'$, the calculations are performed on each stage, in accordance with the composition of the DGHT. On the first stage, all basic transformations T_{φ_k}, $k = 1 : N/2$, are applied as follows

$$\begin{bmatrix} t_{2(k-1)} \\ t_{2k-1} \end{bmatrix} = \begin{bmatrix} \cos \varphi_k & -\sin \varphi_k \\ \sin \varphi_k & \cos \varphi_k \end{bmatrix} \begin{bmatrix} z_{2(k-1)} \\ z_{2k-1} \end{bmatrix}.$$

Then, the first part of the result

$$\mathbf{t}^{(1)} = \left(\underbrace{t_0, t_2, t_4, ..., t_{N-2}}, t_1, t_3, t_5, ..., t_{N-1} \right)$$

is processed similarly by the basic transformations T_{ψ_n}, $n = 1 : N/4$. As a result, other $N/4$ heaps are calculated in the output. On this stage, the input $\mathbf{z}$ is transformed as

$$\mathbf{z} \to \left(\underbrace{t_0^{(1)}, t_2^{(1)}, ..., t_{N/2-2}^{(1)}, t_1^{(1)}, t_3^{(1)}, ..., t_{N/2-1}^{(1)}}, t_1, t_3, t_5, ..., t_{N-1} \right).$$

The process of calculation of the DGHT of $\mathbf{z}$ is continued ($\log_2 N - 2$) times more, and for each stage of calculation, the transform is performed over the half of the output of the transform which has been obtained on the previous stage. Thus, the 2^r-point DGHT of $\mathbf{z}$, when $r > 3$, is defined as

$$H[\mathbf{z}] = \left(\underbrace{\underbrace{\underbrace{t_0^{(r)}, ..., t_{N/16-1}^{(4)}, t_1^{(3)}, ..., y_{N/8-1}^{(3)}, t_1^{(2)}, t_3^{(2)}, ..., t_{N/4-1}^{(2)}, t_1^{(1)}, t_3^{(1)}, ..., t_{N/2-1}^{(1)}}}}, \right.$$

$$\left. t_1, t_3, ..., t_{N-1} \right).$$

As examples, consider the following three matrices of the four-point DGHTs generated by vectors $\mathbf{x} = (1, 1, 1, 1)'$, $(-1, 1, -1, 1)'$, and $(1, 2, 2, 1)'$:

$$\mathbf{H}_1 = \mathbf{H}_{[1,1,1,1]} = \begin{bmatrix} 0.5000 & 0.5000 & 0.5000 & 0.5000 \\ -0.5000 & -0.5000 & 0.5000 & 0.5000 \\ -0.7071 & 0.7071 & 0 & 0 \\ 0 & 0 & -0.7071 & 0.7071 \end{bmatrix},$$

$$\mathbf{H}_2 = \mathbf{H}_{[-1,1,-1,1]} = \begin{bmatrix} 0.5000 & -0.5000 & 0.5000 & -0.5000 \\ -0.5000 & 0.5000 & 0.5000 & -0.5000 \\ 0.7071 & 0.7071 & 0 & 0 \\ 0 & 0 & 0.7071 & 0.7071 \end{bmatrix},$$

$$\mathbf{H}_3 = \mathbf{H}_{[1,2,2,1]} = \begin{bmatrix} 0.3162 & 0.6325 & 0.6325 & 0.3162 \\ -0.3162 & -0.6325 & 0.6325 & 0.3162 \\ -0.8944 & 0.4472 & 0 & 0 \\ 0 & 0 & -0.4472 & 0.8944 \end{bmatrix}.$$

The matrix $\mathbf{H}_1$ coincides with the matrix of the discrete Haar transformation. It is interesting to note that the first basic functions of these three matrices are equal to the corresponding normalized generators with sign ± 1, depending on the sign of x_1. This property holds in the general $N > 4$ case, too. For instance, we consider the generator $\mathbf{x} = (2, 1, 1, 3, 2, 1, 3, 2)$. The matrix $\mathbf{H}$ of the eight-point DGHT generated by this vector can be written in the form $\mathbf{H} = \mathbf{DM}$, where $\mathbf{M}$ is the integer orthogonal matrix

$$\mathbf{M} = \begin{bmatrix} 2 & 1 & 1 & 3 & 2 & 1 & 3 & 2 \\ 12 & 6 & 6 & 18 & -10 & -5 & -15 & -10 \\ 4 & 2 & -1 & -3 & 0 & 0 & 0 & 0 \\ 0 & 0 & 0 & 0 & -26 & -13 & 15 & 10 \\ 1 & -2 & 0 & 0 & 0 & 0 & 0 & 0 \\ 0 & 0 & -3 & 1 & 0 & 0 & 0 & 0 \\ 0 & 0 & 0 & 0 & 1 & -2 & 0 & 0 \\ 0 & 0 & 0 & 0 & 0 & 0 & 2 & -3 \end{bmatrix},$$

and the diagonal matrix $\mathbf{D} = \mathrm{diag}\{0.1741, -0.0318, -0.1826, 0.0292, -0.4472, 0.3162, -0.4472, -0.2774\}$.

The sixteen basic functions of the 16-point DGHT are given in Figure 6.32(a), when the generator is the sampled cosine function defined at 16 equidistant points in the interval $[0, \pi]$. For comparison, the basic functions of the 16-point DHT are also shown in part b. Inside each series, the basic functions of the Haar transforms are the exact shifted waves. The waves of the DGHT are referred to as moving waves which change during the movement.

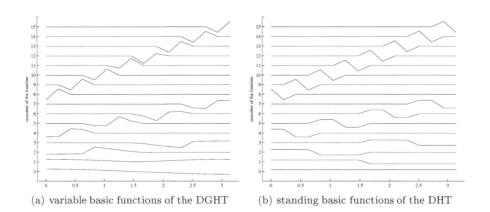

(a) variable basic functions of the DGHT   (b) standing basic functions of the DHT

FIGURE 6.32
(a) The basic functions of the 16-point DGHT generated by the sampled cosine wave $\cos(t)$, $t \in [0, \pi]$, and (b) the basic functions of the 16-point DHT.

We now illustrate these transforms. Figure 6.33 shows the discrete signal $\mathbf{x}$ of length 512 in part a, along with its Givens-Haar transform in b. The vector-generator $\mathbf{x}$ is the sampled cosine wave $\cos(t)$ in the time interval $[0, 4\pi]$. For comparison, the discrete Haar transform of $\mathbf{x}$ is shown in c. These two transforms have the same range of amplitude and vary similarly.

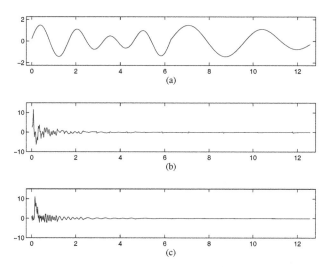

FIGURE 6.33
(a) The original signal of length 512, (b) the DGHT of the signal, when the generator is the sampled cosine wave $\cos(t)$, $t \in [0, 4\pi]$, and (c) the Haar transform of the signal.

The concept of the DGHT can be extended to any order N of the transform. Indeed, there are many ways to compose a few heaps by reducing the process of transform composition to the $N = 2^r$ case. For instance, when $N = 5$, we can propose the following process:

$$\mathcal{H}_5 : \{x_0, x_1, x_2, \underbrace{x_3, x_4}\} \xrightarrow{I_3 \oplus H_2} \{x_0, x_1, x_2, \underbrace{y_0^{(3)}, y_1^{(3)}}\} \xrightarrow{H_4 \oplus 1} \{y_0, y_2, y_1, y_3, y_1^{(3)}\}.$$

The last two components are rotated first, and then the first three components together with the obtained heap $y_0^{(3)}$ are processed as for the four-point DGHT. We here denote by I_m the unique diagonal matrix $(m \times m)$. The orthogonal matrix of this five-point DGHT, when the generator is $(1, 1, 1, 1, 1)$, equals

$$\mathbf{H}_5 = (\mathbf{H}_4 \oplus 1)(\mathbf{I}_3 \oplus \mathbf{H}_2) = \begin{bmatrix} \frac{1}{2} & \frac{1}{2} & \frac{1}{2} & \frac{1}{2\sqrt{2}} & \frac{1}{2\sqrt{2}} \\ -\frac{1}{2} & -\frac{1}{2} & \frac{1}{2} & \frac{1}{2\sqrt{2}} & \frac{1}{2\sqrt{2}} \\ -\frac{1}{\sqrt{2}} & \frac{1}{\sqrt{2}} & 0 & 0 & 0 \\ 0 & 0 & -\frac{1}{\sqrt{2}} & \frac{1}{2} & \frac{1}{2} \\ 0 & 0 & 0 & -\frac{1}{\sqrt{2}} & \frac{1}{\sqrt{2}} \end{bmatrix}.$$

In the $N = 6$ case, we can first process the last four components of the input similarly to the four-point DGHT, and then the first two components of the input with two obtained heaps $y_0^{(2)}$ and $y_2^{(2)}$ as follows:

$$\mathcal{H}_6 : \{x_0, x_1, \underbrace{x_2, x_3, x_4, x_5}\} \xrightarrow{I_2 \oplus H_4} \{x_0, x_1, y_0^{(2)}, y_2^{(2)}, y_1^{(2)}, y_3^{(2)}\} \rightarrow$$

$$\xrightarrow{H_4 \oplus I_2} \{y_0, y_2, y_1, y_3, y_1^{(2)}, y_3^{(2)}\}.$$

The orthogonal matrix of this 6-point DGHT, when the generator is $[1, 1, 1, 1, 1, 1]$, equals

$$\mathbf{H}_6 = (\mathbf{H}_4 \oplus \mathbf{I}_2)(\mathbf{I}_2 \oplus \mathbf{H}_4) = \begin{bmatrix} \frac{1}{2} & \frac{1}{2} & 0 & 0 & \frac{1}{2} & \frac{1}{2} \\ -\frac{1}{2} & -\frac{1}{2} & 0 & 0 & \frac{1}{2} & \frac{1}{2} \\ -\frac{1}{\sqrt{2}} & \frac{1}{\sqrt{2}} & 0 & 0 & 0 & 0 \\ 0 & 0 & -\frac{1}{\sqrt{2}} & -\frac{1}{\sqrt{2}} & 0 & 0 \\ 0 & 0 & -\frac{1}{\sqrt{2}} & \frac{1}{\sqrt{2}} & 0 & 0 \\ 0 & 0 & 0 & 0 & -\frac{1}{\sqrt{2}} & \frac{1}{\sqrt{2}} \end{bmatrix}.$$

The seven-point DGHT of the vector $\mathbf{x}$ can be defined as the eight-point DGHT of the extended vector $[\mathbf{x}, 0]$.

$$\mathbf{H}_7 = \begin{bmatrix} 0.3780 & 0.3780 & 0.3780 & 0.3780 & 0.3780 & 0.3780 & 0.3780 \\ -0.3273 & -0.3273 & -0.3273 & -0.3273 & 0.4364 & 0.4364 & 0.4364 \\ -0.5000 & -0.5000 & 0.5000 & 0.5000 & 0 & 0 & 0 \\ 0 & 0 & 0 & 0 & -0.4082 & -0.4082 & 0.8165 \\ -0.7071 & 0.7071 & 0 & 0 & 0 & 0 & 0 \\ 0 & 0 & -0.7071 & 0.7071 & 0 & 0 & 0 \\ 0 & 0 & 0 & 0 & -0.7071 & 0.7071 & 0 \end{bmatrix}.$$

Such a transformation can also be composed as $\mathbf{H}_7 = (\mathbf{H}_6 \oplus 1)(\mathbf{I}_5 \oplus \mathbf{H}_2)$, or $\mathbf{H}_7 = (\mathbf{H}_5 \oplus \mathbf{I}_2)(\mathbf{I}_3 \oplus \mathbf{H}_4)$.

6.4.2 Experimental results

The above considered examples illustrate that the discrete Givens-Haar transforms generated by vectors may have effective applications, as the traditional Haar transform. Such transforms can be used, for instance, for signal compression. Here the generator plays the essential role, and it is expected that many generators may lead to effective compression, even better than the Haar transform does. As an example, we consider the discrete signal $\mathbf{z}$ of length 512, which is given in Figure 6.34a. This signal is composed from random triangles. The vector-generator is the sampled cosine wave $\cos(t)$ calculated at 512 equidistant time-points of the interval $[0, 4\pi]$. The values of means of the DGHT and DHT equal respectively 1.1038 and 0.6497, and the variances 15.9170 and 15.9420. The DHT provides the small mean, but big variance.

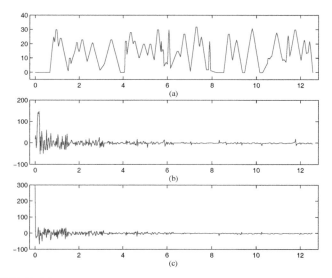

FIGURE 6.34
(a) Original signal, (b) the DGHT of the signal (the generator is the sampled cosine wave $\cos(t)$, $t \in [0, 4\pi]$), and (c) the Haar transform of the signal.

We here consider the simple method of signal compression, when truncating (or filtering) a certain number L of the last coefficients of the transform. Figure 6.35 shows the 512-point DGHT of the signal with the last 256 truncated to zero coefficients in part a, along with the 512-point DHT with the same number of zero coefficients in b, and the reconstructions of the signal by their inverse transforms in c and d, respectively. The mean-square-root error (MSR) for the reconstruction of the signal by the DGHT equals 0.0593, and 0.0594 by the DHT.

We now consider the $L = 384$ case, or 75% of compression. Figure 6.36 shows the 512-point DGHT of the signal with the last 384 zero coefficients in part a, along with the similarly filtered 512-point DHT in b, and the reconstructions of the signal by the inverse DGHT and DHT in c and d, respectively. The MSR error for signal reconstruction by the DGHT equals 0.1158, and 0.1164 for the DHT.

In both $L = 256$ and 384 cases, the compression by truncating the DGHT coefficients provides better results, when compared with the DHT. The MSR errors of approximation, when filtering L last coefficients of these transforms, when $L = 256 : 480$, are shown in Figure 6.37 part a, and separately for $L = 364 : 405$ in b. It can be seen that these two error curves are almost the same for $L = 256 : 384$, but they differ in the next interval $[385, 480]$, where the DGHT leads to a better approximation of the signal than the DHT.

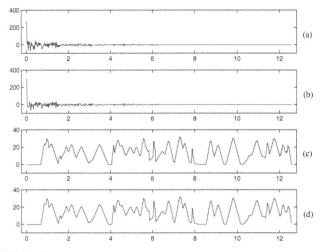

FIGURE 6.35
(a) The filtered DGHT of the signal, (b) the filtered DHT of the signal, (c) the inverse DGHT, and (d) the inverse DHT, when truncating 50% of the coefficients of these transforms.

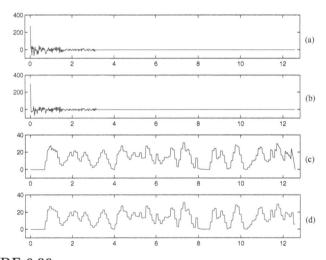

FIGURE 6.36
(a) The filtered DGHT of the signal, (b) the filtered DHT of the signal, (c) the inverse DGHT and (d) the inverse DHT, when truncating 75% of the coefficients of these transforms.

6.4.3 Characteristics of basic waves

We consider briefly the main characteristics of the basic functions of the Givens-Haar transformations. The complete system of basic functions of the

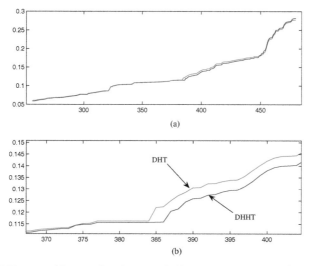

FIGURE 6.37 (See color insert following page 242.)
(a) The mean-square-root error curves for signal reconstruction after truncating L coefficients of the 512-point DGHT and DHT, where $L \in [256, 480]$,
and (b) the same errors in the interval $(365, 405)$.

Givens-Haar transformation can be described in the form of moving waves
originated from the generator. Given $N > 1$, we consider the matrix of the
N-point discrete DGHT as a set of moving waves

$$
\mathbf{H} = \begin{bmatrix} x(n) \\ h_{t=1}(n) \\ h_{t=2}(n) \\ ... \\ h_{t=N-1}(n) \end{bmatrix}, \tag{6.28}
$$

where $x(n)$, $n = 0 : (N - 1)$, are normalized components of the vector-
generator $\mathbf{x}$.

Let $\mathbf{z} = (z_0, z_1, ..., z_{N-1})'$ be a real vector, which is defined at time-points
$t_0, t_1, ..., t_{N-1}$ of an interval $[1, b]$, say $[t_0, t_{N-1}]$. We assume that $||\mathbf{z}|| = 1$.
Then, the numbers $p_k = z_k^2$ can be considered as probabilities of z_k, $k = 0 :$
$(N - 1)$. Each basic function, or row $m_n(k)$ of the matrix of the DGHT, is
referred to as a wave moving in the field of the generator $\mathbf{x}$ of this transform.
Therefore, we can apply for these waves the concepts of the centroid and mean
width of the wave,

$$
\bar{m}_n = \sum_{k=0}^{N-1} t_k p_k = \sum_{k=0}^{N-1} t_k m_n^2(k), \qquad d(m_n) = \sqrt{\sum_{k=0}^{N-1} (t_k - \bar{m}_n)^2 m_n^2(k)},
$$

where $n = 0 : (N-1)$. As an example, we first consider the $N = 2^r = 32$ case, and the generator $(1, 1, ..., 1)$, which leads to the Haar transform. Figure 6.38 shows the centroids of the basic waves of the 32-point DHT in part a, along with the mean widths of these waves in b. For waves of numbers 17 through

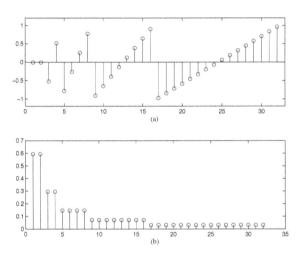

(a)

(b)

FIGURE 6.38
(a) Centroids of waves of the 32-point DHT and (b) mean widths of the waves.

32, the centroid moves uniformly with a step $\Delta = 1/8$ from left to right. Indeed, the centroids for this DHT are defined as

$$\bar{m}_n = \sum_{k=2n,2n+1} t_k m_n^2(k) = (t_{2n}+t_{2n+1})\frac{1}{2} = \frac{1}{2^r}(2n+2n+1)\frac{1}{2} = \frac{1}{2^{r-1}}n + \frac{1}{2^{r+1}},$$

and therefore $\Delta = \bar{m}_n - \bar{m}_{n-1} = 1/2^{r-1}$, for $n = 1 : (N-1)$.

For the basic waves of numbers 9 to 16, 5 to 8, and 3 to 4, centroids move with steps 2Δ, 4Δ, and 8Δ, respectively. The step of movement of centroids of waves of one series thus increases twice when compared with the previous series of waves. The first two waves are immovable. This movement of centroids of the waves explains why the Haar basic functions (as well as basic functions of other following wavelet transforms) are defined in the form $\psi(t) = \psi((t - b)/a)$, $b = \Delta, 2\Delta, 4\Delta, ...$, where the value of the constant-parameter a is connected with the width of the wave. As an illustration, Figure 6.38(b) shows widths of waves of each series, which are values of a.

We consider the generator being the Gaussian function $g(t) = \exp(-t^2/\sigma^2)$ sampled at 32 equidistant points in the time-interval $[-1, 1]$. The value of variance equals $\sigma^2 = 1/16$. Figure 6.39 shows the centroids of the basic waves of the 32-point DGHT in part a, along with the mean widths of these waves in b. We see the same picture of the centroids of the waves as for the Haar

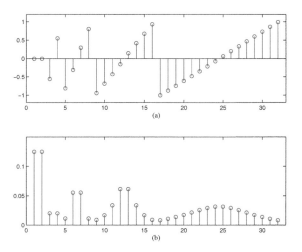

FIGURE 6.39
(a) Centroids of the basic waves of the 32-point DGHT generated by the
Gaussian vector with zero mean and variance one, and (b) mean widths of
the waves.

transform, but the widths of these waves change and have a form similar to
the Gaussian function. This fact is also reflected in the wavelet theory, when
instead of the function $\psi(t)$ defined above, the modified function is taken,
$\psi(t) = k(a)\psi((t-b)/a)$, $b = \Delta, 2\Delta, 4\Delta, \dots$.

If we consider other generators, we will have a similar picture of waves. For
instance, let the generator be the cosine wave $x(t) = \cos(8t)$ sampled at 32
equidistant points in the time-interval $[-1, 1]$. Figure 6.40 shows the centroids
of the basic waves of the 32-point DGHT in part a, along with the mean
widths of these waves in b. The widths of the waves in this scale also change
with accordance with the generator, but not the centroids.

In the case of large N, we observe similar results. As an example, Figure 6.41
shows the centroids and mean widths of the basic waves of two 128-point
DGHTs. The first transform has been generated by the Gaussian function
$g(t)$ sampled at 128 equidistant points in the time-interval $[-1, 1]$. The second
DGHT has been generated by the cosine function $\cos(8t)$ sampled at the
same 128 time-points. In part a, the centroids of waves of these transforms
are plotted together. The mean widths of the waves of the first and second
DGHTs are shown in b and c, respectively. The centroids of basic waves for
both transforms are almost the same, but the widths are very different as
their corresponding generators.

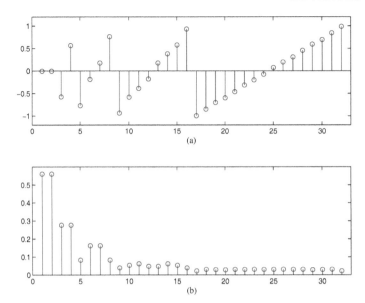

FIGURE 6.40

(a) Centroids of waves of the 32-point DGHT generated by cosine wave, and
(b) mean widths of the waves.

6.4.4   Givens-Haar transforms of any order

It was shown above that the order of the Givens-Haar transformation does
not tie with the power of two N. In general, we can divide the sequence of
length N by parts, for instance two parts of integer length $[N/2]$. If there is
a component that lies beyond these parts, we will not process it on the first
stage of the transformation, but use it later. As an example, we can use the
following MATLAB-based codes for computing the matrices of the N-point
discrete Givens-Haar transform, for any integer $N > 1$.

```
% ---------------------------------------------------------------
% call: GivenHaar_transforms.m (library of codes of Grigoryans)
% to compute the N-point Givens-Haar transform and its matrix
  function y=mer_ghaar(t)
    N=length(t);
    a=sqrt(2);
    if N==1
        y=t;
    else
    y=zeros(1,N);
    sign_mod=mod(N,2);
    t1=t(1+sign_mod:2:N);
    t2=t(2+sign_mod:2:N);
    for i1=1:(N-sign_mod)/2
```

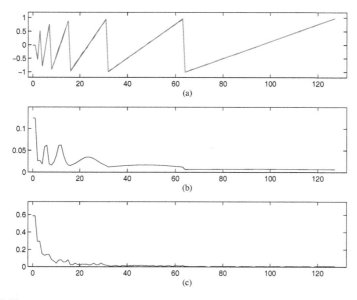

FIGURE 6.41

(a) Centroids of waves of the 128-point DGHT generated by Gaussian and cosine waves, and mean widths of the waves of the DGHT generated by (b) Gaussian and (c) cosine functions.

```
        bi1=t1(i1)-t2(i1);
        b1=t1(i1)+t2(i1);
        t1(i1)=b1/a;
        t2(i1)=bi1/a;
    end;
    if sign_mod
        t1=[t(1) t1];
    end
        y=[mer_ghaar(t1) t2];
    end;
% ------------------------------
  function T=mat_GH(N)
    T=zeros(N);
    for i1=1:N
        y=zeros(1,N);
        y(i1)=1;
        a=mer_ghaar(y);
        T(:,i1)=a(:);
    end;
% -----------------------------------------------------------
```

According to this code, the matrix of the five-point Givens-Haar transformation equals

$$H_5 = \begin{bmatrix} \frac{1}{\sqrt{2}} & \frac{1}{2\sqrt{2}} & \frac{1}{2\sqrt{2}} & \frac{1}{2\sqrt{2}} & \frac{1}{2\sqrt{2}} \\ \frac{1}{\sqrt{2}} & -\frac{1}{2\sqrt{2}} & \frac{1}{2\sqrt{2}} & -\frac{1}{2\sqrt{2}} & \frac{1}{2\sqrt{2}} \\ 0 & \frac{1}{2} & \frac{1}{2} & -\frac{1}{2} & -\frac{1}{2} \\ 0 & \frac{1}{\sqrt{2}} & -\frac{1}{\sqrt{2}} & 0 & 0 \\ 0 & 0 & 0 & \frac{1}{\sqrt{2}} & -\frac{1}{\sqrt{2}} \end{bmatrix},$$

and the matrix of the six-point Givens-Haar transformation equals

$$H_6 = \begin{bmatrix} \frac{1}{2} & \frac{1}{2} & \frac{1}{2\sqrt{2}} & \frac{1}{2\sqrt{2}} & \frac{1}{2\sqrt{2}} & \frac{1}{2\sqrt{2}} \\ \frac{1}{2} & \frac{1}{2} & -\frac{1}{2\sqrt{2}} & -\frac{1}{2\sqrt{2}} & -\frac{1}{2\sqrt{2}} & -\frac{1}{2\sqrt{2}} \\ 0 & 0 & \frac{1}{2} & \frac{1}{2} & -\frac{1}{2} & -\frac{1}{2} \\ \frac{1}{\sqrt{2}} & -\frac{1}{\sqrt{2}} & 0 & 0 & 0 & 0 \\ 0 & 0 & \frac{1}{\sqrt{2}} & -\frac{1}{\sqrt{2}} & 0 & 0 \\ 0 & 0 & 0 & 0 & \frac{1}{\sqrt{2}} & -\frac{1}{\sqrt{2}} \end{bmatrix}.$$

The determinants of these matrices equal $\det(H_5) = -1$ and $\det(H_6) = 1$.

Figure 6.42 shows the gray-scale images of the Givens-Haar transformations for $N = 33$ and 52 in parts a and b, respectively.

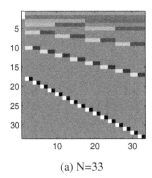
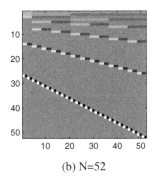

(a) N=33   (b) N=52

FIGURE 6.42
Matrices of the (a) 33-point DGHT and (b) 52-point DGHT.

It is interesting to know if the complete system of functions of the N-point DGHT has the same structure as in the case of the transforms of orders N equal powers of two. Namely, we would like to see if the centroids and the mean widths of the basic-waves of the DGHTs change similarly to the 2^r-point Givens-Haar transforms. As an example, we consider the $N = 33$ case, which is expected to be similar to the $N = 32$ case. We can notice from Figure 6.43 that the the centroids and mean widths of the basic-waves of the 33-point DGHT do change almost as in Figure 6.40.

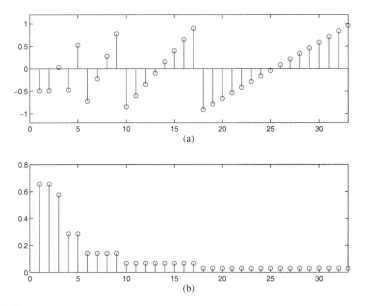

FIGURE 6.43

(a) Centroids and (b) mean widths of the waves of the 33-point DGHT.

Thus the movement of the basic-waves of the DGHT in the $N = 33$ case changes a little when compared with the $N = 32$ case. We observe a similar property even in the case when N differs much from the power of two. As an example, Figure 6.44 shows the centroids of basic-waves of the 52-point DGHT in part a, along with the mean widths of these waves in b.

Below is a simple example of MATLAB-based code for computing the centers and widths of the waves of the N-point discrete Givens-Haar transform, for any integer $N > 1$.

```
% ----------------------------------------------------------------
% call: waveGH_property.m (library of codes of Grigoryans)
% to compute and plot the centers and widths of the waves of DGHT
    N=33;
    t=linspace(-1,1,N);
    H1=mat_GH(N);
    m_center=zeros(1,N);
    d_width=zeros(1,N);
    for n=1:N
        y=H1(n,:);
        f=y.*y;        % the density function
        m_center(n)=t*f';
        x=t-m_center(n);
        x2=x.*x;
        d_width(n)=sqrt(sum(f.*x2));
    end;
```

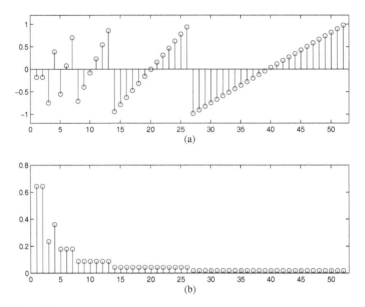

FIGURE 6.44
(a) Centroids and (b) mean widths of the waves of the 52-point DGHT.

```
% to plot the graphs:
figure;
subplot(2,1,1);
stem(m_center);
subplot(2,1,2);
stem(d_width);
axis([0,33,0,.8]);
% --------------------------------------------------
```

Thus, we described the class of the discrete unitary transformations that use the Haar path but are generated by any discrete signal. This class of Givens-Haar transformations generalizes the concept of the Haar transformation, which was used in wavelet theory to compose unitary transforms but in the space of continuous-time functions. The main differences of the discrete Haar transformation, or wavelet transformations, and the Givens-Haar transformations are in the following:

1. In wavelet theory, the functions $k(a)\psi((t-b)/a)$ are used, which are referred to as plane waves. In the case of the Givens-Haar transforms, the decomposition of the signal and its reconstruction are performed by functions $k(a,t)\psi((t-b)/a)$, which are not planar waves. The system of such functions is generated by inputs.

2. The discrete Givens-Haar transforms are unitary; they are generated by real generators of any length, and complex generators as well. In the wavelet theory, the problem of developing the unitary discrete transforms has not been

solved.

3. The discrete Givens-Haar transformations are fast and can be used together with the traditional DHT in different areas of signal and image processing. The method of selection of generators plays an important role in the application of the proposed Givens-Haar transforms. Different generators will allow for tuning the proposed transforms for better processing different classes of signals and images as well. Therefore, the method of finding the best generators is considered to be the next stage of development of the theory of Givens-Haar transformations.

We have presented above only one subclass of discrete signal-induced heap transformations, DsiHT, which are defined by the Haar path and by only one generator. There are many other interesting paths that can be used as well, for instance, the path of the Hadamard transformation or paired transformation, paths which describe the processes of synthesis and decay, whirlwind and chain reaction, and so on. The DsiHT can be generated by rotations or other basic transformations in 2-D and 3-D space with two and more generators. All these subclasses of one-, two-, and multidimensional DsiHTs are unitary and they are characterized by specific and in many cases complex forms of movement and interaction of the basic waves.

Problems

Problem 6.1 Show that the operation of the separation of the even part of the function is (a) the particular case of the mixed Fourier transformation and (b) singular and find the matrix $\mathbf{A}$ of this transformation.

Problem 6.2 Show that the operation of the separation of the odd part of the function is (a) the particular case of the mixed Fourier transformation and (b) singular and find the matrix $\mathbf{A}$ of this transformation.

Problem 6.3 Show that the following transformation of the functions:

$$f(t) \rightarrow \frac{1}{\sqrt{2\pi}} \int f(t) \cos(tp) dt$$

is (a) the particular case of the mixed Fourier transform and (b) singular.

Problem 6.4 Show that the following transformation of the functions:

$$f(t) \rightarrow \frac{1}{\sqrt{2\pi}} \int f(t) \sin(tp) dt$$

is (a) the particular case of the mixed Fourier transform and (b) singular.

Problem 6.5 Compute the A-wavelet transform of the signal $f(t) = 2\cos(2t) + \sin(3t)$.

Problem 6.6 Compute the S-wavelet transform of the signal $f(t) = 2\cos(1.3t)$.

Problem 6.7 Compute the B-wavelet transform of the signal $f(t) = \cos(\pi t/4)$.

Problem 6.8 It is interesting to note that by changing the central point t of the intervals $I_n(\lambda, t)$, we can observe that the main change in the transform $\Psi(\lambda, t, n)$ occurs at points of discontinuity of the signal, i.e., at locations of peaks. To show that, consider the signal defined in (6.23) and compute the cosine wavelet transform $\Psi(\lambda, t_n)$ calculated for the points $t = 2, 0$, and -4.

Problem 6.9 For the signal in (6.23), compute the cosine wavelet transform $\Psi(\lambda, t_n)$ at the time-points t_n with centers t at -4.225. Verify if the maximums of the transform are achieved at points of the peaks as expected.

Problem 6.10 Show that the Fourier correlation function is actually the product of two vectors,

$$F(\lambda, t) = (\cos(\lambda t), \sin(\lambda t)) \begin{pmatrix} \mathrm{Re}\,\hat{f}(\lambda) \\ \mathrm{Im}\,\hat{f}(\lambda) \end{pmatrix}.$$

Problem 6.11 Given vector $\mathbf{x} = (x_1, x_2)'$, consider in the real plane R^2 the rotation by angle ω anti-clockwise and on the X-axis,

$$\mathbf{x} \to R(\omega)\mathbf{x} = \begin{pmatrix} \cos(\omega) & -\sin(\omega) \\ \sin(\omega) & \cos(\omega) \end{pmatrix} \begin{pmatrix} x_1 \\ x_2 \end{pmatrix} = \begin{pmatrix} y \\ 0 \end{pmatrix}.$$

Note that the value of the first component is not unique, i.e., $y = \pm||\mathbf{x}||$. To remove this ambiguity, develop a program for uniquely computing this component through arctan function.

Problem 6.12 Compute the DsiHT transform of the signal $\mathbf{z} = (1, 2, 1, 3, 2, 1, 5, 2, 1, 7, 5, 4, 6, 6, 2)$ of length 15 when the generator is $\mathbf{x} = (1, -1, 1, 2, 1, 3, 1, 1, 1, -2, 1, -3, 1, 1, 2)$, and the path equals $P = (0, 14)(0, 13)(0, 12)...(0, 1)$.

Problem 6.13 Compute the matrix of the DsiHT considered in Problem 6.12.

Problem 6.14* Show the matrix of the DsiHT is always orthogonal, regardless of the path of the transform.

Problem 6.15 Find the path of the Haar transformation of order $N = 16$, when the transform is calculated by Givens rotations. We will call this path *the Haar path*.

Problem 6.16 Calculate the matrix of the Givens-Haar generated by the vector $\mathbf{x} = (1, 1, 1, 1, ..., 1)$ of length 16 with the Haar path. Verify if the obtained matrix is the matrix of the Haar transform.

Problem 6.17 Calculate and plot the centers and mean widths of the basis waves of the 128-point Haar transformation.

Problem 6.18 Consider the 32-point generator sampled from the Gaussian function $x(t) = e^{-t}$ in the interval $(-\pi, \pi)$.

A. Calculate and plot the centers and mean widths of the basis waves of the Givens-Haar transformation.

B. Show that the matrix of the Givens-Haar transformation is orthogonal.

Problem 6.19 Consider the 32-point generator sampled from the Mexican hat function $x(t) = (1 - t^2)e^{-t^2/2}$ in the interval $(-\pi, \pi)$.

A. Calculate and plot the centers and mean widths of the basis waves of the Givens-Haar transformation.

B. Show that the matrix of the Givens-Haar transformation is orthogonal.

Problem 6.20 Consider the 32-point generator sampled from the triangle function $x(t) = 1 - |t|/\pi$ in the interval $(-\pi, \pi)$.

A. Calculate and plot the centers and mean widths of the basis waves of the Givens-Haar transformation.

B. Show that the matrix of the Givens-Haar transformation is orthogonal.

Problem 6.21 Calculate the matrix of the DsiHT generated by the vector $\mathbf{x} = (1, 1, 1, 1, ..., 1)$ of length 16 with the path of the fast paired transform.

A. Calculate and plot the centers and mean widths of the basis waves of the DsiHT .

B. Show that the matrix of the DsiHT is orthogonal.

Problem 6.22* Consider a periodical plane wave on the real line $\psi(x + N) = \psi(x)$.

A. Show that the movement of the plane wave can be represented by powers of the matrix $T = T_N$ as

$$\psi(x - t) = T^t \psi(x),$$

where the coefficients of the matrix T are calculated by

$$t_{n,m} = \delta_{n-1 \bmod N, m}, \quad n, m = 0 : (N - 1).$$

B. Show the movement of such a wave on the vector $\mathbf{x} = (0, 2, 1, 0, 1, 1, 0)'$.

C. Show that the matrix T is orthogonal and $T^N = I$.

Problem 6.23* For $N = 11$, consider the path $P = (N - 1, N - 2), (N - 2, N - 3), (N - 3, N - 4), ..., (1, 0)$ and the matrix T_2 of the basic movement, which is equal to the opposite identity matrix 2×2.

A. Find the matrix T_N that describes the movement of the waves in this case.

B. Find the generator of the transformation with matrix T.

Problem 6.24 Let $\mathbf{x} = (1, 1, 2, 1, 1, -1, 1, 1, 2, -1, 1, 3)'$ be the generator of the DsiHT. Find the generator of the inverse DsiHT.

Problem 6.25 Find the path of the inverse 16-point Givens-Haar transformation.

References

[1] J.W. Cooley and J.W. Tukey, "An algorithm the machine computation of complex Fourier series," *Math. Comput,* vol. 9, no. 2, pp. 297-301, 1965.

[2] N. Ahmed and R. Rao, *Orthogonal Transforms for Digital Signal Processing,* Springer-Verlag, New York, 1975.

[3] O. Ersoy, *Fourier-Related Transform, Fast Algorithms and Applications,* Prentice Hall, New Jersey, 1997.

[4] D.F. Elliot and K.R. Rao, *Fast transforms - Algorithms, Analyzes, and Applications,* Academic, New York, 1982.

[5] R.E. Blahut, *Fast Algorithms for Digital Signal Processing,* Addison-Wesley, Reading, Mass, 1985.

[6] I.C. Chan and K.L. Ho, "Split vector-radix fast Fourier transform," *IEEE Trans. on Signal Processing,* vol. 40, no. 8, pp. 2029-2040, Aug. 1992.

[7] L. Rabiner and R.W. Schaefer, *Speech Signal Processing,* Prentice Hall, Englewood Cliffs, New Jersey, 1983.

[8] A.M. Grigoryan, "2-D and 1-D multi-paired transforms: Frequency-time type wavelets," *IEEE Trans. on Signal Processing,* vol. 49, no. 2, pp. 344-353, Feb. 2001.

[9] A.M. Grigoryan and S.S. Agaian, *Multidimensional Discrete Unitary Transforms: Representation, Partitioning and Algorithms,* Marcel Dekker Inc., New York, 2003.

[10] A.M. Grigoryan, "A novel algorithm for computing the 1-D discrete Hartley transform," *IEEE Signal Processing Letters,* vol. 11, no. 2, pp. 156-159, Feb. 2004.

[11] A.M. Grigoryan, "An algorithm for calculation of the discrete cosine transform by paired transform," *IEEE Trans. on Signal Processing,* vol. 53, no. 1, pp. 265-273, Jan. 2005.

[12] A.M. Grigoryan and V.S. Bhamidipati, "Method of flow graph simplification for the 16-point discrete Fourier transform," *IEEE Trans. on Signal Processing,* vol. 53, no. 1, pp. 384-389, Jan. 2005.

[13] A.M. Grigoryan and S.S. Agaian, "Split manageable efficient algorithm for Fourier and Hadamard transforms," *IEEE Trans. on Signal Processing,* vol. 48, no. 1, pp. 172-183, Jan. 2000.

[14] V.S. Bhamidipati and A.M. Grigoryan, "Parameterized reversible integer disc-rete cosine transforms," *Proc. SPIE Int. Conf. Image Processing: Algorithms and Systems III,* vol. 5298, pp. 13-24, San Jose, CA, 18-22 January 2004.

[15] L.B. Almeida, "The fractional Fourier transform and time frequency repre-sentations," *IEEE Trans. on Signal Processing,* vol. 42, no. 11, pp. 3084-3090, Nov. 1994.

[16] I.C. Chan and K.L. Ho, "Split vector-radix fast Fourier transform," *IEEE Trans. on Signal Processing,* vol. 40, no 8, pp. 2029-2040, Aug. 1992.

[17] P. Duhamel and M. Vetterli, "Fast Fourier transforms: A tutorial review and state of the art," *Signal Processing,* vol. 19, pp. 259-299, 1990.

[18] *Handbook for Digital Signal Processing,* Edited by J.F. Kaiser and S.K. Mitra, Wiley-Interscience, New-York, 1993.

[19] P. Duhamel, "Implementation of "Split-radix" FFT algorithms for complex, real, and real-symmetric data," *IEEE Trans.* ASSP-34, no. 2, pp. 285 - 295, 1986.

[20] A.M. Grigoryan, P.A. Patel, and N. Ranganadh, "Novel algorithm of the Haar transform," *Proc. ISPC & GSPx,* Dallas, March 31-April 3, 2003.

[21] A. Haar, "Zur theorie der orthogonalen funktionensysteme," *Math. Ann,* vol. 69, pp. 331-371, 1910.

[22] C.K. Chui, *An Introduction to Wavelets,* Academic, New York, 1992.

[23] Y. Meyer, *Wavelets. Algorithms and Applications,* Society for Industrial and Applied Mathematics, Philadelphia, PA, 1993.

[24] P.P. Vaidynathan, *Multirate Systems and Filter Banks,* Englewood Cliffs, NJ: Prentice Hall, 1993.

[25] G. Kaiser, "The fast Haar transform," *IEEE Potentials,* vol. 17, no. 2, pp. 34-37, April-May 1998.

[26] J.C. Goswami and A.K. Chan, *Fundamentals of Wavelets: Theory, Algo-rithms, and Applications,* New York: Wiley, 1999.

[27] K. Hwang, *Advanced Computed Architecture: Parallelism, Scalability, Pro-grammability,* Mcgraw-Hill College Division, New York, 1993.

[28] S.C. Pei and J.J. Ding, "The integer transforms analogous to discrete trigono-metric transforms," *IEEE Trans. on Signal Processing,* vol. 48, no. 12, pp. 3345-3364, Dec. 2000.

[29] S. Oraintara, Y.-J. Chen, and T.Q. Nguyen, "Integer fast Fourier algorithms," *IEEE Trans. on Signal Processing,* vol. 50, no. 3, pp. 607-618, Mar. 2002.

[30] W. Sweldens, "The lifting scheme: A custom-design construction of biorthog-onal wavelets," *Appl. Comput. Harmon. Anal.,* vol. 3, no. 2, pp. 186-200, 1996.

[31] W. Sweldens, "The lifting scheme: A construction of second generation wavelets," *SIAM J. Math. Anal.,* vol. 29, no. 2, pp. 511-546, 1998.

[32] V.K. Goyal, "Transform coding with integer-to-integer transforms," *IEEE Trans. Information Theory,* vol. 46, no. 2, pp. 465-473, Mar. 2000.

[33] G. Bi and Y. Zeng, *Transforms and Fast Algorithms for Signal Analysis and and Representations,* Birkhauser Verlag, Oct. 2003.

[34] K. Komatsu and K. Sezaki, "Reversible subband coding of images," *Proc. SPIE VCIP,* vol. 2501, pp. 676-684, May 1995.

[35] L.Z. Cheng, H. Xu, and Y. Luo, "Integer discrete cosine transform and its fast algorithm," *Electronics Letters,* vol. 37, no. 1, pp. 64-65, Jan. 2001.

[36] R. Calderbank, I. Daubechies, W. Sweldens, and B.-L. Yeo, "Wavelet transforms that map integers to integers," *Appl. Comput. Harmon. Anal.,* vol. 5, no. 3, pp. 332-369, 1998.

[37] W.K. Cham and P.C. Yip, "Integer sinusoidal transforms for image processing," *Int. J. Electron.,* vol. 70, no. 6, pp. 1015-1030, 1991.

[38] S.C. Pei and J.J. Ding, "Integer discrete Fourier transform and its extension to integer trigonometric transforms," *Proc. IEEE Int. Symp. Circuits Syst.,* vol. 5, pp. 513-516, May 28-31, 2000.

[39] Y. Zeng, G. Bi, and Z. Lin, "Integer sinusoidal transforms based on lifting factorization," *IEEE Intl. Conf. Acoust., Speech, Signal Processing,* vol. 2, pp. 1181-1184, May 2001.

[40] Y. Zeng, L. Cheng, G. Bi, and A.C. Kot, "Integer DCTs and fast algorithms," *IEEE Trans. Acoust., Speech, Signal Processing,* vol. 49, no. 11, pp. 2774-2782, Nov. 2001.

[41] K. Komatsu and K. Sezaki, "Reversible discrete cosine transform," *IEEE Trans. Acoust., Speech, Signal Processing,* vol. 3, pp. 1769-1772, 12-15 May 1998.

[42] V.K. Goyal, "Transform coding with integer-to-integer transforms," *IEEE Trans. Information Theory,* vol. 46, no. 2, pp. 465-473, Mar. 2000.

[43] Y.J. Chen, "Integer discrete cosine transform (IntDCT)," *The Second Int. Conf. Inform. Commun. Signal Process.,* Singapore, Dec. 1999.

[44] W. Philips, "Lossless DCT for combined lossy/lossless image coding," *Proc. IEEE Int. Conf. Image Process.,* vol. 3, pp. 871-875, 1998.

[45] K.R. Rao and P. Yip, *Discrete Cosine Transform—Algorithms, Advantages, Applications,* Academic Press, Boston, 1990.

[46] W.B. Pennebaker and J.L. Mitchell, *JPEG Still Image Compression Standard,* Van Nostrand Reinhold, New York, 1993.

[47] G. Mandyam, N. Ahmed, and N. Magotra, "Lossless image compression using the discrete cosine transform," *JVCIR,* vol. 8, no. 1, pp. 21-26, Mar. 1997.

[48] M. Iwahashi, N. Kambayashi, and H. Kiya, "Bit reduction of DCT basis for transform coding," *ECJ,* vol. 80, no. 8, pp. 81-91, Aug. 1997.

[49] M.S. Moellenhoff and M.W. Maier, "DCT transform coding of stereo images for multimedia applications," *IndEle,* vol. 45, no. 1, pp. 38-43, Feb. 1998.

[50] M. Hamidi and J. Pearl, "Comparision of the cosine and Fourier transforms of Markov-1 signals," *IEEE Trans. Acoust., Speech, Signal Processing,* vol. ASSP-24, pp. 428-429, Oct. 1976.

[51] M. Uenohara and T. Kanade, "Optimal approximation of uniformly rotated images: relationship between Karhunen-Loeve expansion and discrete cosine transform," *IEEE Trans. on Image Processing,* vol. 7, no. 1, pp. 116-119, Jan. 1998.

[52] H.S. Hou, "A fast recursive algorithm for computing discrete cosine transform," *IEEE Trans. Acoust., Speech, Signal Processing,* vol. ASSP-35, pp. 1455-1461, Oct. 1987.

[53] Y. Morlkawa, H. Hamada, and N. Yamane, "A fast algorithm for the cosine transform based on successive order reduction of the Tchebycheff polynomial," *Trans. Ints. Elec. Communic. Eng. (Japan),* vol. J68-A, pp. 173-180, 1985.

[54] N. Suehiro and M. Hatori, "Fast algorithms for the DFT and other sinusoidal transforms," *IEEE Trans. Acoust. Speech, Signal Processing,* vol. ASSP-34, no. 3, pp. 642-644, June 1986.

[55] Z. Wang, "Pruning the fast discrete cosine transform," *IEEE Trans. Communications,* vol. 39, no. 5, pp. 640-643, May 1991.

[56] C. Loeffler, A. Ligtenberg, and G. Moschytz, "Practical fast 1-D DCT algorithms with 11 multiplications," *Proc. of the IEEE ICASSP '89,* Glasgow, Scotland, pp. 988-991, May 1989.

[57] Z. Cvetković and M.V. Popović, "New fast recursive algorithms for the computation of the discrete cosine and sine transforms," *Proc. of the IEEE ICASSP '91,* Toronto, Canada, p. 2201, May 1991.

[58] Z. Zhijin and Q. Huisheng, "Recursive algorithms for discrete cosine transform," *Proc. of the IEEE ICASSP '96,* Atlanta, GA, pp. 115-118, May 1996.

[59] Z. Wang, "On computing the discrete Fourier and cosine transforms," *IEEE Trans. Acoust. Speech, Signal Processing,* vol. ASSP-33, pp. 1341-1344, Oct. 1985.

[60] Z. Wang, "Fast algorithms for the discrete W transform and for the discrete Fourier transform," *IEEE Trans. Acoust. Speech, Signal Processing,* vol. ASSP-32, no. 2, pp. 803-816, Aug. 1984.

[61] B. G. Lee, "A new algorithm to compute the discrete cosine transform," *IEEE Trans. Acoust. Speech, Signal Processing,* vol. ASSP-35, pp. 1243-1245, Dec. 1984.

[62] G. Bi and L.W. Yu, "DCT algorithms for composite sequence lengths," *IEEE Trans. on Signal Processing,* vol. 46, no. 3, pp. 554-562, Mar. 1998.

[63] S. Mitra, O. Shentov, and M. Petraglia, "A method for fast approximate computation of discrete cosine transforms," *Proc. of the IEEE ICASSP'90,* Albuquerque, NM, pp. 2025-2028.

[64] L.R. Welch, "Walsh functions and Hadamard matrices. Application of Walsh functions," *Proc. 1970. Symp.* Washington. DC, 1970, pp. 163-165.

[65] S.S. Agaian, *Hadamard Matrices and Their Applications,* Lecture Notes in Mathematics 1168. Springer–Verlag, 1985.

[66] W.K. Pratt and J. Kane, "Hadamard transform image coding," *Proc. IEEE,* vol. 57, no. 1, pp. 58-69, 1969.

[67] S.C. Knauer, "Real-time video compression algorithm for Hadamard transform processing," *IEEE Trans. Electromag. Compat.,* vol. 18, no. 1, pp. 28-36, 1976.

[68] M.H. Lee and M. Kaveh, "Fast Hadamard transform based on a simple matrix decomposition," *IEEE Trans. Acoust., Speech, Signal Processing,* vol. 34, no. 6, pp. 1666-1667, 1986.

[69] R. Gordon, "A tutorial on ART (algebraic reconstruction techniques)," *IEEE Trans. Nuclear Science,* vol. 21, pp. 78-93, 1974.

[70] G.T. Herman and A. Lent, "Iterative reconstruction algorithms," *Comput. Biol. Med.,* vol. 6, pp. 273-294, 1976.

[71] Y. Censor, "Finite series-expansion reconstruction methods," *Proc. of IEEE,* vol. 71, no. 3, pp. 409-419, 1983.

[72] A.H. Andersen and A.C. Kak, "Simultaneous algebraic reconstruction technique (SART): a superior implementation of the ART algorithm," *Ultrasonic Imaging,* vol. 6, pp. 81-94, 1984.

[73] R.C. Gonzalez and R.E. Woods, *Digital Image Processing,* Prentice Hall, 2nd Edition, New Jersey, 2001.

[74] J.H. McClellan, "Artifacts in alpha-rooting of images," *Proc. IEEE Int. Conf. Acoustics, Speech, and Signal Processing,* pp. 449-452, Apr. 1980.

[75] A. Beghcladi and A.L. Negrate, "Contrast enhancement technique based on local detection of edges," *Comput. Vision, Graphic, Image Processing,* vol. 46, pp. 162-274, 1989.

[76] S. Aghagolzadeh and O.K. Ersoy, "Transform image enhancement," *Optical Engineering,* vol. 31, no. 3, pp. 614-626, 1992.

[77] S. Liu, W. Smith, L. Zheng, A. Chan, and Hoyler R. "Directional clutter removal of aerial digital images using X-ray wavelet transform and Markov random field," *IEEE Trans. on Remote Sensing,* vol. 49, no. 2, pp. 344-453, 1999.

[78] M. Malfait and D. Roose, "Wavelet-based image denoising using a Markov random field a priori model," *IEEE Trans. on Image Processing,* vol. 6, pp. 549-565, Apr. 1997.

[79] Y. Xu, J.B. Weaver, D.M. Healy, and J. Lu, "Wavelet transform domain filters: A spatially selective noise filtration technique," *IEEE Trans. on Image Processing,* vol. 3, pp. 747-758, 1994.

[80] A. Akansu and R. Haddad, *Multiresolution Signal Decomposition, Transforms Subbands and Wavelets*, Academic Press, 1992.

[81] D. Gabor, "Theory of communication," *J. Inst. Elec. Eng.*, vol. 93, pp. 429-457, 1946.

[82] J.G. Daugman, "Complete discrete 2-D Gabor transforms by neural networks for image analysis and compression," *IEEE Trans. Acoust., Speech, Signal Processing,* vol. 36, no. 7, pp. 1169-1179, 1988.

[83] R.L. De Valois and K.K. De Valois, *Spatial Vision,* New York: Oxford Univ. Press, 1988.

[84] O. Rioul and M. Vetterli, "Wavelets and signal processing," *IEEE Signal Processing Magazine,* vol. 8, pp. 11-38, Oct. 1991.

[85] A. Aldroubi and M. Unser, *Wavelet in Medicine and Biology,* Boca Raton: CRC Press, 1996.

[86] S.G. Mallat, *A Wavelet Tour of Signal Processing,* San Diego: Academic Press, second edition, 1998.

[87] A.N. Kolmogorov and S.V. Fomin, *Elements of the Theory of Functions and Functional Analysis,* Moscow: Nauka, 1972.

[88] S.G. Mallat, "A theory for multiresolution signal decomposition: The wavelet representation," *IEEE Trans. on Pattern Anal. and Machine Intell.,* vol. 11, no. 7, pp. 674-693, 1989.

[89] L. Prasad, *Wavelet Analysis With Applications to Image Processing,* Boca Raton: CRC Press, 1997.

[90] A.K. Louis, D. Maass, and A. Rieder, *Wavelets: Theory and Applications,* New York: Wiley, 1997.

[91] A. Mertins, *Signal Analysis: Wavelets, Filter Banks, Time-Frequency Transforms and Applications,* New York: John Wiley, 1999.

[92] J.H. Halberstein, "Recursive complex Fourier analysis for real time applications," *Proc. IEEE,* vol. 54, p. 903, 1966.

[93] R.R. Bitmead, "On recursive discrete Fourier transformation," *IEEE Trans. Acoust., Speech, Signal Processing,* vol. 30, pp. 319-322, 1982.

[94] M. Unser, "Recursion in short time signal analysis," *Signal Processing,* vol. 5, pp. 229-240, 1983.

[95] P. Flandrin, "Some aspects of nonstationary signal processing with emphasis on time-frequency and time-scale methods," in *Wavelets: Time-Frequency Methods and Phase Space* (J.M. Combes, A. Grossmann, and P. Tchamichian, Eds), New York: Springer-Verlag, pp. 68-98, 1989.

[96] W. Chen, N. Kehtarnavaz, and T.W. Spencer, "An efficient recursive algorithm for time-varying Fourier transform," *IEEE Trans. on Signal Processing,* vol. 41, no. 7, pp. 2488-2490, 1993.

[97] G. Kaiser, *A Friendly Guide to Wavelets,* Birkhauser, 1994.

[98] T. Edward, *Discrete Wavelet Transforms: Theory and Implementation,* Stanford University, Sep. 1991.

[99] A.M. Grigoryan, "Method of paired transforms for reconstruction of images from projections: Discrete model," *IEEE Transactions on Image Processing,* vol. 12, no. 9, pp. 985-994, September 2003.

[100] A.M. Grigoryan and S.S. Agaian, "Transform-based image enhancement algorithms with performance measure," *Advances in Imaging and Electron Physics,* vol. 130, pp. 165-242, May 2004.

[101] F.T. Arslan and A.M. Grigoryan, "Method of image enhancement by splitting-signals," *Proceedings of the IEEE Int. Conf. on Acoustics, Speech, and Signal Processing (ICASSP '05),* vol. 4, pp. 177180, March 18-23, 2005.

[102] F.T. Arslan and A.M. Grigoryan, "Fast splitting α-rooting method of image enhancement: Tensor representation," *IEEE Trans. on Image Processing,* vol. 15, no. 11, pp. 3375-3384, Nov. 2006.

[103] A.M. Grigoryan and M.M. Grigoryan, "Nonlinear approach of construction of fast unitary transforms," *Proc. of the 40th Annual Conference CISS 2006,* Princeton University, pp. 1073-1078, March 22-24, 2006.

[104] A.M. Grigoryan and M.M. Grigoryan, "Discrete unitary transforms generated by moving waves," *Proc. of the International Conference: Wavelets XII, SPIE: Optics+Photonics 2007,* San Diego, CA, August 27-29, 2007.

[105] A.M. Grigoryan and M.M. Grigoryan, "New discrete unitary Haar-type heap transforms," *Proc. of the International Conference: Wavelets XII, SPIE: Optics+Photonics 2007,* San Diego, CA, August 27-29, 2007.

Index

Printed and bound by CPI Group (UK) Ltd, Croydon, CR0 4YY

29/10/2024

01780639-0001